Enzyme Engineering

Methods and Protocols

Edited by

James C. Samuelson

New England Biolabs, Inc., Ipswich, MA, USA

Humana Press

Editor
James C. Samuelson
New England Biolabs, Inc.
Ipswich, MA, USA

ISSN 1064-3745 ISSN 1940-6029 (electronic)
ISBN 978-1-4939-6287-7 ISBN 978-1-62703-293-3 (eBook)
DOI 10.1007/978-1-62703-293-3
Springer New York Heidelberg Dordrecht London

Printed on acid-free paper

Humana Press is a brand of Springer
Springer is part of Springer Science+Business Media (www.springer.com)

METHODS IN MOLECULAR BIOLOGY™

Series Editor
John M. Walker
School of Life Sciences
University of Hertfordshire
Hatfield, Hertfordshire, AL10 9AB, UK

For further volumes:
http://www.springer.com/series/7651

Preface

The following collection of chapters is intended to provide guidance to investigators wishing to create enzyme variants with desired properties. Whether the pursuit is commercially motivated or purely academic, engineering a novel biological catalyst is an enticing challenge. High-resolution protein structure analysis allows for rational alteration of enzyme function, yet many useful enzyme variants are the product of well-designed selection schemes or screening strategies. Accordingly, this volume contains examples where bugs are employed as workhorses in order to evolve enzyme function or to isolate enzyme variants with improved solubility or stability.

One step away from cell-based selection and screening is the use of in vitro compartmentalization to isolate enzyme variants and their respective nucleotide codes, as the most powerful in vitro methods for enzyme evolution link gene sequence to gene product function. In Chapter 6, Golynskiy et al. present a comprehensive review of the methods used for the in vitro evolution of protein enzymes. The general principles of ribosome display, mRNA display, and DNA display are outlined, and the advantages of each of these approaches are highlighted. For many years, in vitro translation systems have offered the opportunity to produce small quantities of protein containing unnatural amino acids. More recently this objective has been realized in genetically modified organisms so protein yield may no longer be a limiting factor. In Chapter 7, Singh-Blom et al. demonstrate that residue-specific incorporation of tryptophan analogs is possible in a Δ*trp*C derivative of BL21(DE3). Additionally, the same researchers present a step-by-step guide to prepare an S30 lysate from a tryptophan auxotroph so that cell-free synthesis of tryptophan-substituted protein may also be accomplished.

The potential for any directed evolution project is dependent upon the type of gene library and the degree of library diversity. Therefore, multiple chapters outline methods for gene mutagenesis, gene and operon assembly, and efficient do-it-yourself gene synthesis. Remarkably, today's researcher also has the option of purchasing mutant gene libraries now that gene synthesis costs have declined significantly. Once the task of gene library construction is completed and after promising enzyme variants are isolated, a further challenge is thorough protein characterization. When a novel enzyme variant is isolated, many concerns must be addressed: For example, what is the true enzyme specificity and is the turnover rate acceptable for the desired application? In Chapter 2, Demarse et al. showcase an underutilized, yet simple and effective technique for analyzing enzyme kinetics. Isothermal titration calorimetry (ITC) is a direct method for determining the basic parameters of an enzyme-catalyzed reaction (i.e. V_{max}, K_m and k_2). Since ITC is a non-destructive method, precious quantities of an enzyme variant are not consumed. But perhaps more importantly, ITC is suitable for many types of assays since substrate(s) do not require labeling and linkage to a secondary-detectable process is not necessary.

Today many engineering efforts are focused on creating protein-based therapeutics. For example, Chapter 3 presents two examples of a screen for evolving amino acid degradation enzymes. Such enzymes show promise in cancer therapy by limiting the nutrient supply for tumor cells. Amino acid auxotrophic host cells were critical in this cell-based screen and a customized host strain is often the enabling element of a cell-based selection or screen. Therefore, this volume includes a simple method for generating site-specific mutations within bacterial chromosomes. In the last few years, Bryan Swingle (Chapter 9) and others have defined the requirements of oligonucleotide recombination and have found that the process is RecA-independent. In short, a bacterial cell is transformed with a single-stranded DNA oligonucleotide containing the desired mutation and the oligonucleotide anneals and becomes incorporated during replication of the host chromosome. This method is certainly a breakthrough and has already yielded impressive biocatalysts.

This volume also highlights the engineering of two different types of rare-cutting endonucleases that show great potential in gene therapy applications: The newest development is the emergence of TAL effector nucleases or TALENs. TALENS are derived from transcription factors (TAL effectors) and the DNA cleavage domain from the FokI restriction endonuclease. In Chapter 5, Li and Yang describe a simple method for creating designer TALENs (dTALENs) from modular motifs with defined DNA-binding specificities. A related chapter describes a method for characterizing the DNA-binding and cleavage properties of LAGLIDADG homing endonucleases. This family of endonuclease is being recruited for gene targeting due to a natural affinity for rare DNA target sites approximately 22 bp in length. Homing endonuclease specificity is especially difficult to characterize and Baxter et al. have developed a high-throughput method where homing endonucleases are expressed on the surface of yeast and specificity is evaluated against synthetic DNA target sequences using flow cytometry.

Finally, the vast amount of genome sequence data and protein structural data has allowed for the development of two new methods that incorporate rational design. First, the REAP method takes advantage of ancient sequences from a phylogenetic tree. Signatures of functional divergence are identified and used to design a library with a high density of viable protein variants. This focused library is then tested for desirable properties such as thermostability. In contrast, the DECAAF method takes advantage of existing protein structure data (PDB files) to search for "promiscuous" active sites that have the potential to catalyze a desired reaction. The DECAAF analysis arrives at a protein scaffold that serves as the basis for rational engineering of residues within the putative active site. Many natural enzymes are thought to have "moonlighting" domains based on the number of encoded proteins versus verified chemical transformations in simple organisms. The DECAAF method may also be useful in identifying these bona fide secondary active sites.

The contents of this book should be valuable for scientists with a budding interest in protein engineering as well as veterans looking for new approaches to apply in established discovery programs. The following chapters describe newly developed technologies in sufficient detail so that each method can be practiced in a standard molecular biology laboratory. Accordingly, I wish to thank each contributor for sharing his/her expertise with the research community. And finally, I thank my colleagues at New England Biolabs for their support and their commitment to the advancement of basic science.

Ipswich, MA, USA *James C. Samuelson*

Contents

Contributors

GIULIA AGNELLO • *Institute for Cellular and Molecular Biology, University of Texas, Austin, TX, USA*

JAMES C.A. BARDWELL • *Howard Hughes Medical Institute, Chevy Chase, MD, USA; Department of Molecular, Cellular and Developmental Biology, University of Michigan, Ann Arbor, MI, USA*

SARAH K. BAXTER • *Department of Immunology, University of Washington, Seattle, WA, USA; Center for Immunity and Immunotherapies, Seattle Children's Research Institute, Seattle, WA, USA; Northwest Genome Engineering Consortium, Seattle, WA, USA*

SWAPAN K. BHATTACHARJEE • *Department of Biological Sciences, Tata Institute of Fundamental Research, Mumbai, India*

JURATE BITINAITE • *New England Biolabs, Inc., Ipswich, MA, USA*

JASON CANTOR • *Department of Chemical Engineering, University of Texas, Austin, TX, USA*

SANDEEP CHAKRABORTY • *Department of Biological Sciences, Tata Institute of Fundamental Research, Mumbai, India*

MEGAN F. COLE • *School of Biology, Georgia Institute of Technology, Atlanta, GA, USA*

VIRGINIA W. CORNISH • *Department of Chemistry, Columbia University, New York, NY, USA*

VANESSA E. COX • *School of Chemistry and Biochemistry, Georgia Institute of Technology, Atlanta, GA, USA*

T. ASHTON CROPP • *Department of Chemistry, Virginia Commonwealth University, Richmond, VA, USA*

ABHAYA M. DANDEKAR • *Plant Sciences Department, University of California, Davis, CA, USA*

NEIL A. DEMARSE • *TA Instruments, Lindon, UT, USA*

ANDREW D. ELLINGTON • *Institute for Cellular and Molecular Biology, The University of Texas at Austin, Austin, TX, USA*

LINDA FOIT • *Howard Hughes Medical Institute, Chevy Chase, MD, USA; Department of Molecular, Cellular and Developmental Biology, University of Michigan, Ann Arbor, MI, USA*

ERIC A. GAUCHER • *School of Biology, Georgia Institute of Technology, Atlanta, GA, USA*

ERIC R. GEERTSMA • *Department of Biochemistry, University of Zurich, Zurich, Switzerland*

GEORGE GEORGIOU • *Department of Chemical Engineering, Institute of Cellular and Molecular Biology, Section of Molecular Genetics and Microbiology, Department of Biomedical Engineering, University of Texas, Austin, TX, USA*

MISHA V. GOLYNSKIY • *Department of Biochemistry, Molecular Biology and Biophysics, University of Minnesota, St. Paul, MN, USA; BioTechnology Institute, University of Minnesota, St. Paul, MN, USA*
KELSEY L. GRATTON • *The Wallace H. Coulter, Department of Biomedical, Engineering, Georgia Institute of Technology and Emory University, Atlanta, GA, USA*
LEE D. HANSEN • *Department of Chemistry, Brigham Young University, Provo, UT, USA*
JOHN C. HAUGNER III • *Department of Biochemistry, Molecular Biology and Biophysics, University of Minnesota, St. Paul, MN, USA; BioTechnology Institute, University of Minnesota, St. Paul, MN, USA*
HIDEHIKO HIRAKAWA • *Department of Chemistry and Biotechnology, School of Engineering, The University of Tokyo, Tokyo, Japan*
PEI-CHUNG HSIEH • *New England Biolabs, Inc., Ipswich, MA, USA*
RANDALL A. HUGHES • *Institute for Cellular and Molecular Biology, The University of Texas at Austin, Austin, TX, USA*
JORDAN JARJOUR • *Northwest Genome Engineering Consortium, Seattle, WA, USA; Pregenen Inc., Seattle, WA, USA*
MARIE C. KILLIAN • *Department of Chemistry, Brigham Young University, Provo, UT, USA*
ABIGAIL R. LAMBERT • *Center for Immunity and Immunotherapies, Seattle Children's Research Institute, Seattle, WA, USA; Northwest Genome Engineering Consortium, Seattle, WA, USA*
TING LI • *Department of Genetics, Development and Cell Biology, Iowa State University, Ames, IA, USA*
JIA LIU • *Department of Chemistry, Virginia Commonwealth University, Richmond, VA, USA*
JIAN-WEI LIU • *CSIRO Ecosystem Sciences, Canberra, Australia*
RENU MINDA • *Department of Biological Sciences, Tata Institute of Fundamental Research, Mumbai, India*
ALEARDO MORELLI • *Department of Biochemistry, Molecular Biology and Biophysics & BioTechnology Institute, University of Minnesota, St. Paul, MN, USA*
DANA MORRONE • *Department of Biochemistry, Molecular Biology and Biophysics & BioTechnology Institute, University of Minnesota, St. Paul, MN, USA*
TERUYUKI NAGAMUNE • *Department of Chemistry and Biotechnology, School of Engineering, The University of Tokyo, Tokyo, Japan*
DAVID L. OLLIS • *Research School of Chemistry, Australian National University, Canberra, Australia*
NILI OSTROV • *Department of Chemistry, Columbia University, New York, NY, USA*
OLGA PALEY • *Department of Chemical Engineering, University of Texas, Austin, TX, USA*
COLETTE F. QUINN • *TA Instruments, Lindon, UT, USA*
BASUTHKAR J. RAO • *Department of Biological Sciences, Tata Institute of Fundamental Research, Mumbai, India*
LIPIKA SALAYE • *Department of Biological Sciences, Tata Institute of Fundamental Research, Mumbai, India*

ANDREW M. SCHARENBERG • *Department of Immunology, University of Washington, Seattle, WA, USA; Center for Immunity and Immunotherapies, Seattle Children's Research Institute, Seattle, WA, USA; Northwest Genome Engineering Consortium, Seattle, WA, USA; Pregenen Inc., Seattle, WA, USA*
BURCKHARD SEELIG • *Department of Biochemistry, Molecular Biology and Biophysics & BioTechnology Institute, University of Minnesota, St. Paul, MN, USA*
AMRITA SINGH-BLOM • *Department of Molecular Genetics and Microbiology, The University of Texas at Austin, Austin, TX, USA*
BRADLEY J. STEVENSON • *Research School of Chemistry, Australian National University, Canberra, ACT, Australia*
EVERETT STONE • *Department of Biomedical Engineering, University of Texas at Austin, Austin, TX, USA*
BRYAN M. SWINGLE • *United States Department of Agriculture, Agricultural Research Service, Ithaca, NY, USA; Department of Plant Pathology and Plant-Microbe Biology, Cornell University, Ithaca, NY, USA*
ROMUALDAS VAISVILA • *New England Biolabs, Inc., Ipswich, MA, USA*
LAURA M. WINGLER • *Department of Chemistry, Columbia University, New York, NY, USA*
BING YANG • *Department of Genetics, Development and Cell Biology, Iowa State University, Ames, IA, USA*
SYLVIA H.-C. YIP • *Research School of Chemistry, Australian National University, Canberra, Australia*
TAE HYUN YOO • *Division of Applied Chemistry and Biological Engineering, Department of Molecular Science and Technology, Ajou University, Suwon, South Korea*

Chapter 1

A Tripartite Fusion System for the Selection of Protein Variants with Increased Stability In Vivo

Linda Foit and James C.A. Bardwell

Abstract

We describe here a genetic selection system that directly links protein stability to antibiotic resistance, allowing one to directly select for mutations that stabilize proteins in vivo. Our technique is based on a tripartite fusion in which the protein to be stabilized is inserted into the middle of the reporter protein β-lactamase via a flexible linker. The gene encoding the inserted protein is then mutagenized using error-prone PCR and the resulting plasmid library plated on media supplemented with increasing concentrations of β-lactam antibiotic. Mutations that stabilize the protein of interest can easily be identified on the basis of their increased antibiotic resistance compared to cells expressing the unmutated tripartite fusion.

Key words: Genetic selection, Protein stability, Protein evolution, Mutagenesis, Reporter protein, Tripartite fusion, Sandwich fusion

1. Introduction

Most soluble, globular proteins exhibit only marginal thermodynamic stabilities between approximately −5 and −10 kcal/mol (1, 2). Such low protein stability imposes significant challenges on the use of these polypeptides in many biotechnological, biomedical, and practical applications, where large amounts of stable and soluble protein are needed. The identification of stabilizing mutations, however, is difficult, since most random amino acid substitutions actually decrease stability (3–5). Computational methods that estimate the effect of mutations on protein stability are available but usually require detailed structural knowledge about the target protein, information that is often not available. Unfortunately, though computational methods are often good at predicting the destabilizing effect of mutations they are generally less accurate at predicting stabilizing mutations (6).

James C. Samuelson (ed.), *Enzyme Engineering: Methods and Protocols*, Methods in Molecular Biology, vol. 978,
DOI 10.1007/978-1-62703-293-3_1,

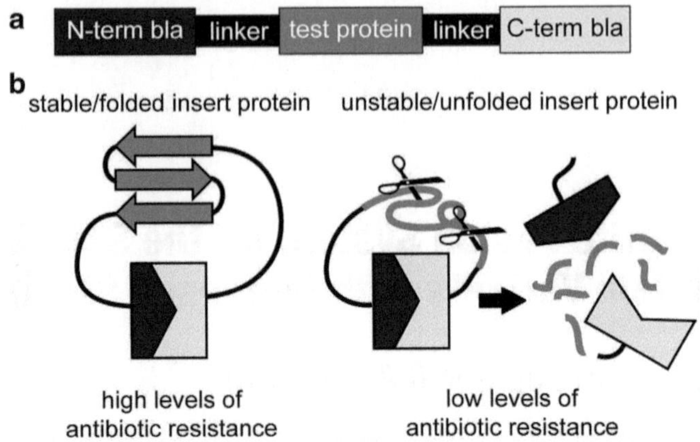

Fig. 1. Schematic representation of the tripartite fusion system. (**a**) The protein of interest is inserted into β-lactamase via a linker. N-term bla = N-terminal part β-lactamase, C-term bla = C-terminal part β-lactamase, linker = flexible glycine-serine linker. (**b**) If the test protein is folded properly (*left*), the two parts of β-lactamase can interact and provide β-lactamase activity. This results in high levels of antibiotic resistance. Unfolded proteins on the other hand (*right*) are subject to degradation by cellular proteases, leading to a large reduction in antibiotic resistance.

Recently, a number of approaches have been developed that utilize reporter proteins to monitor or increase the stability of proteins in the cellular environment (6–12). The genetic system we established (12) allows for monitoring and increasing protein stability and combines the following attractive features: (1) it directly links the in vivo stability of proteins to antibiotic resistance, a quantitative readout, (2) it is a *selection* rather than a *screen*, omitting laborious testing of individual protein mutants, (3) it does not require prior knowledge about structure or function of the insert protein, and (4) it can be used with a variety of different proteins.

Our approach is based on the reporter protein TEM1-β-lactamase. This enzyme is located in the periplasm of Gram-negative bacteria and confers resistance towards β-lactam antibiotics like ampicillin or penicillin (13). TEM1-β-lactamase is tolerant of insertions or deletions that occur in a solvent-exposed loop around residue 196 (14, 15). Based on this knowledge, we generated a tripartite fusion in which a test protein is inserted between residues 196 and 197 of the enzyme via flexible glycine-serine linkers (Fig. 1a) (12). Cleavage of the enzyme at this position results in two fragments that are catalytically inactive when expressed separately (16). However, when these fragments are fused to interacting partner protein as part of a protein complementation assay, activity will be restored (16).

The underlying principle for the use of this tripartite fusion as a readout for protein stability in our genetic selection system is the following: If the insert test protein is folded properly and is stable, the two fragments of β-lactamase can interact with each other, providing enzymatic activity. Cells expressing such a fusion construct exhibit high levels of resistance towards β-lactam antibiotics (Fig. 1b, left). Unstable insert proteins on the other hand are subject to degradation by cellular proteases. Proteolysis of such unstable and unfolded insert protein leads to separation of the two β-lactamase fragments. The result is a substantial decrease in antibiotic resistance levels (Fig. 1b, right).

To select for protein variants with increased in vivo stability, the gene encoding the target protein is randomly mutated and the resulting plasmid library transformed into *Escherichia coli* cells. The cells are then spread on plates containing increasing concentrations of a β-lactam antibiotic. Colonies showing increased levels of antibiotic resistance compared to cells expressing a construct containing the wild-type protein are selected and the protein sequence is determined.

For the model protein immunity protein 7 (Im7), we found that the in vivo steady-state levels of tripartite fusions containing different Im7 variants correlated well with the resulting level of antibiotic resistance (12). Moreover, the vast majority of constructs selected for their increased antibiotic resistance encoded protein variants that were both thermodynamically and kinetically more stable in vitro when expressed in absence of the fusion partner.

Our tripartite fusion system is not just useful for identifying stabilized protein variants. It can also be utilized to select for mutant proteins with other improved properties that lead to increased steady-state levels of the protein in the periplasm. Examples are increased solubility (12), the elimination of disulfide bonds or kinetic traps that are problematic for protein folding in vivo (17), decreased proteolytic susceptibility, and possibly increased translocation efficiency or a combination of these factors.

In addition, alterations that improve the folding of proteins are in principle not limited to mutagenesis of the gene encoding the protein inserted into β-lactamase. The system could possibly also be used to select for increased activity and specificity of chaperones or other folding factors that are co-expressed with a tripartite fusion containing a target protein. For instance, we have shown that it is possible to randomly mutate the host chromosome to generate bacterial strains that enhance expression of target proteins. In doing so we have identified a novel molecular chaperone called Spy (18).

Our selection works in the bacterial periplasm, a very oxidizing environment (19). To avoid nonnative and unwanted disulfide bond formation within the protein of interest, we strongly recommend the use of insert proteins that (1) do not contain unpaired cysteines that are normally part of a disulfide and (2) If they possess disulfides that the number of disulfides present is small, zero is ideal, but up to a maximum of two disulfide bonds can be tolerated, especially if they occur between consecutive cysteines.

In this chapter, we will focus on the selection of protein variants with increased in vivo stability.

2. Materials

2.1. Biological and Chemical Materials

Prepare all stock solutions using double-distilled water or deionized and then distilled water (ddH_2O).

1. Plasmids.
   (a) pBR322, e.g., from New England Biolabs (NEB) (20, 21) (see Note 1).
   (b) pBAD33 (22) (see Note 2).
   (c) Plasmid or chromosomal DNA encoding the insert protein (for amplification with PCR).
2. Primer solutions, 10 or 100 μM, stored at −20°C.
   (a) Primer 1 (polyacrylamide gel electrophoresis purified): 5′-CCGCTCCCGGATCCTGAGCTCGAGCCACCA CCACCAGAACCACCACCACCTAGTTCGCCA GTTAATAGTTTGCGCAACGTTGTTGCC-3′.
   (b) Primer 2 (polyacrylamide gel electrophoresis purified): 5′-TTCCGGAAGCGGAGGAGGTGGTTCAGG CGGAGGTGGAAGCCTTACTCTAGCTTCCCGG CAACAATTAATAGACTGGATGGAGGCG-3′.
   (c) Primer 3: 5′-ATAGGTACCAGGAGGAATTCATGAGTA TTCAACATTTCCGTGTCGC-3′.
   (d) Primer 4: 5′-GGTGGCAGTCTAGATTACCAATGCTTA ATCAGTGAGGCACC-3′.
   (e) Primer 5: 5′-TATCGTGCGGCCGCTCATGTTTGACA GCTTATCATCG-3′.
   (f) Primer 6: 5′-AGCTAGTCTAGACCGCGGGAAGATCC TTTTTGATAATCTC-3′.
   (g) Primer 7: 5′-GCTATACTAGTTCTTCCCCATCGGTGA TGTCGGCG-3′.
   (h) Primer 8: 5′-ATCGATGCGGCCGCATGTATTTAGAA AAATAAACAAAAGAG-3′.
   (i) Primers 9 + 10: Forward and reverse primers containing appropriate restriction sites for cloning of the gene encoding the protein of interest into the tripartite fusion expression plasmid of choice (see Note 3).
   (j) Primers 11 + 12: Forward and reverse primers for the random mutagenesis of the gene encoding for the protein of interest (see Note 4).
3. Individual stock solutions of dNTPs (10 mM), stored at −20°C.

4. Highly competent *E. coli* cells, e.g., NEB10-beta electrocompetent *E. coli,* transformation efficiency: $2–4 \times 10^{10}$ colony forming units (cfu)/μg pUC19 (New England Biolabs, Ipswich, MA, USA).
5. SOC outgrowth medium: 2% Vegetable Peptone, 0.5% Yeast Extract, 10 mM NaCl, 2.5 mM KCl, 10 mM $MgCl_2$, 10 mM $MgSO_4$, 20 mM Glucose.
6. Culture tubes with 5 ml Luria Broth (LB) medium.
7. LB medium: 1% Vegetable Peptone, 0.5% Yeast Extract, 1% NaCl.
8. LB plates containing 34 μg/ml chloramphenicol or 15 μg/ml of tetracycline.
9. Plates with medium of choice with appropriate concentrations of β-lactam antibiotic, e.g., ampicillin, penicillin V (see Note 5).
10. DNA polymerase with proofreading ability, 2 U/μl, e.g., Phusion® High Fidelity DNA Polymerase (New England Biolabs, Ipswich, MA, USA).
11. T4 Polynucleotide kinase, 10 U/μl, e.g., from New England Biolabs (Ipswich, MA, USA).
12. T4 DNA ligase, 400 cohesive end units/μl, e.g., from New England Biolabs (Ipswich, MA, USA).
13. GeneMorph II Random Mutagenesis Kit (Agilent Technologies, Santa Clara, CA, USA).
14. Pfu Turbo, 2.5 U/μl (Agilent Technologies, Santa Clara, CA, USA).
15. Restriction enzymes, e.g., from New England Biolabs (Ipswich, MA, USA) stored at −20°C.
    (a) *Dpn*I, 20 U/μl.
    (b) *Kpn*I, 10 U/μl.
    (c) *Not*I, 10 U/μl.
    (d) *Spe*I, 10 U/μl.
    (e) *Xba*I, 20 U/μl.
    (f) Restriction enzymes needed for cloning the gene encoding the insert protein into the tripartite fusion expression plasmid.
16. DNA gel extraction kit, e.g., QIAquick Gel extraction kit (Qiagen, Valencia, CA, USA).
17. PCR purification kit, e.g., QIAquick PCR purification kit (Qiagen, Valencia, CA, USA).
18. Plasmid DNA extraction kit, e.g., QIAprep Spin Miniprep kit (Qiagen, Valencia, CA, USA).

19. 70% ethanol.
20. 100% ethanol.
21. Phosphate buffered saline (PBS): 1.35 mM NaCl, 2.7 mM KCl, 10 mM Na_2HPO_4, 2 mM KH_2PO_4, adjusted to pH 7.4 with HCl.
22. L-(+)-arabinose.
23. Agarose gels: 1% electrophoresis grade agarose in TAE buffer: 40 mM Tris–acetate, 1 mM ethylene diamine tetraacetic acid (EDTA), 0.5 μg/ml ethidium bromide.
24. Pellet Paint® co-precipitant and supplied 3 M sodium acetate, pH 5.2 (Merck KGaA, Darmstadt, Germany).

2.2. Equipment and Consumables

1. Agarose gel running system.
2. Electroporator, e.g., Electroporator 2510 (Eppendorf, Hauppauge, NY, USA).
3. Incubator, capable of shaking.
4. Microcentrifuge, e.g., Centrifuge 5414 R (Eppendorf, Hauppauge, NY, USA).
5. Thermocycler, e.g., Veriti® Thermal Cycler (Life Technologies Corporation, Carlsbad, CA, USA).
6. Thermomixer, e.g., Thermomixer R (Eppendorf, Hauppauge, NY, USA).
7. UV-Spectrometer, e.g., Genesys 10vis (Thermo Scientific, Waltham, MA, USA).
8. 8- or 12-channel pipette, pipetting range 2–20 μl.
9. 8- or 12-channel pipette, pipetting range 20–200 μl.
10. Sterile electroporation cuvettes, 1 mm gap.
11. Sterile, thin-walled PCR tubes.
12. Sterile 17 × 100 mm round-bottom tubes.
13. Sterile 96-well plate (well volume >250 μl).

3. Methods

Perform all procedures at room temperature unless noted otherwise. When using DNA gel extraction, PCR purification, or plasmid DNA extraction kits, elute DNA from spin column with ddH_2O.

3.1. Construction of a Tripartite Fusion Expression Plasmid (see Note 6)

3.1.1. Construction of a Plasmid for the Expression of the Tripartite Fusion Under Its Native, Constitutive Promoter (pBR322-bla-link)

1. Mix 0.5 μM primer 1, 0.5 μM primer 2, 10 ng of *dam*-methylated pBR322 plasmid DNA, 200 μM of each dNTP, and 1 U Phusion® DNA polymerase in the supplied high fidelity (HF) buffer in a total volume of 50 μl in a thin-walled PCR tube. Keep mixture on ice (see Note 7).
2. Perform PCR in a thermocycler using the following program: 98°C for 5 min (initial denaturation); 25 cycles of: 98°C for 10 s (denaturation), 68°C for 30 s (annealing), 72°C for 3 min (elongation). Following these cycles perform a final elongation at 72°C for 10 min and then hold at 4°C (see Note 8).
3. Run sample of PCR product on an analytical agarose gel to assess yield of the full-length PCR product (see Note 9).
4. Remove the *dam*-methylated template DNA by adding 20 U *Dpn*I. Incubate at 37°C for 2 h. Heat inactivate the restriction enzyme by incubation at 80°C for 20 min.
5. Purify the linear PCR product using a PCR purification kit.
6. Phosphorylate the linear PCR product by combining 0.2–1 μg of the DNA and 10 U of T4 Polynucleotide Kinase in 1× T4 DNA ligase buffer in a total volume of 50 μl. Incubate at 37°C for 30 min. Heat inactivate the enzyme by incubation at 65°C for 20 min.
7. Add 400 cohesive end units T4 DNA ligase. Incubate at 16°C for 12 h. Heat inactivate the enzyme by incubation at 65°C for 10 min.
8. Transform 0.5 μl of ligation reaction into 25 μl of electrocompetent NEB10-beta cells in a chilled electroporation cuvette using the following conditions: 1.8 kV, 200 Omega, and 25 μF. Typical time constants are 4.8–5.1 ms. Immediately after electroporation, add 975 μl of pre-warmed SOC medium to the cuvette and transfer cells into a 17×100 mm round-bottom culture tube. Shake at 250 rpm at 37°C for 1 h (see Note 10).
9. Spread different cell dilutions on pre-warmed LB plates containing the appropriate antibiotic. Incubate plates overnight at 37°C.
10. Select a single colony and isolate plasmid DNA using a plasmid DNA extraction kit.
11. Verify accuracy of the nucleotide sequence by DNA sequencing. The resulting vector is called pBR322-bla-link (see Note 6 for availability of plasmid pBR322-bla-link).

3.1.2. Construction of a Plasmid for the Expression of the Tripartite Under an Arabinose-Inducible Promoter (pMB1-ara-bla-link)

1. For the amplification of the *bla-link* gene from pBR322-bla-link, mix 0.5 μM primer 3, 0.5 μM primer 4, 10 ng pBR322-bla-link plasmid DNA, 200 μM of each dNTP, and 1 U Phusion® DNA polymerase in the supplied high fidelity (HF) buffer in a total volume of 50 μl in a thin-walled PCR tube. Keep mixture on ice (see Note 8).

2. Perform PCR in a thermocycler using the following program: 98°C for 30 s (initial denaturation); 30 cycles of: 98°C for 10 s (denaturation), 69°C for 30 s (annealing), 72°C for 20 s (elongation). Following these cycles perform a final elongation at 72°C for 5 min and then hold at 4°C.
3. Run sample of PCR product on an analytical agarose gel to assess yield of the full-length PCR product (see Note 11).
4. Purify the PCR product using a PCR purification kit.
5. Digest 0.2–1 μg PCR product as well as 0.2–1 μg pBAD33 plasmid with restriction enzymes *Kpn*I and *Xba*I according to the manufacturer's instructions.
6. Heat inactivate the restriction enzymes at 65°C for 20 min. Purify the PCR product and the vector using a PCR purification kit.
7. Mix 50 ng of vector with a threefold molar excess of insert and 400 cohesive end units T4 DNA ligase in T4 DNA ligase buffer. Incubate at 16°C for 12 h. Heat inactivate the enzyme by incubation at 65°C for 10 min.
8. Perform transformation and verification of the vector sequence as described in Subheading 3.1.1, steps 8–11. The resulting vector is called pBAD33-bla-link.
9. Mix 200 μM of each dNTP, 1 U Phusion® DNA polymerase in the supplied high fidelity (HF) buffer in a total volume of 50 μl in a thin-walled PCR tube. Add either 0.5 μM primer 5, 0.5 μM primer 6, and 10 ng of pBR322 (PCR 1) or 0.5 μM primer 7, 0.5 μM primer 8, and 10 ng of pBAD33-bla-link (PCR 2). Keep mixtures on ice.
10. Perform the following PCR in a thermocycler using the following program: 98°C for 30 s (initial denaturation); 30 cycles of: 98°C for 10 s (denaturation), 67°C (PCR 1) or 65.3°C (PCR 2) for 30 s (annealing), 72°C for 1 min (elongation). Following these cycles perform a final elongation at 72°C for 5 min and then hold at 4°C.
11. Purify both full-length PCR products on a preparative agarose gel. Extract DNA from agarose gel using gel DNA gel extraction kit.
12. Digest PCR product 1 with *Not*I and *Xba*I according to the manufacturer's instructions (see Note 12).
13. Digest PCR product 2 with *Spe*I and *Not*I according to the manufacturer's instructions (see Note 12).
14. Heat inactivate the restriction enzymes according to the manufacturer's instructions. Purify the PCR products using PCR purification kit.

15. Mix 100 ng of PCR product 1 and 400 cohesive end units T4 DNA ligase with PCR product 2 in molar ratios of 1:3, 1:1, and 3:1 in supplemented T4 ligase buffer. Incubate at 16°C for 12 h. Heat inactivate the enzyme by incubation at 65°C for 10 min.
16. Perform transformation and verification of the vector sequence as described in Subheading 3.1.1, steps 8–11. The resulting vector is called pMB1-ara-bla-link (see Note 6 for availability of plasmid pMB1-ara-bla-link).

3.1.3. Insertion of the Target Protein Gene into a Tripartite Fusion Expression Plasmid

1. Perform a PCR to amplify the gene of interest with primers carrying appropriate restriction sites using a proofreading polymerase (see Note 3).
2. Clone the gene encoding the target protein into a tripartite fusion expression plasmid of choice, e.g., pMB1-ara-bla-link. Adapt steps 3–8, Subheading 3.1.2, followed by steps 8–11, Subheading 3.1.1. The resulting plasmid is called, e.g., pMB1-ara-bla-link-insert.

3.2. Construction of Expression Libraries

The construction of a library of tripartite fusion plasmids in which only the target gene (but not the *bla* gene or the vector backbone) is mutated is based on the MEGAWHOP technique (23). First, the target gene is amplified in an error-prone PCR, generating a pool of mutated target genes, also termed a megaprimer (23, 24) (see Note 13). This megaprimer is then used to substitute the wild-type version of the target gene in the tripartite fusion expression plasmid, involving a whole-plasmid PCR with a proofreading polymerase.

1. For the random mutagenesis of the gene encoding the target gene, combine 0.1–1,000 ng of target gene, 125 ng of each mutagenesis primer, 200 μM of each dNTP, and 2.5 U Mutazyme II DNA polymerase (from the GeneMorph® II Random Mutagenesis kit) in supplemented Mutazyme II reaction buffer in a total volume of 50 μl in a thin-walled PCR tube. Keep mixture on ice (see Note 4).
2. Perform PCR in a thermocycler using the following program: 95°C for 2 min (initial denaturation); 25–35 cycles of: 95°C for 30 s (denaturation), melting temperature of primers −5°C for 30 s (annealing), 72°C for 1 min for target DNA ≤1 kb or 1 min/kb for target DNA >1 kb (elongation). Following these cycles perform a final elongation at 72°C for 10 min and then hold at 4°C.
3. Purify pool of mutated PCR products on a preparative agarose gel. Extract amplicon band using a DNA gel extraction kit.
4. Combine 200–500 ng of the megaprimer, 50 ng of *dam*-methylated expression plasmid, 200 μM of each dNTP, and

5 U of high-fidelity polymerase Pfu Turbo® in the supplemented buffer in a total volume of 50 μl in a thin-walled PCR tube (see Note 14).

5. Perform PCR in a thermocycler using the following program: 95°C for 3 min (initial denaturation); 25–30 cycles of: 95°C for 2 min 30 s (denaturation), 55°C for 1 min (annealing), 72°C for 2 min/kb (elongation). Following these cycles perform a final elongation at 72°C for 20 min and then hold at 4°C (see Note 15).
6. Remove the *dam*-methylated template DNA by adding 20 U *Dpn*I. Incubate at 37°C for 2 h. Heat inactivate the restriction enzyme by incubation at 80°C for 20 min.
7. To concentrate the undigested PCR product and remove salts, add 2–4 μl Pellet Paint® Co-Precipitant and 0.1 volume of 3 M sodium acetate buffer to the PCR. Mix briefly.
8. Add one volume of isopropanol. Vortex briefly. Incubate at room temperature for 5 min.
9. Spin at 16,000 × *g* for 20 min in a microcentrifuge (see Note 16).
10. Discard supernatant without disturbing the pellet. Add 500 μl of 70% ethanol. Spin at 16,000 × *g* for 5 min in a microcentrifuge.
11. Discard supernatant without disturbing the pellet. Add 500 μl of 100% ethanol. Spin at 16,000 × *g* for 5 min in a microcentrifuge.
12. Remove supernatant completely without disturbing pellet. Dry pellet in thermomixer at 55°C for 5–10 min, leaving the lid of the tube open.
13. Resuspend DNA pellet in 3 μl ddH_2O.
14. Transform 0.5–2 μl of the library into 25–100 μl of highly electrocompetent cells as described in Subheading 3.1.1, step 8 (see Note 17).
15. Spread different various cell dilutions (e.g., 1:1,000, 1:500, 1:100) on pre-warmed LB plates containing an antibacterial agent selecting for the chosen expression plasmid, not the antibiotic degraded by the tripartite fusion (any β-lactam antibiotic). Incubate plates at 37°C for 16 h.
16. Count the number of colonies on each plate and calculate the total library size (see Note 18).
17. Randomly select at least 20, better 40 single colonies, and isolate the plasmid DNA using a plasmid DNA extraction kit.
18. Sequence the target protein encoding sequence and determine the mutagenesis rate per kilobase (see Note 19).

3.3. Selection for Increased Levels of Resistance

1. Transform the library generated above into an *E. coli* strain of choice, using optimized conditions (see Note 18). As a control, transform the expression plasmid that was used as a template for the library construction (containing the wild-type version of the target protein gene).
2. Spread cells on pre-warmed LB plates spanning a range of different concentrations of β-lactam antibiotic (e.g., 0, 500, 1,000, ... 3,000 μg/ml ampicillin). If expressing the tripartite fusion under an arabinose-inducible promoter, include arabinose in selection plates. Incubate plates at 37°C for 16–20 h (see Note 20).
3. Select single colonies from plates containing concentrations of the β-lactam antibiotic that did not permit growth of cell containing the control plasmid. Isolate plasmid DNA using a plasmid DNA extraction kit.
4. Sequence the target protein encoding sequence (see Note 21).
5. Optional: Repeat the mutagenesis, this time using a mixture of mutant plasmids (e.g., different plasmids encoding for various single point mutations in the protein of interest that have been selected for increased resistance) (see Note 22).

3.4. Determining the Level of Antibiotic Resistance Using Spot Titers

3.4.1. Spot Titer Experiment

1. Re-transform mutant DNA isolated in Subheading 3.3, step 3, into fresh competent cells using the same bacterial strain (e.g., commercially available NEB10-beta) that was originally used for the selection (see Note 23).
2. Inoculate 5 ml LB medium with a single colony expressing the mutated tripartite fusion. Cells expressing tripartite fusions containing the wild-type insert protein serve as a control (see Note 24).
3. Grow cells at 37°C until they reach A_{600nm} of about 0.5–0.7.
4. Harvest 1 ml culture by spinning at 16,000 × *g* for 5 min in a microcentrifuge. Discard supernatant. Keep cells on ice.
5. Adjust cells in 1× PBS to $A_{600nm} = 1$. Keep cells on ice.
6. Prepare tenfold dilutions of the cell suspension with 1× PBS (up to 10^{-6}) in a 96-well plate using a multichannel pipette. Keep cells on ice (see Note 25).
7. Using a multichannel pipette, spot 2 μl of each cell dilution onto pre-warmed LB plates containing increasing concentrations of β-lactam antibiotic. If expressing the tripartite fusion under an arabinose-inducible promoter, include arabinose in plates (see Note 20). Wait until spots are dry. Incubate at 37°C for 16–20 h. An example plate is shown in Fig. 3a (see Note 26).

3.4.2. Calculation of the Average Minimal Inhibitory Concentration

1. For each strain and each concentration of β-lactam antibiotic, score cell growth at each cell dilution with "growth" or "no growth" (Fig. 3a) (see Note 27).
2. Plot the maximal cell dilution allowing cell growth versus the antibiotic concentration for each strain (Fig. 3b).
3. For each strain and each cell dilution, assess the smallest concentration of β-lactam antibiotic that inhibits cell growth for this particular dilution, this is the minimal inhibitory concentration or MIC (see Table 1) (see Note 28).
4. For each cell dilution, normalize the MIC for each strain (e.g., expressing the tripartite fusion containing the mutant protein) to the MIC of the reference strain (e.g., expressing the tripartite fusion containing the wild-type (WT) protein) by calculating the value for MIC (mutant)/MIC (WT) (see Table 1) (see Note 29).
5. Average the MIC (mutant)/MIC (WT) ratios for at least three cell dilutions. Standard deviations should be calculated for independent experiments, not for different cell dilutions within one experiment (see Note 30).

Table 1
Example MIC values for three different strains

	MIC						
Cell dilution [10^{-x}]	**6**	**5**	**4**	**3**	**2**	**1**	**0**
Im7 WT	1,400	1,400	1,800	2,600	2,850	3,100	n.d.
Im7 L34A	700	900	1,000	1,400	1,900	3,100	n.d.
Im7 I54V	500	600	600	900	1,400	2,850	n.d.
	MIC_{mut}/MIC_{WT}						
Cell dilution [10^{-x}]	**6**	**5**	**4**	**3**	**2**	**1**	**0**
Im7 WT	1.00	1.00	1.00	1.00	1.00	1.00	n.d.
Im7 L34A	0.50	0.64	0.56	0.54	0.67	1.00	n.d.
Im7 I54V	0.36	0.43	0.33	0.35	0.49	0.92	n.d.

Upper part: Minimal inhibitory concentrations (MIC) for serial dilutions of NEB10β cells expressing tripartite fusions containing Im7 WT, Im7 L34A, or L54A, respectively. *Lower part*: MIC(mutant)/MIC(WT) ratios for serial dilutions of cells expressing tripartite fusions containing Im7 WT, Im7 L34A, or L54A, respectively

4. Notes

1. This plasmid is both ampicillin and tetracycline resistant.
2. This plasmid is chloramphenicol resistant.
3. The primers should be designed to have a similar melting temperature. Appropriate restriction sites (1) are present in the linker-encoding region of the bla-link gene (Fig. 2), (2) are unique within the chosen tripartite fusion expression plasmid, and (3) are not present in the target gene.
4. Mutagenesis primers should be designed to have a similar melting temperature (ranging from 55 to 72°C). The forward primer should cover the 5′ end of the target gene, the reverse primer the 3′ end of the target gene. Since the mutagenesis rate is decreased in the region covered by the primers, short primers are preferred. The amount of template DNA is determined by the amount of actual target DNA (length of the gene encoding for the protein of interest), not by the total amount of DNA added to the reaction in form of the tripartite fusion expression plasmid. The mutation frequency can be increased by lowering the amount of target DNA result and/or by increasing the number of cycles in the PCR program. Note that PCR yields might decrease at target DNA amounts below 0.1 ng. The mutation frequency can further be increased by using a pool of already mutated sequences as a template for the error-prone PCR.
5. Square plates are preferred over round plates, because they allow one to compare more strains on a single plate.
6. The tripartite fusion can be expressed from a variety of different plasmids. For smaller, nontoxic *E. coli* proteins (smaller than about 15 kDa), expression of the tripartite fusion can often be achieved simply by using the constitutive, native β-lactamase promoter (e.g., expression plasmid pBR322-bla-link, see Subheading 3.1.1). For heterologous, larger or more toxic *E. coli* proteins (larger than about 15 kDa), we found it

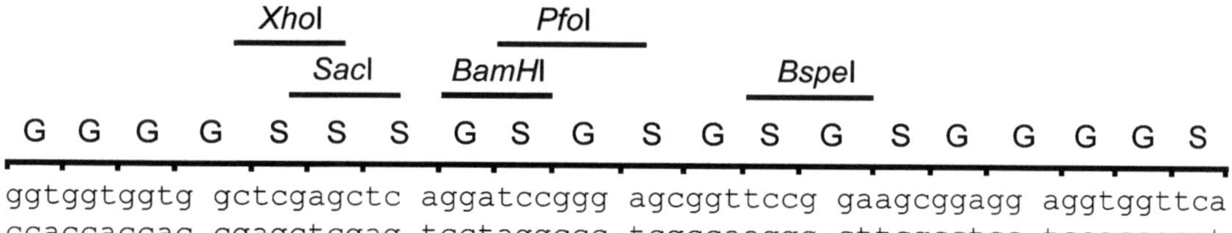

Fig. 2. Glycine-serine linker. Shown is the nucleotide sequence encoding for residues 6–25 of the glycine-serine linker. Restriction sites are indicated. The amino acid sequence of the entire linker is $(GGGGS)_2SSGSGSGSG(GGGGS)_2$.

beneficial to be able to fine-tune expression of the tripartite fusion and thus optimize the basal level of antibiotic resistance by using a regulatable promoter such as the arabinose promoter (e.g., expression plasmid pMB1-ara-bla-link, see Subheading 3.1.2). Choosing a pMB1 origin of replication (about 15–20 plasmid copies per cell, e.g., on pBR322) over lower copy number origins like pSC101 (about 5 copies per cell, e.g., on pBAD43) or p15A (10–12 copies per cells, on, e.g., pBAD33) will increase the basal level of antibiotic resistance conferred by the tripartite fusion and is preferred. Plasmids that already carry the intact β-lactamase (*bla*) gene absolutely need to be avoided since the wild-type *bla* gene will overwhelm any antibiotic resistance encoded by the tripartite fusion. Additionally, plasmids that even contain fragments of the *bla* gene should also be avoided to prevent unwanted recombination with the *bla* portions of the tripartite fusion encoding sequence. The standard length of the linker in the tripartite fusion length is 30 amino acids (aa), but can be adjusted to accommodate the insertion of larger test proteins into β-lactamase. We calculated that for a theoretical, perfectly spherical protein of 50 kDa with N- and C-termini at opposite sites, a 30 aa long linker should be more than long enough to allow interaction of the two β-lactamase fragments and therefore activity. For non-spherical proteins that have their termini far apart, the use of a longer linker, e.g., with 60 residues, is suggested. This however may lead to lower levels of antibiotic resistance.

Both plasmids, pBR322-bla-link and pMB1-ara-bla-link, can be requested from the authors. The description of their derivation is included simply to facilitate the reader to prepare similar constructs that are customized for their own purposes.

7. In this PCR, a linker-encoding sequence is inserted into the *bla* gene present on pBR322. The 5′ end of each primer encodes either for the first 17 residues (primer 1) or for the last 13 residues (primer 2) of the linker. The remaining nucleotide sequences of the two primers are complementary to the regions directly upstream (primer 1) or downstream (primer 2) of the insertion site for the linker-encoding sequence within the *bla* gene. The linker-encoding region contains restriction sites, allowing the insertion of a guest protein into approximately the middle of the linker (Fig. 2).
8. High fidelity (proofreading) DNA polymerases other than Phusion® can be used in this step. In this case, follow the instructions of the manufacturer.
9. If more than one PCR product is observed, gel-purification of the full-length PCR product after step 4 is recommended. In this case, omit step 5 and instead extract and purify the appropriately

sized DNA fragment from a preparative agarose gel using a DNA gel extraction kit.

10. Electrocompetent or chemically competent strains other than NEB10-beta can alternatively be used in this step as long as they are not tetracycline resistant.
11. If more than one PCR product is observed, gel-purification of the full-length PCR product after step 4 is recommended. In this case, omit step 4 and instead purify the appropriately sized DNA fragment from a preparative agarose gel using a DNA gel extraction kit.
12. The restriction enzymes *Xba*I and *Spe*I produce compatible ends.
13. In theory, the mutagenesis of the target gene can also be achieved by alternative mutagenesis techniques such as chemical mutagenesis, random insertion and deletion mutagenesis, random oligonucleotide mutagenesis, and so on. For reviews about mutagenesis methods see ref. 25–28.
14. Estimate megaprimer concentration by measuring absorption at 260 nm or by comparing the intensity of the corresponding band on an ethidium bromide stained agarose gel to the intensity of a band representing a DNA fragment of similar size and known concentration.
15. At this time, a PCR product may or may not be visible on an analytical agarose gel. Proceed either way with step 6.
16. At this point, a DNA pellet should have formed, easily visible due to co-precipitation of the dye. If no pellet can be observed, spin for additional 10 min at $16{,}000 \times g$ in a microcentrifuge and/or add more isopropanol.
17. A variety of different strains can be used for the selection. To ensure large library sizes, however, highly competent cells should be used, e.g., commercially available NEB10-beta electrocompetent cells with a transformation efficiency of $2\text{–}4 \times 10^{10}$ cfu/μg pUC19.
18. Typical library sizes are 10^5 colonies and more for about 100 μl competent cells. If the yield of the synthesized plasmid is unsatisfactory, increase the amount of megaprimer used in the PCR and/or optimize the reaction by trying different ratios of megaprimer: plasmid. The total number of transformed cells can further be optimized by using different DNA amounts for the transformation, more cells and/or simply by performing multiple transformations.
19. It is crucial to determine the mutagenesis rate prior to any selection for increased levels of resistance to β-lactam antibiotics. The desired mutagenesis rate depends on the application. A rate that results in single point mutations can be useful for

initial experiments. Proceed with the selection described in Subheading 3.3 only when satisfied with the mutagenesis rate (see Note 4).

20. We found this genetic selection system to work with a variety of different β-lactam antibiotics (e.g., ampicillin, penicillin V), media (e.g., LB, minimal medium, MacConkey medium), *E. coli* strains (e.g., MG1655, NEB10-beta, BW25113), and incubation temperatures (30–42°C). Choose conditions suitable to address the scientific question of interest. We found that including an additional antibiotic (e.g., for maintenance of the expression plasmid) in the selection medium causes additional stress to the cells and is neither recommended nor necessary. When expressing the tripartite fusion under an arabinose-inducible promoter, it is further advisable to optimize the arabinose concentration used for induction prior any selection experiment. For this, perform spot titer experiments (see Subheading 3.4) using plates supplemented with different concentrations of arabinose (e.g., 0, 0.1, 0.2, 0.5, 0.75, 1, 1.5, and 2%) and one to three fixed concentrations of β-lactam antibiotic that prevent growth for some but not all cell dilutions. The arabinose concentration allowing the highest level of resistance (without causing any cell sickness on plates containing no β-lactam antibiotic) should be chosen for downstream experiments.
21. Although less common, mutations within the reporter protein can occur and could theoretically be selected for if the specific activity of β-lactamase was increased as a result of the mutation. To exclude this possibility, sequencing of the entire fusion protein gene and its promoter region, not only of the gene for the protein of interest, is recommended. We have not observed any mutations on the plasmid that should increase its copy number. However, if this is a concern, the rest of the plasmid can be sequenced as well.
22. This additional PCR step can result in recombination of different single point mutants in form of multiple mutations, as well as in the introduction of additional mutations. The consequence can be even higher levels of antibiotic resistance.
23. This step serves to exclude mutations in the host chromosome that could have occurred sporadically and caused an increased level of resistance.
24. If the overall level of antibiotic resistance is extremely low, a plasmid-free strain should be used as an additional control to monitor the basal level of antibiotic resistance this strain exhibits.
25. Depending on insert protein, strain, incubation temperature, and medium, cell dilutions of 10^{-5} or 10^{-6} might or might not

show cell growth. More important than absolute growth is relative growth compared to the control strain expressing a tripartite fusion containing the wild-type insert protein.

26. Include the control strain (strain expressing tripartite fusion containing the wild-type version of the target protein gene) on every plate.

27. For example, in Fig. 3a depicting a LB plate containing 0.9 mg/ml penicillin V, cells expressing a tripartite fusion containing the insert protein Im7 L34A grow at dilutions of 10^0 to 10^{-4}, but not at dilutions of 10^{-5} or 10^{-6}. The maximal cell dilution allowing growth for this mutant at this concentration of β-lactam antibiotic is therefore 10^{-4}. The corresponding data point in Fig. 3b is marked with an asterisk symbol.

28. For example, in Fig. 3b, inspect the graph representing cells expressing a tripartite fusion containing the insert protein Im7 L34A. Cell dilutions of 10^{-3} show growth on plates containing 0–1.3 mg/ml penicillin V, but not on plates containing ≥1.4 mg/ml penicillin V. The MIC is the smallest concentration of β-lactam antibiotic tested that prevents cell growth (in this example, 1.4 mg/ml is the MIC for cell dilution of 10^{-3}). The smaller the concentration differences between plates, the more precise the calculated MIC will be. As an alternative to using the first concentration tested that prevents cell growth for a given dilution, extrapolation of the MIC value from the graphs in Fig. 3b is generally acceptable, too. For example, we often extrapolated the MIC to be, e.g., 0.1 mg/ml higher than the last concentration that showed cell growth for a particular dilution. In the case of cells expressing a tripartite fusion containing the insert protein Im7 L34A, for instance, cells show growth at dilutions of 10^{-1} up to 2.75 mg/ml. We estimated the MIC to be $2.75 + 0.1 = 2.85$ mg/ml (although the next tested concentration was 3 mg/ml). In our experience, extrapolation of MIC values does not significantly influence the ratio of MIC (mutant)/MIC (WT).

29. MIC (mutant)/MIC (WT) values smaller than one indicate decreased levels of antibiotic resistance compared to WT. MIC (mutant)/MIC (WT) values larger than one indicate increased levels of antibiotic resistance compared to WT.

30. Which cell dilutions show the highest phenotypic reproducibility depends on insert protein, strain, incubation temperature, and medium. We frequently average MIC (mutant)/MIC (WT) ratios for cell dilutions of 10^{-4} to 10^{-2}, 10^{-3} to 10^{-1}, or 10^{-4} to 10^{-1}. Phenotypes for cell dilutions of 10^0, 10^{-5}, and 10^{-6} are in general less reproducible and their MIC values should be excluded from the MIC average if possible.

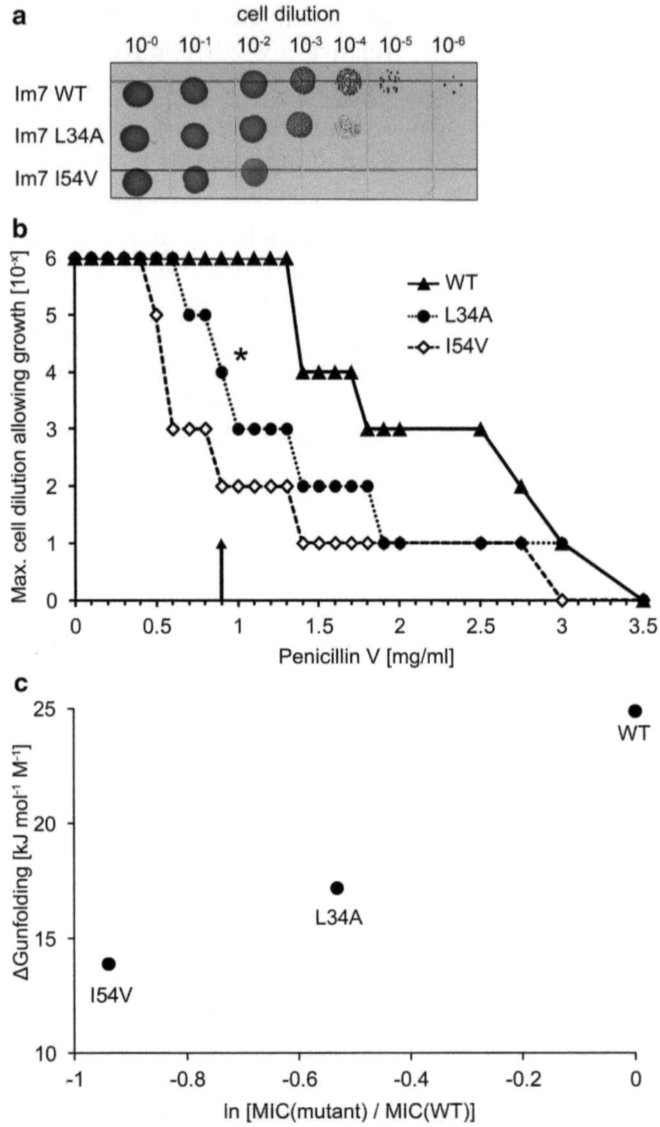

Fig. 3. Spot titer experiments and their analysis. (**a**) Mid-log phase cells of *E. coli* NEB10β expressing tripartite fusions with Im7 WT, Im7 L34A, or Im7 I54V, respectively, were adjusted to $A_{600nm} = 1$ with PBS. Serial cell dilutions of 10^0 to 10^{-6} were spotted on a LB plate containing 0.9 mg/ml penicillin V. (**b**) The maximal cell dilution allowing growth is plotted against the concentration of penicillin V used in LB plates. The *arrow* indicates the penicillin V concentration used in (**a**) (for *, see Note 27). (**c**) The free energy of unfolding ($\Delta G_{unfolding}$) is plotted against the natural logarithm of the average ratio MIC (Im7 mutant)/ MIC (Im7 WT) (28) (see Note 31).

31. At constant temperature in vitro, the free energy of unfolding $\Delta G_{unfolding}$ is directly correlated to the equilibrium constant K of the unfolding reaction through the following equation: $\Delta G_{unfolding}(T) = -R \ln K$. K is defined as the number of molecules occupying the unfolded state divided by the number of

molecules occupying the folded state. In vivo, this ratio is reflected by the steady-state expression level of the tripartite fusion in the periplasm: the more insert proteins occupy the folded state, the higher is the level of the tripartite fusion. Since we found the steady-state expression level to be directly proportional to the level of antibiotic resistance, we plot the natural logarithm of MIC here, analogous to ln K.

References

1. Privalov PL, Khechinashvili NN (1974) A thermodynamic approach to the problem of stabilization of globular protein structure: a calorimetric study. J Mol Biol 86:665–684
2. Giver L, Gershenson A, Freskgard PO, Arnold FH (1998) Directed evolution of a thermostable esterase. Proc Natl Acad Sci U S A 95:12809–12813
3. Taverna DM, Goldstein RA (2002) Why are proteins marginally stable? Proteins 46:105–109
4. Guo HH, Choe J, Loeb LA (2004) Protein tolerance to random amino acid change. Proc Natl Acad Sci U S A 101:9205–9210
5. Soskine M, Tawfik DS (2010) Mutational effects and the evolution of new protein functions. Nat Rev Genet 11:572–582
6. Ghaemmaghami S, Oas TG (2001) Quantitative protein stability measurement in vivo. Nat Struct Biol 8:879–882
7. Ignatova Z, Gierasch LM (2004) Monitoring protein stability and aggregation in vivo by real-time fluorescent labeling. Proc Natl Acad Sci U S A 101:523–528
8. Mayer S, Rudiger S, Ang HC, Joerger AC, Fersht AR (2007) Correlation of levels of folded recombinant p53 in Escherichia coli with thermodynamic stability in vitro. J Mol Biol 372:268–276
9. Barakat NH, Carmody LJ, Love JJ (2007) Exploiting elements of transcriptional machinery to enhance protein stability. J Mol Biol 366:103–116
10. Chautard H, Blas-Galindo E, Menguy T, Grand'Moursel L, Cava F, Berenguer J, Delcourt M (2007) An activity-independent selection system of thermostable protein variants. Nat Methods 4:919–921
11. Waldo GS (2003) Improving protein folding efficiency by directed evolution using the GFP folding reporter. Methods Mol Biol 230:343–359
12. Foit L, Morgan GJ, Kern MJ, Steimer LR, von Hacht AA, Titchmarsh J, Warriner SL, Radford SE, Bardwell JC (2009) Optimizing protein stability in vivo. Mol Cell 36:861–871
13. Delaire M, Lenfant F, Labia R, Masson JM (1991) Site-directed mutagenesis on TEM-1 beta-lactamase: role of Glu166 in catalysis and substrate binding. Protein Eng 4:805–810
14. Hallet B, Sherratt DJ, Hayes F (1997) Pentapeptide scanning mutagenesis: random insertion of a variable five amino acid cassette in a target protein. Nucleic Acids Res 25:1866–1867
15. Zebala J, Barany F (1991) Mapping catalytically important regions of an enzyme using two-codon insertion mutagenesis: a case study correlating beta-lactamase mutants with the three-dimensional structure. Gene 100:51–57
16. Galarneau A, Primeau M, Trudeau L-E, Michnick SW (2002) [beta]-Lactamase protein fragment complementation assays as in vivo and in vitro sensors of protein-protein interactions. Nat Biotechnol 20:619–622
17. Foit L, Mueller-Schickert A, Mamathambika BS, Gleiter S, Klaska CL, Ren G, Bardwell JC (2011) Genetic selection for enhanced folding in vivo targets the Cys14-Cys38 disulfide bond in bovine pancreatic trypsin inhibitor. Antioxid Redox Signal 14:973–984
18. Quan S, Koldewey P, Tapley T, Kirsch N, Ruane KM, Pfizenmaier J, Shi R, Hofmann S, Foit L, Ren G, Jakob U, Xu Z, Cygler M, Bardwell JC (2011) Genetic selection designed to stabilize proteins uncovers a chaperone called Spy. Nat Struct Mol Biol 18:262–269
19. Gleiter S, Bardwell JC (2008) Disulfide bond isomerization in prokaryotes. Biochim Biophys Acta 1783:530–534
20. Watson N (1988) A new revision of the sequence of plasmid pBR322. Gene 70: 399–403
21. Bolivar F, Rodriguez RL, Greene PJ, Betlach MC, Heyneker HL, Boyer HW, Crosa JH, Falkow S (1977) Construction and characterization of new cloning vehicles. II. A multipurpose cloning system. Gene 2:95–113

22. Guzman LM, Belin D, Carson MJ, Beckwith J (1995) Tight regulation, modulation, and high-level expression by vectors containing the arabinose PBAD promoter. J Bacteriol 177: 4121–4130
23. Miyazaki K (2003) Creating random mutagenesis libraries by megaprimer PCR of whole plasmid (MEGAWHOP). Methods Mol Biol 231:23–28
24. Miyazaki K (2011) MEGAWHOP cloning: a method of creating random mutagenesis libraries via megaprimer PCR of whole plasmids. Methods Enzymol 498:399–406
25. Labrou NE (2010) Random mutagenesis methods for in vitro directed enzyme evolution. Curr Protein Pept Sci 11:91–100
26. Rasila TS, Pajunen MI, Savilahti H (2009) Critical evaluation of random mutagenesis by error-prone polymerase chain reaction protocols, Escherichia coli mutator strain, and hydroxylamine treatment. Anal Biochem 388:71–80
27. Wong TS, Roccatano D, Zacharias M, Schwaneberg U (2006) A statistical analysis of random mutagenesis methods used for directed protein evolution. J Mol Biol 355: 858–871
28. Chusacultanachai S, Yuthavong Y (2004) Random mutagenesis strategies for construction of large and diverse clone libraries of mutated DNA fragments. Methods Mol Biol 270:319–334

Chapter 2

Determining Enzyme Kinetics via Isothermal Titration Calorimetry

Neil A. Demarse, Marie C. Killian, Lee D. Hansen, and Colette F. Quinn

Abstract

Isothermal titration calorimetry (ITC) has emerged as a powerful tool for determining the thermodynamic properties of chemical or physical equilibria such as protein–protein, ligand–receptor, and protein–DNA binding interactions. The utility of ITC for determining kinetic information, however, has not been fully recognized. Methods for collecting and analyzing data on enzyme kinetics are discussed here. The step-by-step process of converting the raw heat output rate into the kinetic parameters of the Michaelis–Menten equation is explicitly stated. The hydrolysis of sucrose by invertase is used to demonstrate the capability of the instrument and method.

Key words: Isothermal titration calorimetry, ITC, Enzyme kinetics, Michaelis–Menten kinetics

1. Introduction

Enzymes are biological macromolecules that catalyze the conversion of chemical precursor molecules (substrate) to essential chemical products. When linked in series, enzyme pathways perform numerous critical functions to maintain organismal life (i.e., cell growth, cell differentiation, breakdown of nutrients for energy, energy storage, etc.). When enzyme function is perturbed, serious disease can result. Thus, studying enzyme kinetics and determining the details of an enzyme's activity is a necessary prerequisite for developing novel therapeutics to treat and understand disease. Isothermal titration calorimetry (ITC) is a straightforward and direct method for determining the basic chemical details of an enzyme catalyzed reaction (i.e., V_{max}, K_m, and k_2). The advantages of ITC over other analytical methods are: (1) Substrate(s) do not require labeling or linkage to a secondary-detectable process.

James C. Samuelson (ed.), *Enzyme Engineering: Methods and Protocols*, Methods in Molecular Biology, vol. 978,
DOI 10.1007/978-1-62703-293-3_2, © Springer Science+Business Media New York 2013

(2) ITC is nondestructive to the enzyme. And (3) ITC is compatible with both physiological and synthetic substrates (1, 2). This study uses a standard experimental system (hydrolysis of sucrose by invertase) to illustrate measurement of the kinetics of a reaction by ITC (3). Because processing of enzyme kinetic data collected by ITC is over-simplified in the literature, details of the calculations to convert the heat generated into the kinetic parameters are given (2). Some of the instructions given here are specific to invertase catalyzed hydrolysis of sucrose, our test system, but these can easily be adapted for other enzymes.

2. Materials

2.1. Laboratory Supplies

1. Lint-free laboratory wipes (e.g., KimWipes®).
2. Analytical balance to weigh samples.
3. 12″ stainless steel tweezers.
4. 10 mL graduated cylinder or 10 mL volumetric flask.
5. 100 mL graduated cylinder or 100 mL volumetric flask.
6. 500 mL graduated cylinder or 500 mL volumetric flask.
7. 1 L graduated cylinder or 1 L volumetric flask.
8. 10 μL pipette.
9. 10 μL pipette tips.
10. Waste beaker.

2.2. Instrument Setup Components

1. Nano ITC Low Volume (part number 601000.901) (TA Instruments) (see Notes 1 and 2).
2. Degassing station (part number 6326) (TA Instruments).
3. 500 μL Hamilton filling syringe.
4. Computer running Windows XP or Windows 7.
5. Nano ITCRun data acquisition software (available at TAInstrument.com).
6. Data analysis software (e.g., MATHCAD).

2.3. Instrument Cleaning Components

1. Nano ITC cleaning tool (part number 601028.901) (TA Instruments).
2. Silicone rubber tubing (1/16 inside diameter).
3. 1 L side-arm vacuum flask with a #8 rubber stopper and tubing to connect to a vacuum source.
4. Cleaning solution: 100 mL 5% w/v SDS solution. Weigh 5 g of sodium dodecylsulfate and transfer to 100 mL graduated

cylinder or 100 mL volumetric flask containing 40 mL deionized water. Make up to 100 mL with deionized water. Store the SDS solution at room temperature.

5. Rinsing solution: 1 L deionized water.

2.4. Sample Components and Sample Preparation

1. Sample buffer: 0.1 mM sodium acetate, pH 5.65. Weigh 4.1 g anhydrous sodium acetate (SigmaUltra, ≥99%, Sigma) and transfer to a 1 L volumetric flask or 1 L graduated cylinder. Add 900 mL deionized water. Mix and adjust pH with HCl to 5.65. Make up to 1 L with deionized water. Store at 4°C.
2. Starting enzyme solution: 1.5 mg/mL invertase solution. Weigh 150 mg invertase (β-D-Fructofuranosidase, Grade VII) from *S. cerevisiae* (Sigma) into a small weigh boat and then transfer to a 100 mL graduated cylinder. Add 100 mL of 0.1 mM sample buffer to the 100 mL graduated cylinder. Store at 4°C (see Note 3).
3. Substrate stock solution: 5 mM sucrose. Weigh 854 mg sucrose into a small weigh boat and then transfer to a 500 mL graduated cylinder. Add sample buffer to the 500 mL graduation. Store at 4°C (see Note 3).
4. Experiment enzyme stock solution: 0.0015 mg/mL (5 nM) invertase. Pipette 10 μL of the enzyme stock solution into a 100 mL graduated cylinder that is partially filled with sample buffer. Add sample buffer to the 100 mL graduation. Store at 4°C (see Note 3).

3. Methods

3.1. Preparing the Reference Cell

1. Clean the 500 μL Hamilton filling syringe with degassed and deionized (DG-DI) water by filling and then emptying to a waste container three times.
2. Carefully remove the reference needle from the reference cell with the tweezers.
3. Flush the reference cell three times with at least 300 μL of DG-DI water (see Note 4). Discard the rinse water into a waste container.
4. After flushing the third time, fill the reference cell with 300 μL DG-DI water (see Note 5). Water is used instead of buffer in the reference cell because the difference in heat capacity is not significant for this experiment (see Note 6).
5. Carefully replace the reference needle into the reference cell with the tweezers.

3.2. Preparing the Sample Cell

1. Attach separate lengths of the silicone tubing (1/16 inside diameter) to the inlet and outlet of the Nano ITC cleaning tool, and then position the stem of cleaning tool so that it is inside the instrument sample cell.
2. Connect one length of silicone tubing to a vacuum source via a side-arm flask. The other length of tubing will go to the 5% SDS cleaning solution.
3. Turn on the vacuum source and the cleaning solution will automatically flush through the sample cell. Flush 50 mL to 100 mL of 5% SDS solution.
4. After cleaning solution is flushed through, rinse the cell by flushing with 0.5–1 L deionized water (see Note 7).
5. When cleaning is complete, remove the cleaning tool from the ITC.
6. Flush the Hamilton filling syringe three times with the substrate stock solution.
7. Flush the sample cell three times with the substrate stock solution (see Note 8).
8. After flushing the third time, fill the Hamilton filling syringe with 300 μL, and then fill the sample cell (see Note 9).

3.3. Preparing the Titration Syringe

1. Rinse the 50 μL titration syringe with the sample buffer by filling and then draining the titrant out the back of the syringe. Repeat three times (see Note 10).
2. After rinsing the third time, fill the titration syringe with enzyme solution, remove the plunger, and allow enzyme solution to flow to the open end of the syringe (see Note 11).
3. When the titrant is at the open end of the syringe, insert the plunger (see Note 12).
4. Depress the plunger until some titrant emerges from the tip of the syringe, and then continue to fill the titration syringe to the 45 μL graduation with enzyme solution.
5. Load an additional 5 μL of sample buffer into the syringe until the plunger is at the 50 μL graduation. This "buffer plug" at the syringe tip prevents diffusion of enzyme into the substrate solution prior to titration, which would cause premature initiation of the reaction.

3.4. Final Instrument Preparation

1. Thread the titration syringe into the buret handle.
2. Gently wipe the syringe tip with a lint-free laboratory wipe.
3. Install the buret handle into the calorimeter and ensure that the buret handle is secure.

3.5. ITC Experiment Parameter Setup

1. Setup experiment parameters in ITCRun, so a single injection of 8 μL (5 μL sample buffer and 3 μL invertase) is delivered. Allow the reaction to run for 7,000 s (see Note 13). Refer to the Nano ITC Run Getting Started Guide manual for specific instructions on software use.
2. Enter sample concentrations for both the enzyme (syringe sample) and substrate (cell sample).
3. Set the stirring rate to 350 rpm and set a 300 s initial baseline.
4. After the calorimetric signal has equilibrated, start data collection (see Note 14).
5. After the 300 s initial baseline is completed, the calorimeter will make the injection (8 μL) of the enzyme solution into the substrate solution. The reaction then proceeds until complete (Fig. 1).
6. Initially, a large increase in heat is observed due to catalysis. This could be preceded by a large spike caused by heat of dilution (see Note 15). As the reaction continues the total heat of catalysis decreases due to a decrease in available substrate, and

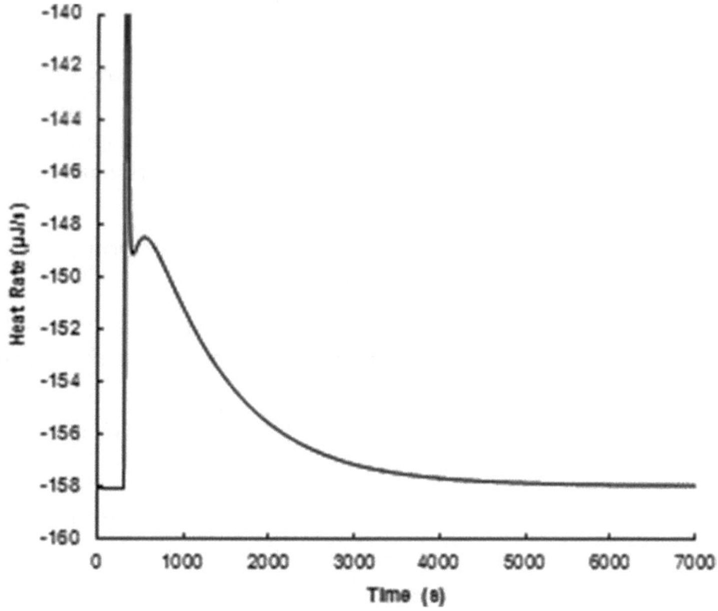

Fig. 1. Raw data for a single injection of invertase (grade VII from baker's yeast, Sigma) into 4.997 mM sucrose in pH 5.65, 0.1 M acetate buffer in a Nano ITC-LV at 25°C. The stirring speed was 350 rpm, and the calorimeter was auto-equilibrated to a slope less than 0.1 μW/h and a peak-to-peak standard deviation less than 10 nW. The initial large peak immediately following the intial baseline period corresponds to the heat of dilution from the injection of enzyme solution into sucrose.

eventually the reaction returns to baseline when all substrate is consumed and the reaction is complete.

7. After the experiment is complete, the instrument should automatically stop stirring and save the data.

3.6. Data Analysis

1. Export the data to a data analysis software package. For the following, analysis was performed in Mathcad.
2. Choose a baseline value (Φ_B). This can be done by taking the average of several points near the end of the run.
3. Calculate the concentration of sucrose ([S]) in the sample cell,

$$[S] = \frac{[S]_{stock}(V_{cell} - V_{inj})}{V_{cell}} \tag{1}$$

where $[S]_{stock}$ is the concentration of sucrose in the stock solution, V_{cell} is the volume of the sample cell, and V_{inj} is the volume injected (see Note 16). Similarly calculate the concentration of the enzyme ([E]) in the sample cell,

$$[E] = \frac{[E]_{stock}(V_{enz})}{V_{cell}} \tag{2}$$

where $[E]_{stock}$ is the concentration of the experimental enzyme stock solution and V_{enz} is the volume of enzyme solution injected into the sample cell (see Note 17).

4. Adjust the time values in the data array so that injection occurs at time = 0 s (see Note 18). The data array consists of time values, t_0, t_1, t_2, ...tN, corresponding to heat rate values Φ_0, Φ_1, Φ_2,...ΦN, where N is the number of data points in the data array.
5. Select the starting data point for the analysis (after the initial injection spike).
6. The object of the data analysis is to find the values of k_2, K_M, $\Delta_R H$, and τ that minimize the following expression.

$$\sum_{j=1}^{M} \left(\Phi_M(t_j) - \Phi_B - \Phi_C(t_j, k_2, K_M, \Delta_R H, \tau)\right)^2 \tag{3}$$

where $\Phi_M(ti)$ is the measured heat rate at time ti, $\Phi_C(ti)$ is the calculated heat rate as a function of k_2 (catalytic rate constant), K_M (Michaelis constant), $\Delta_R H$ (reaction enthalpy), τ (instrument time constant), and Φ_B (the baseline value chosen in part 1 of Data Analysis). The calculated heat rate includes the time delay of the instrument (4).

7. The expression for $\Phi_C(ti)$ for a reaction with Michaelis–Menten kinetics cannot be calculated explicitly. Therefore, it is calculated as follows:

(a) Generate a sequence of equally spaced discrete values for the sucrose concentration, starting with the initial concentration and ending with zero ($S_0 > S_1 > S_2 > \ldots > S_M = 0$) with $M \geq N$.

(b) Use the following equation to calculate a time (t_{calc}) value corresponding to each of the sucrose concentrations, Si in the sequential array.

$$t_{calc}(S) = -\frac{1}{k_2 E}\left[(S - S_0) + K_M \ln\left(\frac{S}{S_0}\right)\right] \quad (4)$$

(c) Using the calculated values for t and S, use a linear interpolation to estimate values of S for each time, t_0, t_1, $t_2 \ldots tN$.

(d) Using the calculated values $S(ti)$, calculate $\Phi_R(ti)$ using the following equation.

$$\Phi_R(t_i) = \Delta_R H V k_2 E \frac{S(t_i)}{K_M + S(t_i)} \quad (5)$$

(e) Use $\Phi_R(ti)$ to calculate the estimated observed heat rate, $\Phi_C(ti)$, according to the following equation.

$$\Phi_C(t_i) = \frac{1}{\tau} e^{-\frac{t_i}{\tau}} \int_0^{t_i} e^{\frac{S}{\tau}} \Phi_R(S)\, dS \quad (6)$$

Use the Nelder–Mead simplex direct search algorithm (5) or another nonlinear minimization routine to find the values extual of k_2, K_M, $\Delta_R H$, and τ that minimize Eq. 3. The initial values for k_2, K_M, and τ should be chosen to be close to the expected values of the parameters. The initial value of $\Delta_R H$ can be estimated by integrating the area under the curve $\Phi_M(ti) - \Phi_B$ using trapezoidal integration and dividing the area by the number of moles of sucrose in the reaction cell (V_{cell} [S]).

8. The k_2 and K_M parameters are those defined by the equation for Michaelis–Menten kinetics, $d[S]/dt = -k_2[E] \times [S]/K_M + [S]$ where [S] is the concentration of substrate, k_2 is the rate constant, and K_M is the equilibrium constant for binding substrate to the enzyme. $\Delta_R H$ is the enthalpy change for the catalyzed reaction, hydrolysis of sucrose in this case. τ is the time constant of the calorimeter.

9. The estimated results from the fit of the data are $K_M = 16.3$ mM, $k_2 = 2.21 \times 10^5$ s^{-1}, $\Delta_R H = -13.4$ kJ/mol, and $\tau = 14.5$ s. The fit of the results to the raw data is shown in Fig. 2.

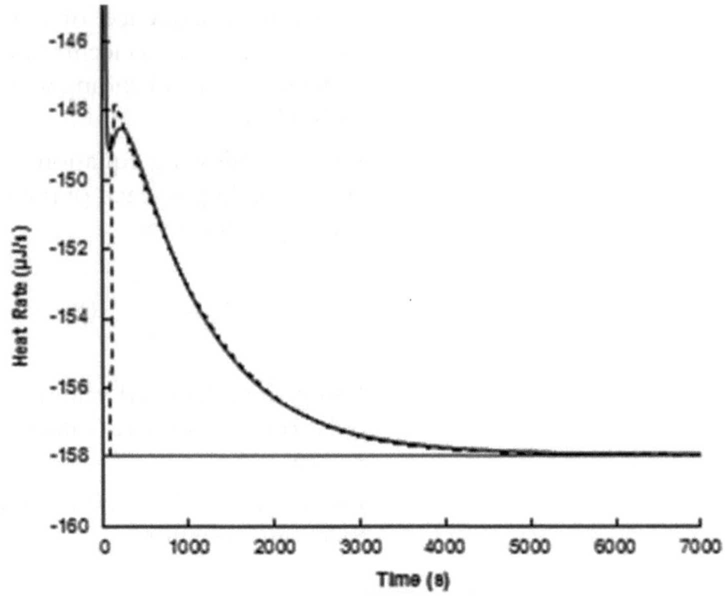

Fig. 2. Raw data (*solid, black line*), fit to the Michaelis–Menten model (*dashed, black line*), and baseline (*solid, gray line*) for hydrolysis of sucrose by invertase.

4. Notes

1. The enzyme kinetics experiments described in this chapter could be performed on other commercially available differential, power compensation isothermal titration calorimeters, not only those supplied by TA Instruments.
2. A Nano ITC Standard Volume could also be used for this test, but it is important to adjust the sample concentrations accordingly (3).
3. Obtaining the exact concentration stated in the directions is not as critically important as knowing the exact concentration of the sample that was prepared. For example, directions to formulate a 5 mM solution of sucrose are provided, but the actual concentration achieved was 4.997 mM sucrose solution. Uncertainty in concentration could lead to increased error in results. Thus, knowing the precise concentration is essential.
4. To properly fill the reference cell, first you must carefully insert the needle into the fill tube and then carefully guide the syringe needle to the bottom of the cell. By filling from the bottom of the cell, all air will be displaced by the reference material and no air bubble will form. Do not forcefully put the syringe in the cell as this could damage the cell and significantly reduce instrument performance.

5. The suggested fill volume for the Nano ITC Low Volume is 300–700 μL.
6. For samples that contain concentrations of organic solvent >20%, it is suggested that this same concentration is added in the reference cell.
7. Other dilute cleaning solutions (i.e., Contrad60 or Micro90) may also be used. To confirm the instrument is clean, perform a titration experiment of water titrated into water at 25°C (3). For additional details on cleaning the ITC, reference the Nano ITC Getting Started Guide that is available at TAInstruments.com.
8. Flushing the sample cell with the substrate stock solution will ensure that sample cell is properly conditioned. When substrate solution is limited, rinse the cell with the buffer that the substrate solution was made. Also, when rinsing and filling the sample cell be very careful not to apply any force onto cell surface with the filling syringe needle. Excessive force to the sample cell could reduce instrument performance.
9. All commercially available differential, power compensation isothermal titration calorimeters utilize an overfill design. For example, the active cell volume on the Nano ITC Low Volume (TA Instruments) is 190 μL, but 300 μL is used for filling. Overfilling improves data quality by minimizing the degree to which evaporation is measured by the instrument, and it also allows for improved temperature equilibration of the titrant.
10. When an air bubble(s) is present in the titration syringe during cleaning, the liquid may not drain easily. In this case lightly tapping, or applying a vacuum, to the butt of the glass–syringe barrel will aid the flow of liquid.
11. Ensure there is no bubble, or air gap, between the titrant solution and the plunger tip. If a bubble or gap is present, the injection volume may not be accurate due to compression of the air bubble.
12. It is very important that contact with the titration syringe needle is minimized in order prevent inadvertent bending. When the titration syringe needle is bent it could rub against the inside of the instrument when stirring is activated. This will create heat and increase drift in the baseline.
13. The recommended injection interval for this particular experiment is 7,000 s, but optimization to shorter or longer times is possible by adjusting concentrations of enzyme and substrate accordingly. The experiment is complete when no additional heat is produced from the reaction (e.g., the signal has come back to baseline).

14. Commercially available differential, power compensation isothermal titration calorimeters are now equipped with software that will allow for autoequilibration of the calorimetric signal. Refer to the respective instrumentation software manual for detailed information on setting autoequilibration settings.
15. The initial peak following the injection represents heat of dilution. This can be minimized or eliminated completely by matching samples precisely. This is most commonly done by dialysis (6). Dialysate should be saved for rinsing instrument cell and syringe and to adjust sample concentrations if necessary.
16. The following conditions were used to calculate [S] in Eq. 1.

    $[S]_{stock} = 4.997$ mM

    $V_{cell} = 190$ μL

    $V_{inj} = 8$ μL
17. The following conditions were used to calculate [E] in Eq. 2.

    $[E]_{stock} = 5.6$ nM

    $V_{enz} = 3$ μL

    $V_{cell} = 190$ μL
18. All times were adjusted so that the maximum of the injection peak occurred at time = 0, $t_0 = 84$ s. The first 420 data points were truncated, which included the 300 s initial baseline.

References

1. Todd MJ, Gomez J (2001) Enzyme kinetics determined using calorimetry: a general assay for enzyme activity? Anal Biochem 296: 179–187
2. Olsen SN (2006) Applications of isothermal titration calorimetry to measure enzyme kinetics and activity in complex solution. Thermochimica Acta 448:12–18
3. Demarse NA, Quinn CF, Eggett DF, Russell DJ, Hansen LD (2011) Calibration of Nanowatt isothermal titration calorimeters with overflow reaction vessels. Anal Biochem 417:247–255
4. Hansen CW, Hansen LD, Nicholson AD, Chilton MC, Thomas N, Clark J, Hansen JC (2011) Correction for instrument time constant and baseline in determination of chemical kinetics. Int J Chem Kinetics 43(2):53–61
5. http://en.wikipedia.org/wiki/Nelder%E2%80%93Mead_method. Accessed 4 Apr 2012.
6. McPhie P (1971) Dialysis. Meth Enzymol 22:23–33

Chapter 3

GFP Reporter Screens for the Engineering of Amino Acid Degrading Enzymes from Libraries Expressed in Bacteria

Olga Paley, Giulia Agnello, Jason Cantor, Tae Hyun Yoo, George Georgiou, and Everett Stone

Abstract

There is significant interest in engineering human amino acid degrading enzymes as non-immunogenic chemotherapeutic agents. We describe a high-throughput fluorescence activated cell sorting (FACS) assay for detecting the catalytic activity of amino acid degrading enzymes in bacteria, at the single cell level. This assay relies on coupling the synthesis of the GFP reporter to the catalytic activity of the desired amino acid degrading enzyme in an appropriate *E. coli* genetic background. The method described here allows facile screening of much larger libraries (10^6–10^7) than was previously possible. We demonstrate the application of this technique in the screening of libraries of bacterial and human asparaginases and also for the catalytic optimization of an engineered human methionine gamma lyase.

Key words: Amino acid auxotrophs, GFP reporter, Enzyme engineering

1. Introduction

The screening of large libraries of enzyme variants for improved or altered catalytic function generally presents a daunting challenge. Unlike the screening of protein libraries for ligand binding, which can be accomplished using the same assay format regardless of the identity of the ligand, enzymes catalyze a plethora of reactions which in turn necessitates the use of specific individualized activity assays. Many, if not most, enzyme-catalyzed reactions are not easily amenable to high-throughput assays. This in turn limits the size of the sequence space that can be practically sampled in directed evolution experiments (1).

James C. Samuelson (ed.), *Enzyme Engineering: Methods and Protocols*, Methods in Molecular Biology, vol. 978, DOI 10.1007/978-1-62703-293-3_3, © Springer Science+Business Media New York 2013

Many tumors are known to exhibit defects either in amino acid biosynthesis or alternatively to have a very high demand for certain amino acids relative to non-malignant tissues. Additionally tumor cells exhibit deregulation of stress responses that normally protect cells from the effects of nutrient starvation (2). The use of enzymes as a means for systemically depleting particular amino acids to restrict tumor growth with minimal toxicities has been under investigation for over 50 years and has resulted in significant clinical successes. Because of the paucity of human enzymes that display sufficient catalytic and pharmacological properties (e.g., serum stability and distribution, catalytic selectivity, ease of expression in heterologous hosts) until very recently amino acid depletion therapies in cancer have relied almost exclusively on the use of heterologous enzymes. However, heterologous enzymes are immunogenic leading to the elicitation of antibody responses that at best result in the neutralization of enzyme therapeutic function and at the worst, induce hypersensitivity and anaphylactic shock.

In recent years we have demonstrated the use of protein engineering strategies to develop human enzymes with the requisite properties for cancer therapy (3–6). Bacterial L-asparaginases so far constitute the only approved enzymes for amino acid depletion therapy and have been shown to dramatically enhance the response rate to combination chemotherapy in childhood acute lymphoblastic leukemia (7). However, immunogenicity of the bacterial enzymes is still an issue for a considerable fraction of patients that display hypersensitivity and for the treatment of patients that relapse, underscoring the need for a non-immunogenic human enzyme alternative. The enzymatic depletion of L-Methionine has been investigated for years but the bacterial enzyme is highly immunogenic in primates (8). As with L-asparaginase the engineering of a human L-methionine-γ-lyase (MGL) is of significant therapeutic interest, yet no human enzymes with these catalytic activities exist.

High-throughput screens for L-asparaginase and MGL activities can greatly aid the engineering of therapeutic enzymes by directed evolution of human enzymes that catalyze similar reactions. The development of such screens necessitates the engineering of unique *E. coli* strains that are auxotrophic for either L-aspartate (L-Asp) or α-ketobutyrate (a metabolite required for the synthesis of L-isoleucine (L-Ile). These auxotrophies can then be rescued by the enzymatic hydrolysis of L-Asn to L-Asp or degradation of L-Met to produce α-ketobutyrate, respectively. In turn, these reactions can complement the auxotrophic strain enabling the synthesis of GFP. Thus a very active enzyme allows GFP to be translated at levels commensurate with nonselective conditions, whereas a less active or inactive enzyme results in reduced GFP expression due to the depleted amino acid pool in the auxotroph. Fluorescence activated cell sorting (FACS) can then be used to isolate individual clones based on their fluorescence profile from large libraries of enzyme variants expressed in *E. coli*. Employing several rounds of FACS sorting we can attain thousands of fold enrichment of clones with desired activity and *then* use a secondary assay method to analyze a handful of clones for the rank ordering of enzyme variants and assessment of kinetics. Figure 1 shows the general scheme and workflow of the method.

A high-throughput FACS screen for the activity of *E. coli* L-Asparaginase was reported recently (9). Here we demonstrate that this assay can be applied to the directed evolution of the human asparaginase-like protein 1 (hASRGL1), an enzyme that has β-aspartyl peptidase activity but also exhibits a low level of L-Asparaginase activity (10). The activity of hASRGL1 for L-Asn hydrolysis is about three orders of magnitude lower than the *E. coli* L-Asparaginase II enzyme approved for therapy. Briefly, *E. coli* BL21(DE3) [*ΔaspCΔtyrBΔansAΔansBΔiaaA*] cells auxotrophic for L-Asn and stably transformed with a plasmid expressing GFP under control of an arabinose-inducible promoter are transformed with either parental hASRGL-1 or a mutagenized library of hASRGL-1 variants. Cells are grown and allowed to express hASRGL-1 or hASRGL-1 variants under nonselective conditions. The cells are then shifted to selective conditions concomitant with induction of GFP synthesis. Thus intracellular fluorescence is dependent on the availability of L-Asp, which in turn is proportional to the activity of the enzyme variant hydrolyzing L-Asn.

As a second example, we describe the application of our method for screening MGL activity. The non-mammalian enzyme MGL catalyses the α,γ-elimination of L-methionine to methanethiol, α-ketobutyrate, and ammonia. MGL from *P. putida* has shown efficacy in controlling tumor growth in a variety of xenograft models (11–18) but unfortunately has proven to be highly immunogenic in primates (8). Recently we employed rational design and scanning saturation mutagenesis to convert the structurally related human enzyme cystathionine-γ-lyase (CGL) into a human L-Methionine γ-lyase (MGL) that catalyzes the α,γ-elimination of L-Methionine to methanethiol, α-ketobutyrate, and ammonia but does not accept L-Met as a substrate. The assay available to screen this class of enzymes is a 96-well plate colorimetric assay for α-ketobutyrate production. *E. coli* produces α-ketobutyrate as an intermediate metabolite in the de novo production of L-isoleucine (L-Ile) by action of threonine deaminase (*ilvA*). We thus deleted the *ilvA* gene to create an L-Ile auxotroph that could be rescued by α-ketobutyrate produced from L-Met degradation by MGL (Fig. 2a). Additionally, because CGL can also produce α-ketobutyrate from L-cystathionine and L-homocysteine which are metabolites in the *E. coli* pathway for L-Met production, we also deleted the *metA* gene (Fig. 2b).

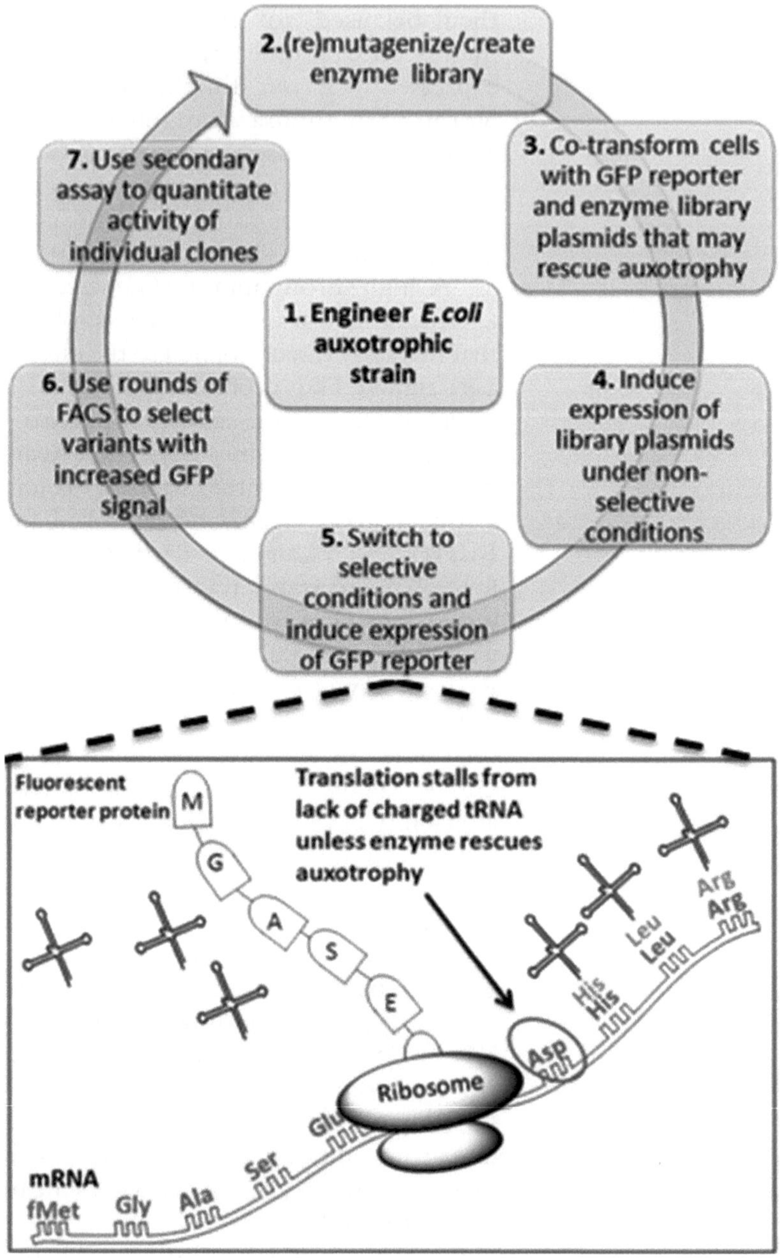

Fig. 1. (1) Engineer *E. coli* strain that can no longer make a metabolic intermediate needed for growth. (2) Create mutagenized libraries of enzymes that have the potential to make the missing metabolic intermediate. (3) Co-transform auxotrophic strain with library and GFP plasmids that have mutually exclusive promoters. (4) Culture under nonselective conditions and induce expression of enzyme plasmids. (5) Shift culture to selective conditions and induce expression of GFP. If an enzyme variant can rescue the auxotrophy, then the cell will express greater levels of GFP. (6) Use rounds of FACS sorting to select cells with high GFP fluorescence. (7) Use secondary screen/assay to rank order clones and if necessary repeat mutagenesis on variants with increased activity.

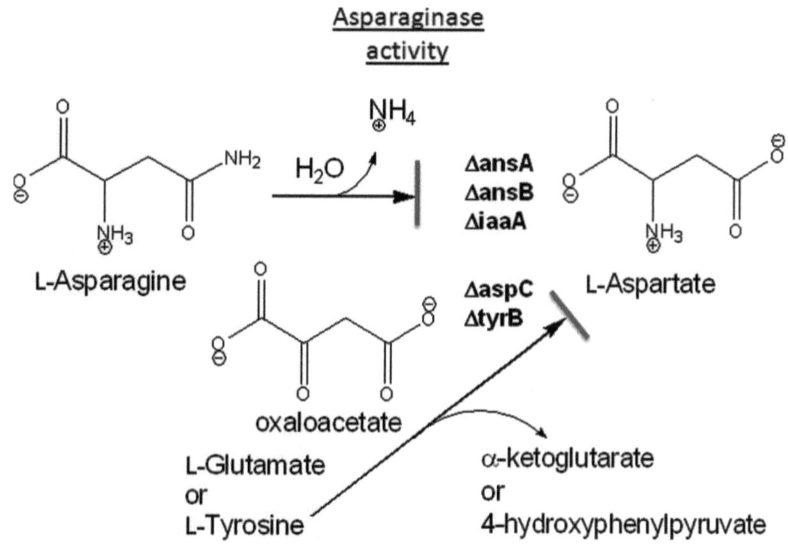

Fig. 2. Summary of genes deleted in *E. coli* to create L-aspartic acid auxotroph strain that can be rescued by recombinant asparaginase activity.

2. Materials

2.1. Asparaginase Screen

Unless otherwise stated *E. coli* plates and cultures are incubated at 37°C.

1. *E. coli* strain BL21(DE3) *ΔaspCΔtyrBΔansAΔansBΔiaaA*. This strain has deletions of the three native asparaginases and two aminotransferases that can produce L-Asp which for the sake of brevity we have named *E. coli D-aux* to reflect L-Asp auxotrophy (Fig. 3).
2. *Competent cells: E. coli D-aux* harboring pBAD-GFP prepared by an electrocompetent method (19).
3. Arabinose-inducible pBAD-GFP construct with a chloramphenicol marker as a reporter plasmid.
4. Isopropyl-β-D-1-thiogalactopyranoside (IPTG) inducible pET28a with a kanamycin marker as an expression plasmid for parental hASRGL-1 and libraries.
5. *Nonselective medium*: Medium supplemented with all amino acids; 1× M9 media supplemented with 0.4% glucose, 3.5 μg/mL thiamine, 1 mM $MgSO_4$, 0.1 mM $CaCl_2$, 160 μg/mL of the amino acid L-Asp, 80 μg/mL of the remaining 19 amino acids, 50 μg/mL kanamycin, and 34 μg/mL chloramphenicol. The nonselective medium can be supplemented with IPTG (see Note 1).

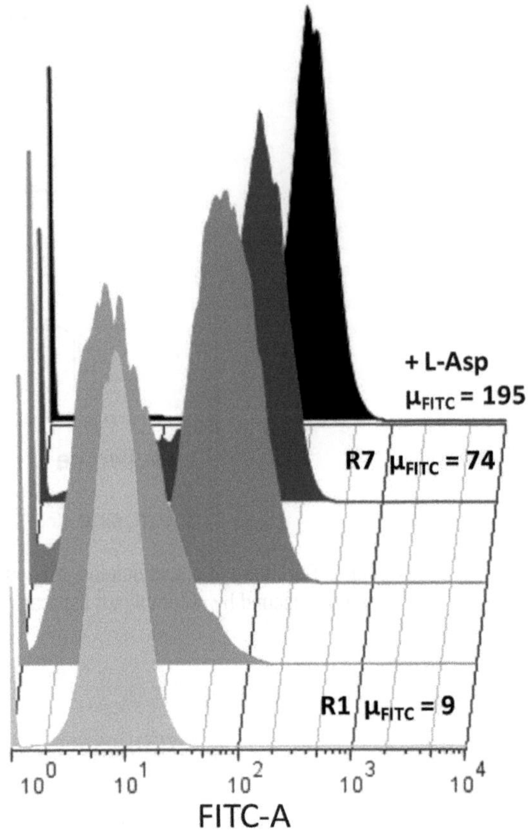

Fig. 3. Flow cytometric analysis of *E. coli D-aux* strain expressing an error-prone library of hASRGL-1 variants. Representative fluorescence histograms are shown from round 1 (R1) of sorting to round 7 (R7), along with a positive control containing L-Asp (see Note 4). Mean of the fluorescence histograms is reported in FACS diagram.

6. *Selective medium*: Medium deprived of L-Aspartate; 1× M9 supplemented with 2% arabinose, 1% glycerol, 3.5 μg/mL thiamine, 1 mM $MgSO_4$, 0.1 mM $CaCl_2$, no L-Asp, 5 μg/mL L-Asn, 80 μg/mL of the remaining 18 amino acids, 50 μg/mL kanamycin, and 34 μg/mL chloramphenicol.
7. *General medium*: 2xYT supplemented with 50 μg/mL kanamycin and 34 μg/mL chloramphenicol.
8. *Plates*: 2xYT agar supplemented with 50 μg/mL kanamycin and 34 μg/mL chloramphenicol.
9. Filter-sterilized 0.9% NaCl (4°C).
10. Filter-sterilized 1× phosphate buffered saline at 4°C (PBS).
11. Filter-sterilized solution of 4 g/L L-Asp in water.
12. Filter-sterilized solution (1 M) of IPTG.

13. Culture tubes (5 mL).
14. Falcon tubes (15 mL).
15. For flow cytometric analyses, a FACSAria instrument (BD Biosciences) with a 488-nm solid state laser for the excitation of GFP. A 530/30 band pass filter is used for the detection.
16. FACS tubes (BD Biosciences).

2.2. Methionine-γ-Lyase Screen

1. *E. coli* strain BL21(DE3) Δ*ilvA*Δ*metA*. For the sake of brevity this strain was named *E. coli MI-aux* to reflect its dual L-Met and L-Ile auxotrophies (Fig. 2).
2. *Competent cells*: *E. coli MI-aux* harboring pBAD-GFP prepared by an electrocompetent method (19).
3. Arabinose-inducible pBAD-GFP with a chloramphenicol acetyltransferase marker as a reporter plasmid.
4. IPTG-inducible pET28a with a kanamycin resistance marker as an expression plasmid for parental hCGL and pMGL.
5. *Nonselective medium*: Enzyme expression medium; M9 minimal medium supplemented with 0.4% glucose, 1 mM $MgSO_4$, 0.1 mM $CaCl_2$, 150 μg/mL of the amino acids L-Met and L-Ile, 75 μg/mL of the remaining amino acids excluding L-Leu and L-Val (see Note 6), 50 μg/mL kanamycin, and 34 μg/mL chloramphenicol.
6. *Selective medium*: GFP Expression Medium: M9 medium supplemented with 0.4% glucose, 1 mM $MgSO_4$, 0.1 mM $CaCl_2$, 150 μg/mL of L-Met, 75 μg/mL of the remaining amino acids (excluding L-Ile, L-Leu and L-Val), 50 μg/mL kanamycin, 34 μg/mL chloramphenicol, and 2% arabinose.
7. *General medium*: LB supplemented with 50 μg/mL kanamycin and 34 μg/mL chloramphenicol.
8. *Plates*: 2xYT agar supplemented with 50 μg/mL kanamycin and 34 μg/mL chloramphenicol.
9. Filter-sterilized 0.9% NaCl (4°C).
10. Filter-sterilized 1× PBS at 4°C.
11. Filter-sterilized solution of L-Ile (250 mM) in PBS.
12. Filter-sterilized solution (1 M) of IPTG.
13. Culture tubes (5 mL).
14. Falcon tubes (15 mL).
15. A FACSAria instrument (BD Biosciences) with a 488-nm solid state laser for the excitation of GFP. A 530/30 band pass filter was used for the detection.
16. FACS tubes (BD Biosciences).

3. Methods

3.1. Asparaginase Screen

Use your laboratories' favorite methods for library construction, competent cells, transformations, etc.

3.1.1. Cell Growth and GFP Expression

1. Grow an overnight culture (5 mL) of *E. coli D-aux* cells co-transformed with pBAD-GFP and either a library or a single mutant in 2xYT (see Note 2).
2. Subculture the overnight culture into 10 mL nonselective medium supplemented with all the amino acids to a final OD_{600} of 0.1. (To perform the FACS analysis of one sample (i.e., a round of a library or a single mutant) 10 mL of nonselective medium is sufficient).
3. Grow cells at 37°C with shaking at 250 rpm until OD_{600} ~0.9–1.1.
4. Harvest cells by centrifugation of 1 mL aliquots of culture in 1.5 Eppendorf tubes at 13,000 × *g* for 5 min at 4°C. For each sample spin down at least 3 mL for subsequent positive control (step 8).
5. Re-suspend the pellets in cold 0.9% NaCl (by vortexing at low speed) and centrifuge at 13,000 × *g* for 5 min at 4°C. Use 1 mL of 0.9% NaCl per mL of culture harvested.
6. Repeat the wash with cold 0.9% NaCl (step 5) another three times (see Note 3).
7. Media shift: re-suspend cell pellets into selective media deprived of L-Aspartate. Use 1 mL of medium per tube.
8. Transfer culture into two separate 5 mL culture tubes (1 mL per tube): one sample tube and one positive control tube. In the positive control tube add L-aspartate to a final concentration of 160 μg/mL.
9. Shake the sample and positive control tubes at 250 rpm and 37°C for 2 h to induce the expression of GFP (see Note 4).
10. Harvest the sample and positive control cultures by centrifugation at 13,000 × *g* for 5 min at 4°C.
11. Re-suspend the sample and positive control pellets in cold PBS (by vortexing at low speed) (use 1 mL of PBS per tube) and centrifuge at 13,000 × *g* for 5 min at 4°C.
12. Repeat the wash with 1 mL cold PBS (step 11) another two times (see Note 5).
13. Re-suspend cell pellets in 1 mL cold PBS then dilute the positive control and the sample by 50–100-fold for flow cytometric analysis and cell sorting.

3.1.2. Cytometric Analysis and Cell Sorting

FACS screening. The instructions reported here describe the procedure for a single sample (either a round of a library or a single mutant). For multiple samples adjust the volumes reported accordingly to the number of samples to analyze.

1. Adjust the throughput rate of cells to 4,000–5,000 events per second. Set a gate in the fluorescence channel to recover the 2% most highly fluorescent cells. Select the purity mode for the initial sort of each library, and then switch to the single cell mode and sort ~10^7 cells for each round.
2. Collect each round of sorting in 0.5 mL 2xYT medium and then spread onto 2xYT agar plates supplemented with 50 μg/mL kanamycin and 34 μg/mL chloramphenicol.
3. After overnight growth at 30°C, pool the clones and store them in 15% glycerol aliquots at −80°C.

3.1.3. Secondary Screening and Analysis

1. Following the final round of sorting use the colorimetric asparaginase 96 wells plate assay using L-aspartic-acid-β-hydroxomate to isolate active hASRGL1 clones, as previously reported (9).
2. After rank ordering, select two or three clones for sequence verification.
3. Express, purify enzyme variants, and analyze rates of L-Asn hydrolysis by HPLC assay to determine steady-state kinetic parameters (10).
4. Performing this asparaginase FACS screening method on an error-prone library of hASRGL-1 variants resulted in a ~eight-fold increase in GFP signal after seven rounds of sorting (Fig. 4). A novel variant isolated from the screen had a corresponding twofold increase in k_{cat}/K_M (M^{-1} s^{-1}) for L-Asn hydrolysis relative to wt hASRGL-1.

3.2. Methionine-γ-Lyase Screen

Use your laboratories favorite methods for library construction, competent cells, transformations, etc.

3.2.1. Cell Growth and GFP Expression

1. Grow an overnight culture (5 mL) of *E. coli* BL21(*MI-aux*) co-transformed with pBAD-GFP and plasmids either a library or a single mutant in 2xYT (see Note 2).
2. Inoculate 1 mL of *nonselective* media in a sterile culture tube with *E. coli* BL21(*MI-aux*) cells harboring plasmids of pBAD-GFP reporter and pET28a containing gene of interest to a final OD_{600} of 0.1. The cell source for the inoculation can be previously frozen aliquots or overnight cultures of either a single mutant or a library of variants. Make sure to include tubes for a positive and negative control (1 mL each) for each round if planning to sort a library.
3. Grow cultures with shaking at 250 rpm at 37°C to an OD_{600} of 0.3–0.4.

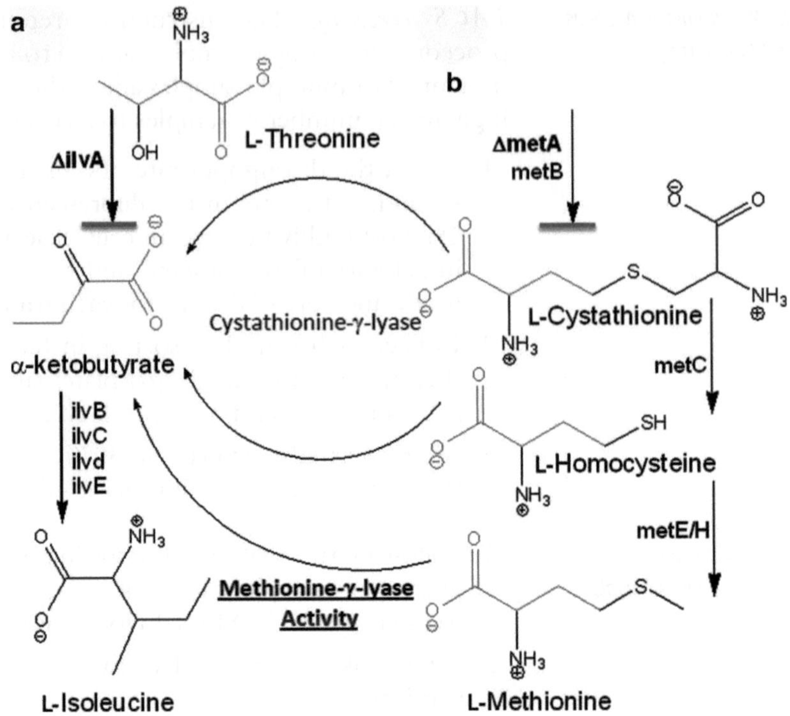

Fig. 4. (**a**) Summary of the pathway and gene deletions in *E. coli* needed to create an α-ketobutyrate auxotrophic strain that can be rescued by recombinant methionine-γ-lyase activity. (**b**) Summary of the pathway and gene deletions needed to create an L-methionine auxotroph to prevent parental cystathionine-γ-lyase activity from rescuing α-ketobutyrate auxotrophy.

4. Add 1 M IPTG solution to a final concentration of 1 mM and shift the cultures to 25°C for 2 h with shaking at 250 rpm for enzyme expression (see Note 7).
5. Harvest the cells by centrifugation (4,000 × *g*, 3 min, 4°C). Wash the cells twice with cold 0.9% NaCl by re-suspending and centrifuging at 4,000 × *g*, for 3 min at 4°C (see Note 8). Use sterile 1.5 mL Eppendorf tubes to harvest the cells and at least 1 mL of the 0.9% NaCl solution for each wash.
6. Re-suspend the washed cells in 1 mL of GFP Expression Media (containing arabinose inducer for GFP expression). Split each 1 mL sample into two small culture tubes (500 μL per tube) and add 250 mM L-Ile solution to a final concentration of 1 mM in one of the two tubes. This will serve as a control for each sample (see Note 9).
7. Continue to grow at 25°C for an additional 2 h (shake at 250 rpm).
8. Harvest the cells by centrifugation (4,000 × *g*, 3 min) and re-suspend in PBS to a final OD_{600} ~0.05–0.1 for flow cytometric analysis and cell sorting.

3.2.2. Cytometric Analysis and Cell Sorting

1. Perform flow cytometric analysis with a FACSAria (BD Biosciences) or an equivalent system. Adjust the throughput rate of cells to 4,000–5,000 events per second.
2. Perform all rounds of sorting in single cell mode, except for the initial sort, which should be in purity mode.
3. Set a gate in the fluorescence channel to recover the 4–5% most highly fluorescent cells. Additionally, set gates based on both the forward- and side-scatter channels to exclude sorting non-single cell events (see Note 10).
4. Sort tenfold the number of cells expected to cover the diversity for each round.
5. Collect the sorted cells in 0.5 mL of 2xYT medium and plate onto pre-warmed 2xYT medium supplemented with 50 μg/mL kanamycin and 34 μg/mL chloramphenicol (see Note 11).
6. Following overnight growth at 30°C, pool the clones and store aliquots in 15% glycerol at −80°C. At this step, you may also use the pool to inoculate the next round.

3.2.3. Enrichment Experiment with Controls

1. Inoculate two culture tubes of *nonselective* medium with *E. coli* BL21(*MI-aux*) containing pBAD-GFP/pET28-pMGL for a positive control and pBAD-GFP/pET28-hCGL for a negative control. Follow the protocols from Subheading 3.2.1, steps 3–7, keeping the controls separate.
2. After GFP expression, measure the OD_{600} of the cultures and make several 100 μL aliquots of negative controls in 1.5 mL Eppendorf tubes.
3. Generate a pool that contains the negative control in a 10,000-fold excess by serial dilution of your positive control into your negative control aliquots.
4. Dilute this mixture in PBS to a final OD_{600} ~0.05–0.1 for flow cytometric analysis and cell sorting.
5. Follow the remaining steps of the protocol described in Subheading 3.2.2.
6. Continue additional rounds of sorting until the fluorescence signal of your pool is either: (a) equivalent to that of the positive control alone or (b) no longer changing from one sort round to the next.
7. Sequence at least 20 random clones from the last round plate to verify identity of gene in the pET28a expression plasmid. Alternately, you can perform a *secondary* 96-well plate screen for MGL activity (20) to determine the makeup of the pooled sample.
8. Calculate the fold of enrichment achieved and note the number of sort rounds necessary to achieve the enrichment.

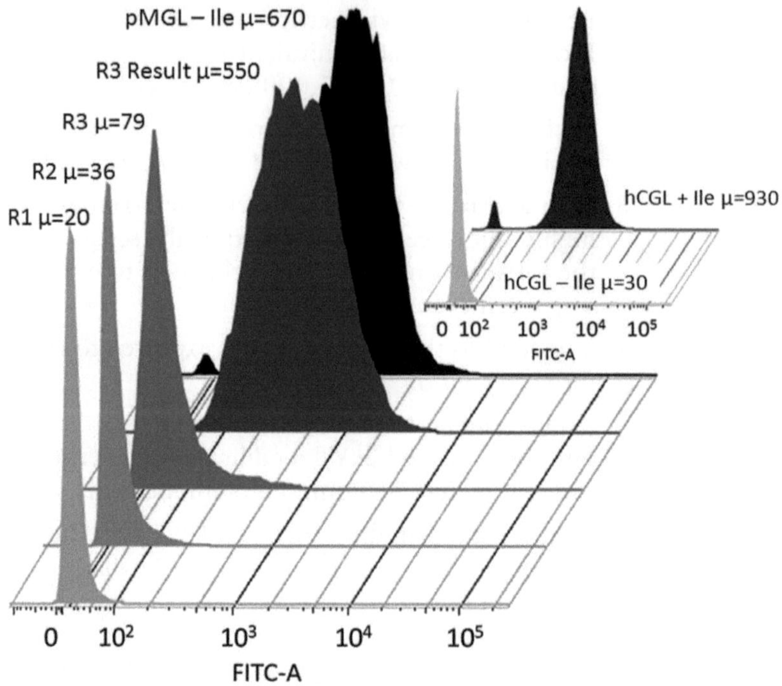

Fig. 5. Histogram showing enrichment of positive enzyme control out of a mixed pool of 10,000:1 hCGL to pMGL over the course of four rounds as well as the signals of a positive enzyme control alone (pMGL -Ile). (*Inset*) A histogram of a positive and negative control experiment using hCGL in the presence or absence of L-Ile (+/– Ile).

9. A successful Enrichment Test should yield results similar to Fig. 5. We achieved a 5,000-fold enrichment within three rounds (Additional rounds may be necessary under different conditions) (see Note 12).

4. Notes

1. The amount of enzyme required to rescue the auxotrophy depends on the catalytic activity and level of "leaky" expression of hASRGL1 and hASRGL1 libraries. In the case of the parental enzyme leaky expression is sufficient for the analyses. For a poorly expressed enzyme one can supplement with IPTG.
2. We find it is easiest to prepare competent cells harboring the reporter plasmid pBAD-GFP and then transform with the expression plasmids of interest. For preparing highly competent *E. coli* cells we use an electrocompetent transformation method described previously (19).
3. The multiple wash steps with 0.9% NaCl ensure that L-Asp is removed from the media.

4. In the positive control the expression of GFP will be independent from the enzymatic activity, as the medium is nonselective and supplemented with L-Aspartate. In the sample the expression of GFP will be dependent on the hASRGL-1 asparaginase activity, as the medium is selective and deprived of L-Aspartate. A negative control can also be added using an inactive enzyme variant to assess background levels of GFP expression.
5. The multiple wash steps with PBS ensure removal of the selective expression media for the subsequent cytometric analysis.
6. L-Val and L-Ile are excluded from the nonselective growth medium and selective growth medium because these branched-chain amino acids are known to inhibit the L-Ile biosynthesis pathway (21, 22). The addition of these two amino acids to the selective growth media without L-Ile in our experiments resulted in undetectable GFP expression in all variants tested.
7. Some enzymes express better at lower temperatures thus this parameter can be adjusted on a case-by-case basis.
8. The multiple wash steps ensure that L-Ile is removed from the media.
9. In the positive control for each sample the expression of GFP will be independent from the enzymatic activity, as the medium is nonselective and supplemented with L-Ile. In the sample the expression of GFP will be dependent on the L-Met degrading activity of the single mutant or of the library, as the medium is selective and deprived of L-Ile.
10. During flow cytometric analysis the applied voltage may be adjusted to improve the signal separation of the controls. A higher voltage will result in higher signals for both the positive and negative samples, but will also improve the signal-to-noise ratio when the positive control signal is at the lower end of the optimal dynamic range for the instrument and the negative control is near 0 (FITC-A 0-10,000).
11. When plating cells collected from a round of sorting, optimal recovery is achieved if the agar plate is pre-warmed for 30 min to 1 h at 37°C.
12. Various parameters may be adjusted to optimize the process. Optimization should maximize the fluorescence signal difference between the positive and negative control variants. The expression of enzyme can be tuned by changing the IPTG inducer concentration, altering the expression temperature and time, as well as by choosing a different OD_{600} reading to initiate expression. GFP expression can be similarly controlled by altering the arabinose concentration in the selective media and varying expression time and temperature.

References

1. Farinas ET (2006) Fluorescence activated cell sorting for enzymatic activity. Comb Chem High Throughput Screen 9:321–328
2. Hay N, Sonenberg N (2004) Upstream and downstream of mTOR. Genes Dev 18:1926
3. Agrawal V, Woo JH, Mauldin JP, Jo C, Stone EM, Georgiou G, Frankel AE (2012) Cytotoxicity of human recombinant arginase I (Co)-PEG5000 in the presence of supplemental L-citrulline is dependent on decreased argininosuccinate synthetase expression in human cells. Anti-Cancer Drugs 23:51–64
4. Cantor JR, Panayiotou V, Agnello G, Georgiou G, Stone EM (2012) Engineering reduced-immunogenicity enzymes for amino acid depletion therapy in cancer. Methods Enzymol 502:291–319
5. Stone E, Chantranupong L, Gonzalez C, O'Neal J, Rani M, VanDenBerg C, Georgiou G (2011) Strategies for optimizing the serum persistence of engineered human arginase I for cancer therapy. J Controll Release 158:171–179
6. Stone EM, Glazer ES, Chantranupong L, Cherukuri P, Breece RM, Tierney DL, Curley SA, Iverson BL, Georgiou G (2010) Replacing Mn2+ with Co2+ in human arginase I enhances cytotoxicity toward l-arginine auxotrophic cancer cell lines. ACS Chem Biol 5:333–342
7. Richards NG, Kilberg MS (2006) Asparagine synthetase chemotherapy. Annu Rev Biochem 75:629–654
8. Yang Z, Wang J, Lu Q, Xu J, Kobayashi Y, Takakura T, Takimoto A, Yoshioka T, Lian C, Chen C (2004) PEGylation confers greatly extended half-life and attenuated immunogenicity to recombinant methioninase in primates, AACR, pp 6673–6678
9. Cantor JR, Yoo TH, Dixit A, Iverson BL, Forsthuber TG, Georgiou G (2011) Therapeutic enzyme deimmunization by combinatorial T-cell epitope removal using neutral drift. Proc Natl Acad Sci USA 108:1272–1277
10. Cantor JR, Stone EM, Chantranupong L, Georgiou G (2009) The human asparaginase-like protein 1 hASRGL1 is an Ntn hydrolase with beta-aspartyl peptidase activity. Biochemistry 48:11026–11031
11. Yang Z, Wang J, Lu Q, Xu J, Kobayashi Y, Takakura T, Takimoto A, Yoshioka T, Lian C, Chen C, Zhang D, Zhang Y, Li S, Sun X, Tan Y, Yagi S, Frenkel EP, Hoffman RM (2004) PEGylation confers greatly extended half-life and attenuated immunogenicity to recombinant methioninase in primates. Cancer Res 64:6673–6678
12. Tan Y, Xu M, Tan X, Wang X, Saikawa Y, Nagahama T, Sun X, Lenz M, Hoffman RM (1997) Overexpression and large-scale production of recombinant L-methionine-a-deamino-g-mercaptomethane-lyase for novel anticancer therapy. Protein Expr Purif 9:233–245
13. Tan Y, Xu M, Guo H, Sun X, Kubota T, Hoffman RM (1996) Anticancer efficacy of methioninase in vivo. Anticancer Res 16: 3931–3936
14. Lishko VK, Lishko OV, Hoffman RM (1993) Depletion of serum methionine by methioninase in mice. Anticancer Res 13:1465–1468.
15. Tan Y, Zavala JSR, Han Q, Xu M, Sun X, Tan X, Magana R, Geller J, Hoffman RM (1997) Recombinant methioninase infusion reduces the biochemical endpoint of serum methionine with minimal toxicity in high-stage cancer patients. Anticancer Res 17:3857–3860
16. Hu J, Cheung NK (2009) Methionine depletion with recombinant methioninase: in vitro and in vivo efficacy against neuroblastoma and its synergism with chemotherapeutic drugs. Int J Cancer 124:1700–1706
17. Yang Z, Wang J, Yoshioka T, Li B, Lu Q, Li S, Sun X, Tan Y, Yagi S, Frenkel EP, Hoffman RM (2004) Pharmacokinetics, methionine depletion, and antigenicity of recombinant methioninase in primates. Clin Cancer Res 10:2131–2138
18. Yoshioka T, Wada T, Uchida N, Maki H, Yoshida H, Ide N, Kasai H, Hojo K, Shono K, Maekawa R, Yagi S, Hoffman RM, Sugita K (1998) Anticancer efficacy in vivo and in vitro, synergy with 5-fluorouracil, and safety of recombinant methioninase. Cancer Res 58: 2583–2587
19. Varadarajan N, Cantor JR, Georgiou G, Iverson BL (2009) Construction and flow cytometric screening of targeted enzyme libraries. Nat Protocols 4:893–901
20. Takakura T, Mitsushima K, Yagi S, Inagaki K, Tanaka H, Esaki N, Soda K, Takimoto A (2004) Assay method for antitumor L-methionine-lyase: comprehensive kinetic analysis of the complex reaction with L-methionine. Anal Biochem 327:233–240
21. Leavitt RI, Umbarger H (1962) Isoleucine and valine metabolism in Escherichia coli XI. K-12: Valine inhibition of the growth of Escherichia coli strain. J Bacteriol 83:624–630
22. Gollop N, Chipman DM, Barak Z (1983) Inhibition of acetohydroxy acid synthase by leucine. Biochimica et Biophysica Acta: Protein Struct Mol Enzymol 748:34–39

Chapter 4

Flow Cytometric Assays for Interrogating LAGLIDADG Homing Endonuclease DNA-Binding and Cleavage Properties

Sarah K. Baxter, Abigail R. Lambert, Andrew M. Scharenberg, and Jordan Jarjour

Abstract

A fast, easy, and scalable method to assess the properties of site-specific nucleases is crucial to understanding their *in cellulo* behavior in genome engineering or population-level gene drive applications. Here we describe an analytical platform that enables high-throughput, semiquantitative interrogation of the DNA-binding and catalytic properties of LAGLIDADG homing endonucleases (LHEs). Using this platform, natural or engineered LHEs are expressed on the surface of *Saccharomyces cerevisiae* yeast where they can be rapidly evaluated against synthetic DNA target sequences using flow cytometry.

Key words: Homing endonuclease, Meganuclease, Yeast surface display, Yeast transformation, Flow cytometry, Binding affinity, Protein–DNA interaction

1. Introduction

LAGLIDADG homing endonucleases (LHEs) are a family of highly specific DNA-cleaving enzymes that have been discovered in diverse unicellular eukaryotes, bacteria, and viruses (1). LHE genes are able to surpass vertical patterns of inheritance by engaging in a horizontal transposition process called "homing." There are two prerequisites for horizontal propagation: (1) recipient genomic locations which can accept the insertion of an autonomous genetic element without incurring a prohibitive fitness cost and (2) biochemical machinery that enables the transfer of genetic information into such a locus. LHE genes are embedded in and transpose to phenotypically neutral genomic locations, such as within self-splicing group I introns or as N- or C-terminal fusions with permissive recipient proteins (2, 3). Furthermore, LHE genes have devised a mechanism

James C. Samuelson (ed.), *Enzyme Engineering: Methods and Protocols*, Methods in Molecular Biology, vol. 978, DOI 10.1007/978-1-62703-293-3_4, © Springer Science+Business Media New York 2013

of transposition that relies on the nearly universal process of DNA repair by homologous recombination (4). LHE genes encode for DNA endonucleases and embed themselves within a cleaved target site; they become self-propagating by creating DNA breaks and using their own coding sequences as repair templates for homology-driven DNA break resolution (5).

The DNA target sequences bound and cleaved by LHEs are approximately 22 base pairs in length. Similar to microRNAs or commonplace PCR primers, this length of DNA surpasses a hypothetical threshold of uniqueness within a genome, thereby enabling genome-specific operations. LHEs are one of two (the other being the recently discovered TAL effector proteins) natural protein scaffolds currently known which are able to recognize DNA with specificity properties at or above this "genomic threshold." Two decades of experimentation with natural LHEs has demonstrated that they can be expressed at high levels in orthogonal cell types, altering genomic loci without causing any overt toxicity, thereby confirming genomic-level DNA cleavage specificity (6–11). Two important overarching questions still dominate the LHE field: (a) what are the more in-depth patterns of their DNA recognition specificity, and (b) can their high level of specificity be redirected to target a broad range of nonnative sequences relevant to the research and/or therapeutic communities.

Here, we present flow cytometric methods that allow for efficient, semiquantitative analysis of homing endonuclease binding and cleavage activities (12, 13). These assays can be performed to develop in-depth characterizations of LHE-DNA recognition properties. If carried out using a wide array of DNA substrate sequences, they can lead to a better understanding of correlations (or lack thereof) between DNA binding and cleavage and also how multiple simultaneous base pair substitutions influence the interaction. The system is fast to set up: following transformation with an episomal plasmid, yeast display on their cell surface a homing endonuclease of interest, and can be rapidly validated by antibody staining of a C-terminal Myc epitope to detect the presence of stable, full-length enzyme. Synthetically generated, fluorescently labeled DNA substrates are then used to perform fast, multi-well-scalable assays to uncover DNA-binding and cleavage properties of interest to the investigator (Fig. 1).

2. Materials

Prepare all solutions using ultrapure RNAse- and DNAse-free water (0.22 μm filtered, deionized water) and analytical grade reagents. Cultures and reagents may be prepared at the bench, but care should be taken to use sterile components and aseptic technique.

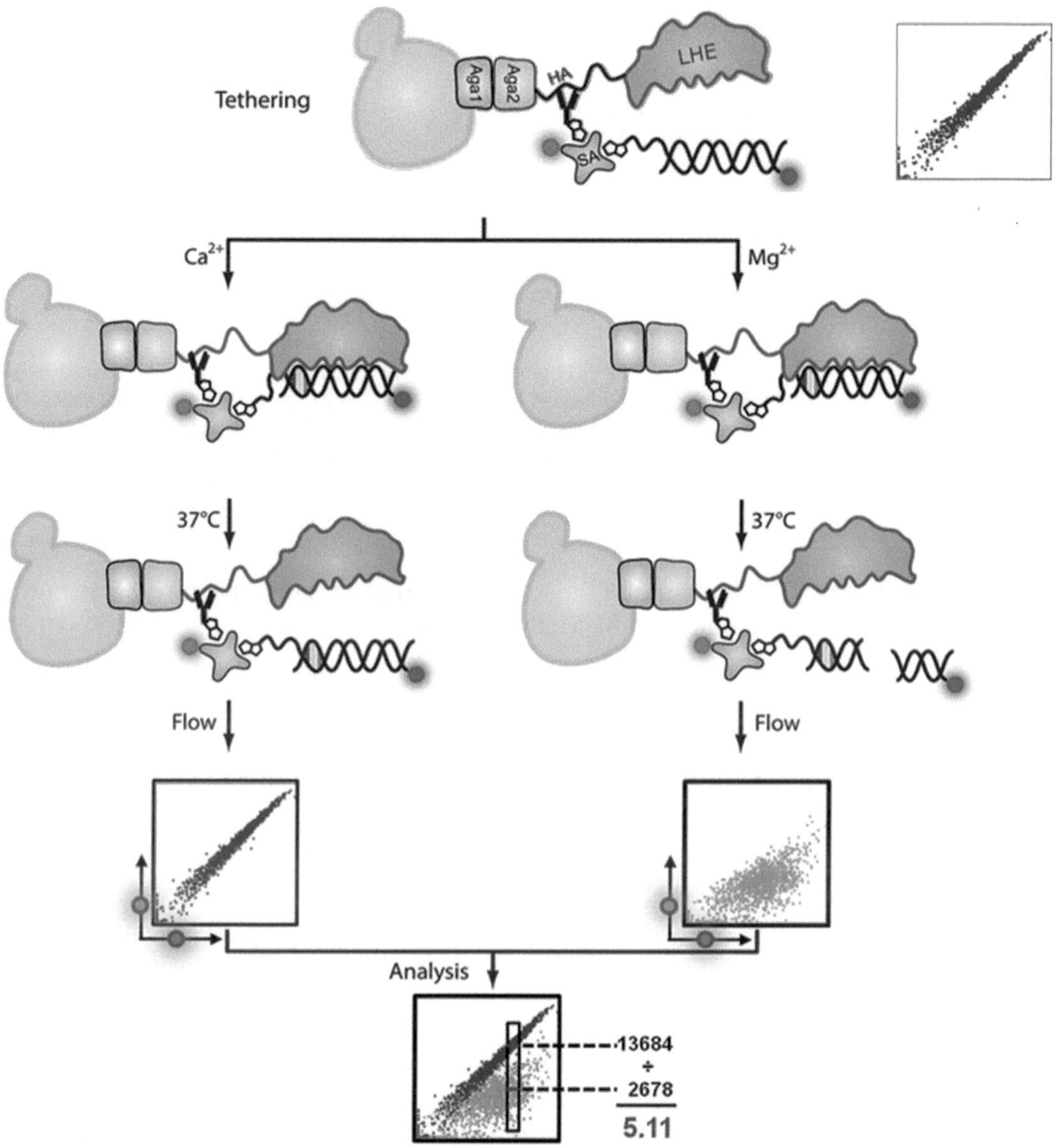

Fig. 1. Schematic representation of a flow cytometric DNA cleavage assay. The homing endonuclease of interest is expressed on the surface of yeast through fusion with the yeast Aga2P protein. Aga2P forms a disulfide linkage with Aga1P, which is co-induced in the EBY100 strain upon galactose induction. An N-terminal hemagglutinin (HA) tag enables cis-tethering of the double-stranded DNA substrate to the enzyme: a streptavidin-PE "bridge" connects the bound anti-HA-biotin with biotinylated and A647-labeled DNA substrate. This tethered target substrate results in the characteristic colinear PE and A647 fluorescence profile observed on the flow cytometer. With the addition of calcium, the enzyme may bind the DNA substrate, but it cannot cleave it; with the addition of magnesium, the enzyme is able to bind and cleave the substrate provided a productive interaction is formed. Cleavage of the tethered substrate leads to a loss of fluorescence in the A647 channel. This loss in A647 fluorescence can be used to quantify cleavage by gating on a population of yeast normalized for their PE signal and comparing the median A647 fluorescence intensities of the corresponding Ca^{2+} and Mg^{2+} samples.

Template for target with universal primers:
TGGACACGACTTGAGCTGAGGAGGTTTCTCTGTAACCGTAGTGCTGTGCTGA

Target sequence:
TGAGGAGGTTTCTCTGTAAC

Universal forward primer conjugated to biotin:
BIOTIN-TGGACACGACTTGAGC

Universal reverse primer conjugated to A647 fluorophore:
A647-TCAGCACAGCACTACG

Fig. 2. Reagents for PCR production of DNA target oligo substrate. The desired target sequence should be flanked by forward and reverse "universal primer" sequences. The primers currently in use in our laboratories are listed; alternative universal primer schemes may be substituted. A647 and biotin moieties included on the primers result in the production of dual-labeled double-stranded DNA substrates for use in both the flow cytometric DNA cleavage and DNA-binding (where the biotin is obsolete) assays.

2.1. Dual-Labeled Double-Stranded DNA Substrates

1. Platinum® Taq DNA Polymerase High Fidelity, with buffer and 50 mM $MgSO_4$ (Invitrogen).
2. Target site oligo template, with flanking universal primer sites, standard desalting purification (Integrated DNA Technologies (IDT)) (Fig. 2).
3. Biotin-labeled universal forward primer (IDT) (Fig. 2) (see Note 1).
4. *A*647-labeled universal reverse primer (IDT) (Fig. 2).
5. 10 mM dNTPs.
6. 0.2 mL PCR strip tubes and 96-well PCR plates (Bio-Rad).
7. Thermal cycler.
8. Exonuclease I (New England Biolabs).
9. MultiScreen HTS-HV filter plate (Fisher Scientific).
10. Illustra sephadex G-100 (GE Healthcare): Make sephadex solution at 1 g/20 mL in water; allow at least 24 h for bead hydration; store at room temperature for a maximum of 4 months.
11. Odyssey infrared imaging system (LI-COR Biosciences) or UV transilluminator.
12. Nanodrop or similar microvolume spectrophotometer (Thermo Scientific).

2.2. Polyacrylamide Gel

1. 30% Acrylamide/*bis* (19:1) solution.
2. 10× Tris/borate/EDTA (TBE) buffer (see Note 2).
3. *N*,*N*,*N*′,*N*′-tetramethylethylenediamine (TEMED).
4. Ammonium persulfate (Fisher Scientific): 10% w/v solution in water.

5. Plastic gel cassette, 1.0 mm.
6. Vertical gel electrophoresis apparatus.
7. 6× Ficoll loading buffer: 18% (w/v) Ficoll-400 (Sigma-Aldrich) in 6× TBE, with NO added dyes.

2.3. Yeast Transformation

1. EBY100 yeast (Invitrogen).
2. 2×YPAD nonselective yeast media: 20 g Bacto yeast extract, 40 g Bacto peptone, 100 mg adenine hemisulfate (Sigma-Aldrich), 50 g glucose, water to 1 L, pH to 6.0. Filter sterilize or autoclave and store at 4°C (see Note 3).
3. Salmon sperm DNA (Sigma-Aldrich), 2 mg/mL solution in 1× TE.
4. 1 M Lithium acetate solution in water.
5. 50% w/v Polyethylene glycol in water, MW 3350 (PEG 3350).
6. Plasmid DNA-encoding endonuclease in pCTCON2 yeast surface display vector (Fig. 3).

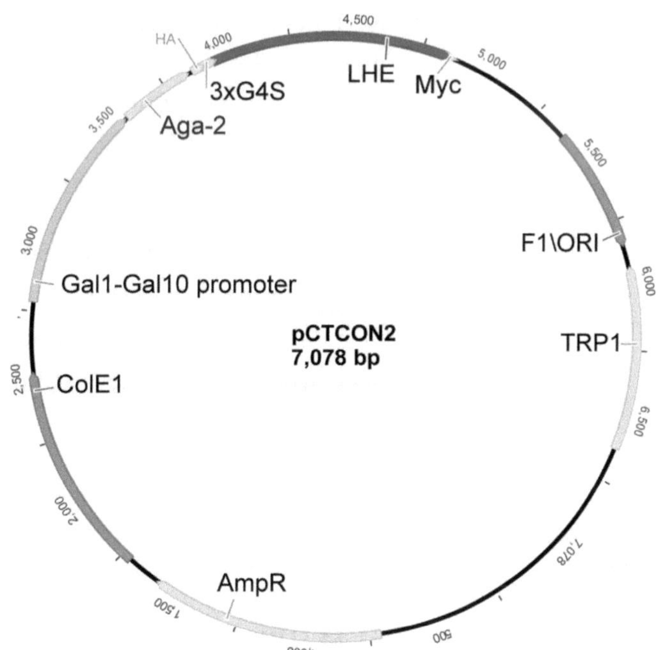

Fig. 3. Schematic representation of pCTCON2 vector for LHE surface display. Gal1-10, galactose inducible promoter; HA, hemagglutinin epitope; 3×G4S, thrice repeated gly–gly–gly–gly–ser linker sequence; LHE, LAGLIDADG homing endonuclease coding sequence; Myc, cMyc epitope. Ampicillin resistance (AmpR) is included for selection in *Escherichia coli*, TRP1 marker used for auxotrophic selection in *Saccharomyces cerevisiae*.

7. 42°C water bath.
8. Yeast-selective growth media "SC–Ura–Trp": 6.7 g yeast nitrogen base without amino acids (Sigma-Aldrich), 1.4 g yeast synthetic dropout media supplement without Trp, Ura, His, Leu (Sigma-Aldrich), 76 mg histidine, 380 mg leucine, 4.34 g MES, and water to 900 mL. Adjust pH to 5.25 with HCl. Sterilize by autoclaving 20 min. Prior to use, add penicillin (100 i.u./mL), streptomycin (100 μg/mL), and kanamycin (25 μg/mL). Store at 4°C.
9. 20% w/v glucose solution, filter sterilized. Store at 4°C.
10. Selective growth media agar plates: Add 20 g of bacteriological agar to 900 mL of SC–Ura–Trp-selective growth media and autoclave for 20 min. Add 100 mL prewarmed (55°C) 20% w/v glucose and penicillin (100 i.u./mL), streptomycin (100 μg/mL), and kanamycin (25 μg/mL). Pour into Petri dishes and let it solidify at room temperature. Store plates at 4°C.
11. Water-jacketed incubator.

2.4. Yeast Growth and Induction

1. EBY100 yeast transformed with surface-expression vector containing homing endonuclease of interest.
2. SC–Ura–Trp-selective growth media (for recipe, see Subheading 2.3).
3. 20% w/v glucose solution, filter sterilized, store at 4°C.
4. 20% w/v D-(+)-raffinose pentahydrate (Sigma-Aldrich) + 0.1% w/v glucose solution, filter sterilized, store at room temperature.
5. 20% w/v D-(+)-galactose solution, filter sterilized, store at 4°C.
6. Baffled Erlenmeyer flask(s).
7. Disposable 15 mL culture tubes.
8. Deep-well 96-well plate (flat bottom).
9. Shaking incubator.
10. Spectrophotometer.

2.5. Yeast Surface Display Flow Cytometric DNA-Binding and Cleavage Assay

1. Induced EBY100 yeast with surface-expressed homing endonuclease.
2. 10× Yeast staining buffer (YSB): 1.8 M KCl, 0.1 M NaCl, 0.1 M HEPES, 2% BSA, 1% w/v D-(+)-galactose, adjust pH to 7.5 with KOH (see Note 4). Filter sterilize and store at 4°C in a light-protected or foil-wrapped container.
3. 1× High-salt yeast staining buffer (YSB + KCl): Dilute 10× YSB to 1× with 466 mM KCl for a final KCl concentration of 600 mM. Store at 4°C in light-protected or foil-wrapped container.

4. 10× In vitro oligo cleavage buffer (IOCB): 1.5 M KCl, 0.1 M NaCl, 0.1 M HEPES, 0.05 M K-Glu (l-glutamic acid potassium salt monohydrate), 0.5% BSA, adjusted to pH 8.25 with KOH (see Note 4). Filter sterilize solution and store at 4°C in a light-protected or foil-wrapped container.
5. 1 M $CaCl_2$ solution. Filter sterilize and store at room temperature.
6. 1 M $MgCl_2$ solution. Filter sterilize and store at room temperature.
7. Biotin-labeled anti-HA antibody (Covance).
8. Streptavidin-PE (BD Biosciences).
9. FITC-conjugated chicken anti-cMyc antibody (Immunology Consultants Laboratory, Inc.).
10. Costar 96-well V-bottom plate (Sigma-Aldrich).
11. BD FACScalibur or LSRII™ cytometer (BD Biosciences) or other cytometer with equivalent optics.
12. FlowJo software (Tree Star Inc.).

3. Methods

Carry out all procedures on ice, unless otherwise specified.

3.1. Preparation of Dual-Labeled Double-Stranded DNA Substrates

1. In 0.2 mL PCR tubes, mix 0.08 μL Platinum High Fidelity Taq, 2 μL 10× Taq buffer, 1.3 μL 50 mM $MgSO_4$, 0.4 μL 10 mM dNTPs, 2 nM (final concentration) target site template oligo, 0.55 nM (final concentration) of each the *A*647 universal FP and biotin universal RP. Add H_2O to a final volume of 20 μL (see Note 5).
2. Thermal cycler program (for PCR amplification of target and incorporation of labels):

   90°C × 1 min.

   40× (86°C × 15 s, 48°C × 15 s, 60°C × 30 s).

   60°C × 15 min.

   40× (70°C × 30 s, decrease by 1°C every cycle) (see Note 6).

   Hold at 4°C.
3. Digest excess single-stranded DNA with exonuclease I: Add 2 units of *Exo*I to each 20 μL PCR reaction in 2 μL total volume of water (see Note 7). Digest 1 h at 37°C. This reaction can be stored at 4°C overnight or at −20°C for extended periods.
4. Load the hydrated sephadex G-100 suspension into the filter plate. For each 20 μL PCR reaction to be purified, add 500 μL

total volume of suspension to a filter plate well. This is best accomplished by loading 320 μL sephadex suspension (using wide bore tips) into each necessary well of the filter plate, centrifuging briefly up to a speed of 500 × *g*, discarding water, and adding the remaining 180 μL sephadex (see Note 8). The plate should then be dehydrated by centrifugation at 2,000 × *g* for 7 min.

5. Load the 22 μL PCR + *Exo*I reaction directly to the center of each sephadex column. Secure a 96-well PCR plate below the filter plate to catch the purified flow-through, using tape if necessary. Centrifuge for 5 min at 2,000 × *g*. Approximately 12–14 μL of flow-through should be present in each recipient well.
6. Determine the concentration of purified target site substrates using a nanodrop spectrophotometer. Target concentration should be approximately 10–25 ng/μL. A concentration greater than 25 ng/μL suggests inadequate *Exo*I digestion or sephadex purification.
7. Run target oligos on a 15% polyacrylamide gel. For a 7.5 mL gel, combine 0.75 mL 10× TBE, 3.75 mL 30% acrylamide/*bis* (19:1), and 2.9 mL water. Add 100 μL 10% APS and 10 μL TEMED, then immediately mix and pipette into the prepared gel cassette. Once the gel is set, load 0.5–1.0 μL purified target, diluted with 1.0 μL 6× Ficoll loading buffer and 4 μL water (see Note 9). Use 0.1 μL of the *A*647-labeled primer as a size standard. Run the gel for 90 min at 120 V.
8. Visualize the gel on a LI-COR Odyssey infrared imager, using the 700 nM laser. The gel should show a prominent single PCR product and minimal contamination by other bands or leftover primers (Fig. 4). Alternatively, the gel can be stained with 1× SybrGold in 1× TBE for 20 min, washed in 1× TBE or water, and visualized on a UV transilluminator.

3.2. Yeast Transformation

(This transformation procedure is based on published protocols by Gietz and Schiestl) (14).

1. Thaw and spin down a frozen aliquot of EBY100 competent yeast cells (see Note 10).
2. Resuspend the pellet in the following transformation mixture: 50 μL denatured 2 mg/mL salmon sperm DNA (heat at 95°C for 5 min, then transfer immediately to ice), 36 μL 1 M LiAc, 260 μL 50% PEG 3350, and 14 μL water plus plasmid DNA (up to 1 μg) (see Notes 11 and 12).
3. Incubate the yeast and transformation mixture at 42°C for 40 min (see Note 13).

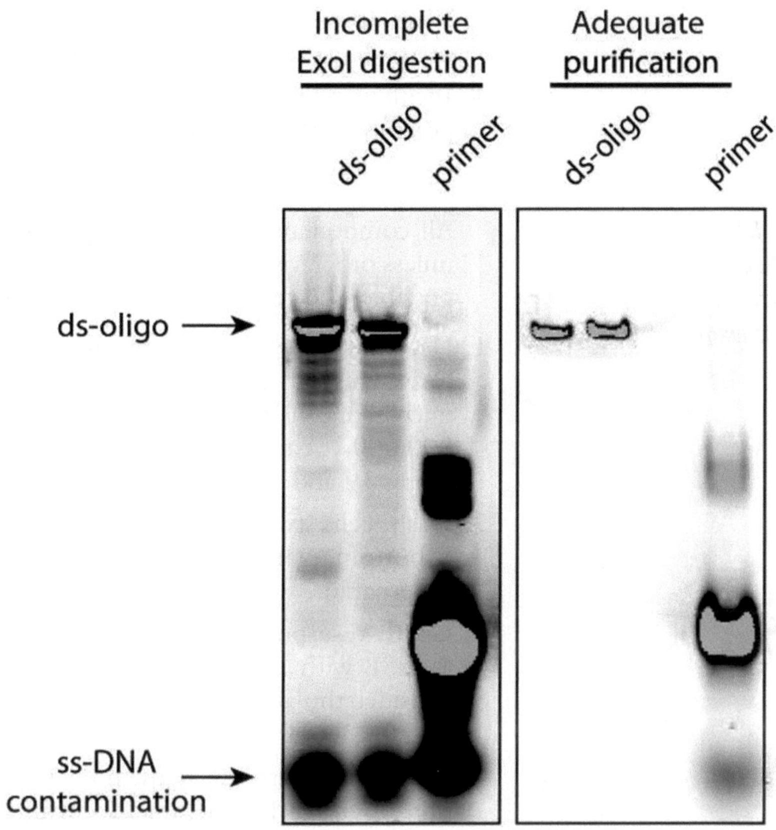

Fig. 4. DNA target production and purification. Following PCR amplification and annealing of double-stranded oligo target, *ExoI* digestion and sephadex G-100 filtration are used to remove contaminating ss-DNA and unincorporated nucleotides. Incomplete digestion with *ExoI* or inadequate G-100 purification, as can occur with cracked or incorrectly packed columns, leaves significant ss-DNA (*left*), which will lead to poor staining with the conjugated target-SAV-PE in the cleavage and binding assays. Correct purification leaves minimal ss-DNA contamination (*right*).

4. Fill the tube with SC–Ura–Trp + 2% glucose media and spin down the cells. Remove supernatant.
5. Resuspend the yeast pellet in 1 mL SC–Ura–Trp + 2% glucose media.
6. Plate 1–10 μL transformed yeast on selective growth media agar plates (SC–Ura–Trp + 2% glucose) and incubate in a 30°C water-jacketed incubator. Colonies of an appropriate size for picking should appear by 48 h.

3.3. Growth and Induction of Yeast

1. Transfer a single colony of transformed yeast into 1.5 mL SC–Ura–Trp + 2% raffinose + 0.1% glucose media (see Note 14).
2. Incubate overnight in a 15 mL culture tube at 30°C with 250 rpm shaking until the cells reach a density of 90–120 million/mL and place on ice for up to 24 h (see Note 15).

3. Wash 30 million cells twice with water and transfer to 1.5 mL of SC–Ura–Trp + 2% galactose media (see Note 16).
4. Incubate the galactose culture on the benchtop (room temperature with no shaking) for 16–18 h for optimal induction (see Note 17).

3.4. Yeast Surface Display Flow Cytometric DNA Cleavage Assay

All components should be kept on ice throughout the assay, unless otherwise specified, including YSB and IOCB buffers. If possible, the centrifuge should also be kept at 4°C.

1. Determine the density of induced yeast in the galactose culture. This can be accomplished using a hemocytometer and microscope or by spectrophotometer (see Note 18). Aliquot 500,000 yeast per sample into a 96-well, V-bottom plate (see Notes 19 and 20).
2. Wash cells twice with 200 μL 1× YSB, centrifuging the V-bottom plate at 3,000 × *g* for 2 min and discarding the supernatant.
3. Gently resuspend cells at a concentration of 50 million/mL in 1× YSB with 1:300 dilution of anti-HA-biotin antibody (i.e., consider the anti-HA-biotin to be a 300× stock). Incubate at 4°C for 30–60 min, mixing gently every 10–15 min.
4. During the anti-HA stain, prepare target oligos for conjugation in either 1.5 mL microcentrifuge tubes or plate format. For 500,000 cells (final cell density of 50 million/mL), use 25 μL total volume per well. SAV-PE should be diluted to 5 nM in the high-salt 1× YSB + KCl buffer. Add the labeled ds-oligo target to a final concentration of 50 nM (see Notes 21 and 22). Aliquot to plates (if necessary) and incubate in the dark and on ice for 20 min.
5. Following the anti-HA incubation, centrifuge cells for 2 min at 3,000 × *g*, and wash yeast cells twice with 200 μL ice-cold high-salt 1× YSB + KCl.
6. Following the second wash, resuspend yeast cells with oligo conjugates. Gently vortex or pipette to resuspend.
7. Incubate 30 min at 4°C, mixing briefly every 5–10 min.
8. During this incubation, make 5 mM $MgCl_2$ and $CaCl_2$ solutions in 1× IOCB, and prewarm to 37°C.
9. Following incubation with oligo, wash yeast with 200 μl ice-cold high-salt 1× YSB + KCl, and centrifuge for 2 min at 3,000 × *g*. Resuspend conjugated yeast in 200 μL ice-cold 1× IOCB (containing no divalent ions).
10. Create duplicate wells (one will contain calcium for no cleavage, and one will contain magnesium to allow cleavage) by transferring half of the resuspended volume into an adjacent

set of wells. Bring volume back to 200 μL with ice-cold 1× IOCB. Centrifuge for 2 min at 3,000 × *g*. Discard the supernatant and blot/tap vigorously on paper towels to remove as much buffer as possible.

11. Add 30 μL 1× IOCB containing either Ca^{2+} or Mg^{2+} (pre-warmed to 37°C) to each set of duplicate wells. Add Ca^{2+} IOCB first to minimize background cleavage events in the negative control, and work as quickly as possible.
12. Incubate 20 min at 37°C.
13. Fill wells with ice-cold 1× YSB to stop the reaction, centrifuge 2 min at 3,000 × *g*, and discard the supernatant.
14. Resuspend in 25 μL 1× YSB containing 1:100 dilution of anti-Myc FITC to a final cell density of 100 million/mL.
15. Incubate 1–2 h at 4°C, with foil wrap or in a refrigerator to protect from light, vortexing occasionally to keep the cells in suspension (see Note 23).
16. Wash cells twice with 1× YSB, and resuspend in a final volume of 60 μL. This will lead to an acquisition rate of 1,000–2,000 events/s depending on the acquisition settings for the cytometer. Acquire data on a BD Biosciences LSRII with HTS. Record FSC-A, FSC-H, SSC-A, SSC-H, APC, PE, and FITC.
17. Analyze the flow cytometry data using FlowJo. Gate live cells (FSC-A by SSC-A), then singlets (FSC-H by FSC-A), then cells staining for both FITC and PE (representing full-length expression of both the C′ and N′ termini) (Fig. 5). Visualize this final subset as APC (*y*-axis) versus PE (*x*-axis). Superimpose the Ca^{2+} and Mg^{2+} samples to observe any cleavage-induced shift in APC signal. Quantitative measurements of cleavage efficiency can be obtained by determining the median APC signal within a small gated subset of live, singlet, expressing cells. A greater Ca^{2+}-to-Mg^{2+} median APC ratio represents increased cleavage efficiency (Fig. 5d).

3.5. Yeast Surface Display Flow Cytometric DNA-Binding Assay

1. Determine the density of induced yeast. This can be done using a hemocytometer or by spectrophotometry, as described above.
2. Aliquot 100,000 yeast to appropriate wells in a 96-well plate (for 384-well format, see Note 24) and wash once with 200 μL 1× IOCB containing 5 mM $CaCl_2$. Centrifuge at 3,000 × *g* for 2 min and discard the supernatant.
3. Resuspend cells in 30 μL 1× IOCB containing 5 mM $CaCl_2$, 1:100 anti-Myc FITC, and the desired concentration of target site oligo (see Note 25). Incubate at 4°C for 2 h, vortexing gently every 30 min.
4. Wash twice with 1× IOCB + 5 mM $CaCl_2$.

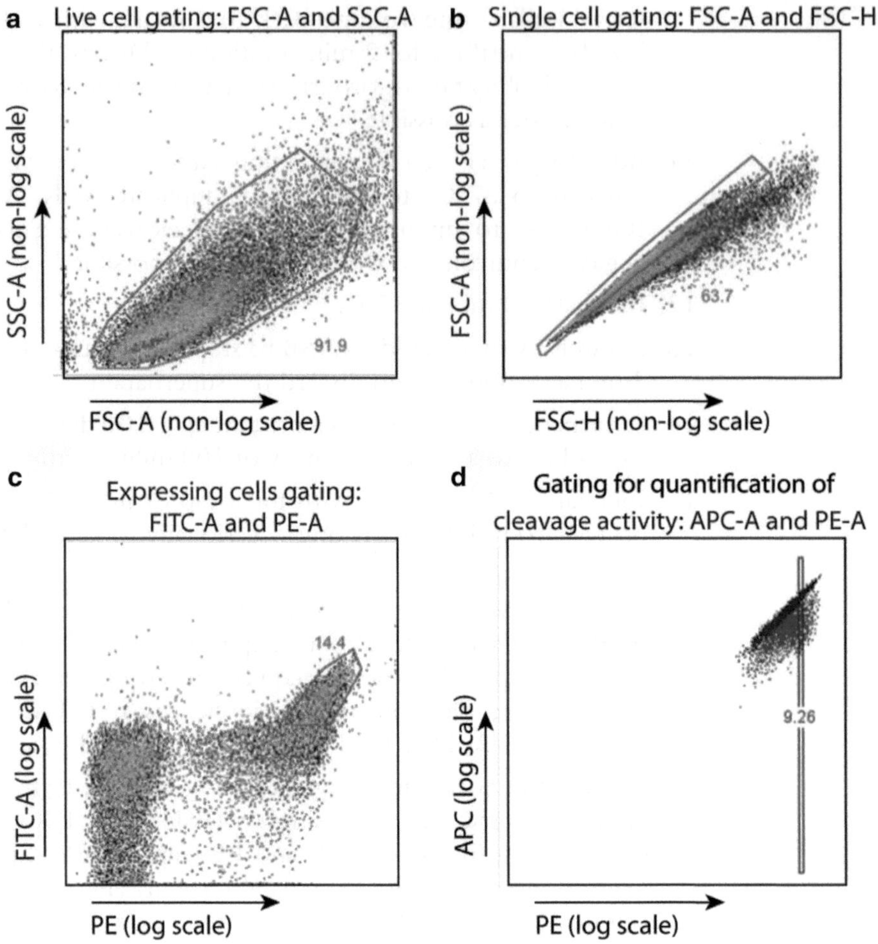

Fig. 5. Analysis of flow cytometry data from on-cell cleavage assay. (**a**) Live cells are gated first, using FSC-A and SSC-A parameters, followed by (**b**) singlet cells using FSC-A and FSC-H. With proper induction conditions, approximately 50% of cells will display the enzyme, and (**c**) the high-expressing population is gated using PE (which is bound to the N-terminal HA tag) and FITC (which is bound by antibody directly to the C-terminal Myc tag). Cleavage activity is visualized within this expressing population in (**d**) a plot of *A*647 fluorescence (in the APC channel, bound to the free end of the conjugated DNA) versus PE. Cleavage activity can be quantified by gating on a PE-normalized subset of the expressing cells and taking the ratio of the median APC fluorescence for the control (Ca^{2+}) and experimental (Mg^{2+}) conditions.

5. Resuspend in 60 μL 1× IOCB + 5 mM Ca^{2+} for flow cytometry analysis. A final volume of 60 μL will lead to 500–1,000 events/s.
6. Acquire data on a BD Biosciences LSRII with HTS. Record FSC-A, FSC-H, SSC-A, SSC-H, APC, and FITC. If using HTS for collecting 96- or 384-well plate samples, set machine to mix samples minimum 3×.
7. Analyze the flow cytometry data using FlowJo. Gate first the live cells, then singlets, then expressing cells, as described above. APC signal represents binding of the *A*647-labeled

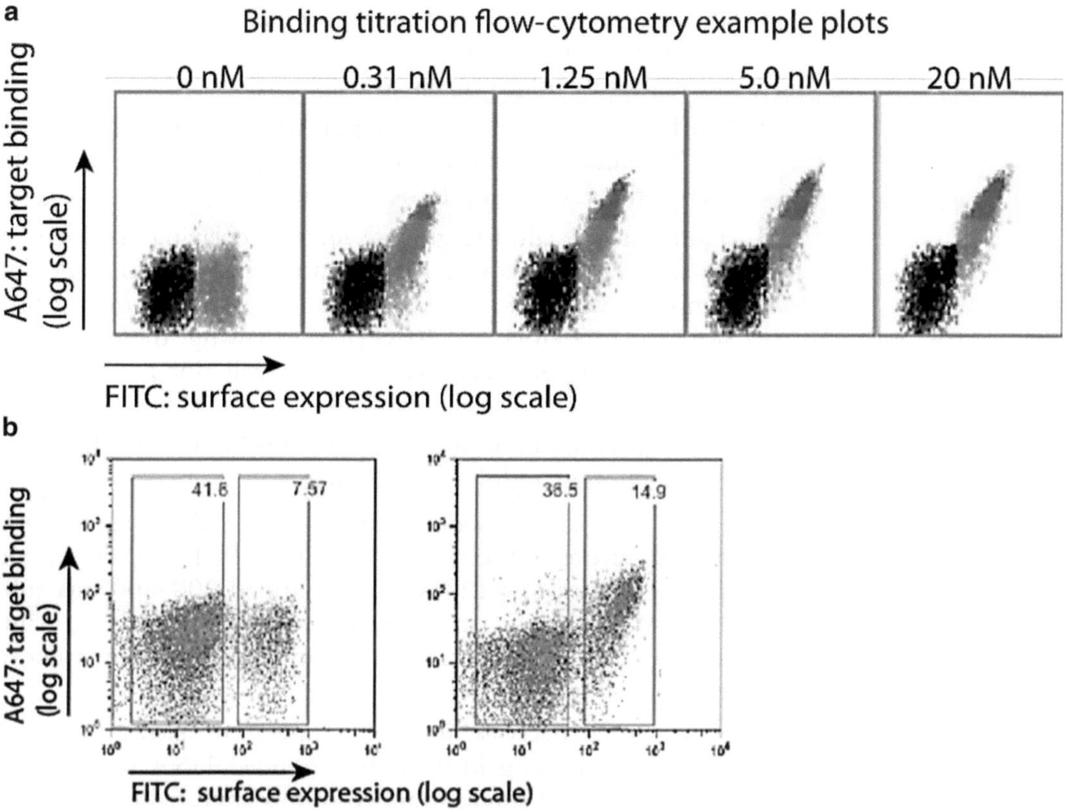

Fig. 6. Example of binding titration and analysis of flow cytometry data from on-cell binding assay. Live and singlet cells are gated as in the cleavage assay. Anti-Myc FITC staining delineates the population of yeast expressing enzyme, and A647 signal represents binding of the *A*647-labeled target oligo substrate. (**a**) Binding affinity on individual clones or sorted population can be quantified using binding titrations where cells expressing enzyme are incubated with a range of target substrate concentrations (0–20 nM as indicated in the depicted experiment). (**b**) Binding affinity can be approximated by gating a FITC-normalized subset of the expressing cells and plotting the median APC fluorescence to a nonlinear regression curve as described previously (12).

oligo to the surface-expressed homing endonuclease. Quantitative measurements of binding can be obtained by determining the ratio of median APC signal of the FITC-positive cells (which are expressing homing endonuclease) to the median APC signal of the FITC-negative cells (no homing endonuclease expression) (Fig. 6).

4. Notes

1. The biotin and *A647* labels can be affixed to either the forward or the reverse primer. Maximum cleavage signal is achieved empirically by testing each endonuclease with both

labeling schemes. For reasons that remain speculative, a more pronounced cleavage shift may occur with the *A*647 label on one end of the oligo versus the other.

2. 10× Tris/acetate/EDTA (TAE) buffer can be used in place of 10× TBE buffer. If this substitution is made, be sure to use 1× TAE as the gel running buffer in place of 1× TBE.
3. If autoclaving, add the glucose *after* autoclaving. To increase the shelf life of this media, make a 50× stock of the adenine hemisulfate and add it at 1× concentration just prior to use. Store the 50× stock at −20°C (upon thawing, there will be a small amount of precipitation which will not go back into solution).
4. Solution should be adjusted to pH 7.5 using KOH. This limits introduction of additional sodium ions.
5. Target oligo can be made in large batches and stored at −80°C in a light-protected container. One 20 μL PCR reaction should yield approximately 12–14 μL of purified 300–500 nM substrate. When scaling up production of these substrates, maintain 20 μL PCR reaction volumes, and increase the number of reactions simultaneously run.
6. This gradual decrease in final temperature allows for high-efficiency annealing into double-stranded DNA target oligo (leaving minimal single-stranded or mis-annealed target).
7. *Exo*I is diluted with water to a total volume of 2 μL per sample for ease and accuracy of transfer to the PCR reaction. Do not add any of the supplied *Exo*I buffer.
8. Special care should be taken to avoid any bubbles in the sephadex suspension when mixing or aliquoting to the filter plate. Bubbles will lead to cracks within the final, centrifuged sephadex columns, and cracked columns should be discarded. We find that careful pipetting, using wide bore tips or standard p200 tips cut at approximately the 50 μL gradation, reduces frequency of cracking. We also find that allowing 30 min between pipetting and centrifugation can significantly reduce column cracking.
9. While it is more difficult to load samples without dye in the loading buffer, this allows for clear visualization of the target oligo and *A*647-labeled primer. Colored dyes will fluoresce under the LI-COR excitation and confound the image.
10. Frozen competent cells are prepared according to the published protocol by Gietz and Schiestl (15). Add 2.5×10^9 EBY100 cells from an overnight 2× YPAD culture to 500 mL fresh 2× YPAD media. Grow at 30°C to a density of at least 20 million/mL. Pellet the cells and wash with sterile water. Resuspend the washed cell pellet in 5 mL of 5% v/v glycerol + 10%

v/v DMSO in water. Aliquot 50 μL volumes to microcentrifuge tubes. Pack the tubes into a Styrofoam rack with lid (or similar form of insulation) and place at −80°C. (The insulation allows for gradual freezing of the cells).

11. When transforming a library of variant homing endonucleases, increase the number of yeast and volume of the transformation mixture according to Gietz and Schiestl (16).
12. When transforming high-quality plasmid DNA, a single frozen aliquot of competent yeast cells in the described volume of transformation mixture can be used for multiple reactions. In this case, divide the resuspended cells (*prior* to addition of DNA) into up to 15 equal volumes and add up to 1 μL total volume of plasmid DNA to each aliquot. Proceed to the incubation step.
13. We have found that an incubation time of 40–42 min at 42°C provides the highest transformation efficiency with lowest cell death. Longer incubation times can lead to significant cell death. If using a high-quality plasmid, a shorter incubation time of 20 min will suffice for the generation of transformed clones.
14. Raffinose cultures can be successfully started using a single colony from a selective media + glucose plate. Alternatively, we have found that an initial overnight incubation in YPAD media (at 30°C with 250 rpm shaking) can substantially increase induction efficiency, and the absence of selective media at this stage does not result in significant plasmid loss.
15. When using vertical tube racks inside a shaking incubator, position the 15 mL culture tubes at a slant to allow for maximum aeration, and do not use more than 1.5 mL of media.
16. Care should be taken to wash yeast from the raffinose culture at least twice before transferring to the galactose media. This limits carry-over of raffinose and/or glucose.
17. Induced cultures should be kept on ice or at 4°C following the galactose induction, in the 2% galactose media. Well-folded homing endonucleases will be stably expressed on the yeast surface for several days, although the total expression levels and catalytic activity may decrease slightly, depending on endonuclease.
18. On our spectrophotometer, the density of a yeast culture can be estimated by mixing a 1:10 dilution of yeast in water and measuring the resulting OD_{600}. A simple calculation of $OD_{600} \times 300$ provides an estimated value for density of the culture in millions of cells per milliliter. The validity of this estimate should be checked when using a different instrument.
19. If running a large number of samples, the assay can be performed in a 384-well conical-bottom plate, with 50,000 cells/well. Volumes for staining and wash steps for a 384-well

plate are 8 μL anti-HA stain, 8 μL conjugated DNA-SAV-PE stain, and 10 μL anti-Myc FITC stain. All washes should use a minimum 100 μL buffer.

20. If running a small number of samples, the assay can be performed using 1.5 mL microcentrifuge tubes. In this format, cells can be spun down in a tabletop centrifuge at speeds up to 10,000 × g for 1 min. Perform 4°C incubation steps on slow rotator, if possible.
21. Include the volume of target oligo in calculations for total conjugation reaction, as the dilution volume should be around 1:8–1:14, and is therefore substantial. The resulting slight decrease in IOCB salt concentration is not problematic at this point and can be disregarded.
22. If using 1.5 mL microcentrifuge tubes, pipette the oligo onto the side of the tube not contacting SAV-PE, then gently vortex to mix SAV-PE and oligo quickly. Likewise, in plate format the oligo should be pipetted onto the side of plate wells not contacting the SAV-PE mixture, if possible, and quickly mixed by vortex or multichannel pipette.
23. Yeast can be left in anti-Myc FITC stain overnight, if necessary. Due to relatively low affinity of this antibody, FITC-stained cells should not be washed or diluted greater than 2× if cells are to sit for more than 4 h prior to acquisition.
24. 25,000 yeast can be assayed in a 384-well plate format. Use a total volume of 20 μL. Low cell number is important to ensure that the effective concentration of ds-oligo substrate is not altered in affinity titration experiments.
25. For the I-OnuI family of homing endonucleases, specific binding can be detected from 100 pM to 50 nM. Higher concentrations of oligo may be nonspecifically bound and lower concentrations can be difficult to detect. This protocol can also be used to determine binding along a titration of various oligo concentrations.

References

1. Takeuchi R, Lambert AR, Mak AN-S, Jacoby K, Dickson RJ, Gloor GB, Scharenberg AM, Edgell DR, Stoddard BL (2011) Tapping natural reservoirs of homing endonucleases for targeted gene modification. Proc Natl Acad Sci U S A 108(32):13077–13082
2. Heath PJ, Stephens KM, Monnat RJ, Stoddard BL (1997) The structure of I-CreI, a group I intron-encoded homing endonuclease. Nat Struct Biol 4(6):468–476
3. Duan X, Gimble FS, Quiocho FA (1997) Crystal structure of PI-SceI, a homing endonuclease with protein splicing activity. Cell 89(4):555–564
4. Gimble FS (2000) Invasion of a multitude of genetic niches by mobile endonuclease genes. FEMS Microbiol Lett 185(2):99–107
5. Jurica MS, Stoddard BL (1999) Homing endonucleases: structure, function and evolution. Cell Mol Life Sci 55(10):1304–1326
6. Thermes V, Grabher C, Ristoratore F, Bourrat F, Choulika A, Wittbrodt J, Joly JS (2002) I-SceI meganuclease mediates highly efficient transgenesis in fish. Mech Dev 118(1–2):91–98
7. Gouble A, Smith J, Bruneau S et al (2006) Efficient in toto targeted recombination in mouse liver

by meganuclease-induced double-strand break. J Gene Med 8(5):616–622

8. Arnould S, Perez C, Cabaniols JP, Smith J, Gouble A, Grizot S, Epinat JC, Duclert A, Duchateau P, Paques F (2007) Engineered I-CreI derivatives cleaving sequences from the human XPC gene can induce highly efficient gene correction in mammalian cells. J Mol Biol 371(1):49–65
9. Gao H, Smith J, Yang M et al (2010) Heritable targeted mutagenesis in maize using a designed endonuclease. Plant J 61(1):176–187
10. Windbichler N, Papathanos PA, Catteruccia F, Ranson H, Burt A, Crisanti A (2007) Homing endonuclease mediated gene targeting in Anopheles gambiae cells and embryos. Nucleic Acids Res 35(17):5922–5933
11. Smith F, Rouet P, Romanienko PJ, Jasin M (1995) Double-strand breaks at the target locus stimulate gene targeting in embryonic stem cells. Nucleic Acids Res 23(24): 5012–5019
12. Jarjour J, West-Foyle H, Certo MT, Hubert CG, Doyle L, Getz MM, Stoddard BL, Scharenberg AM (2009) High-resolution profiling of homing endonuclease binding and catalytic specificity using yeast surface display. Nucleic Acids Res 37(20):6871–6880
13. Volná P, Jarjour J, Baxter S, Roffler SR, Monnat RJ, Stoddard BL, Scharenberg AM (2007) Flow cytometric analysis of DNA binding and cleavage by cell surface-displayed homing endonucleases. Nucleic Acids Res 35(8):2748–2758
14. Gietz RD, Schiestl RH (2007) High-efficiency yeast transformation using the LiAc/SS carrier DNA/PEG method. Nat Protoc 2(1):31–34
15. Gietz RD, Schiestl RH (2007) Frozen competent yeast cells that can be transformed with high efficiency using the LiAc/SS carrier DNA/PEG method. Nat Protoc 2(1):1–4
16. Gietz RD, Schiestl RH (2007) Large-scale high-efficiency yeast transformation using the LiAc/SS carrier DNA/PEG method. Nat Protoc 2(1):38–41

Chapter 5

TAL Effector Nuclease (TALEN) Engineering

Ting Li and Bing Yang

Abstract

TALENs, fusion proteins of DNA binding domains of TAL (transcription activator-like) effectors and the DNA cleavage domains of endonuclease FokI, have emerged as genetic tools for targeted gene modification, holding great potential for basic and applied research, even for gene therapy. Here we present a simple and efficient approach to custom-engineering TALEN genes with four basic TAL repeats and their DNA recognition cipher. The "modular assembly" method also involves the "Golden Gate" cloning strategy, using 53 ready-to-use plasmids in just two rounds of restriction and ligation to assemble TALENs with up to 24 repeat units that recognize up to 24 bp of target DNA.

Key words: TAL effector, FokI, TAL effector nuclease, TALEN, Enzyme engineering, Golden gate cloning, Genetic engineering

1. Introduction

TAL effector nucleases (TALENs) are a type of fusion protein comprised of full-length or truncated TAL effector and the DNA cleavage domain of endonuclease FokI (1, 2). The newly merged TALENs have been used as a promising genetic tool to create site-specific gene modification in plant cells (3, 4), yeast (5), animals (6–9), and even human pluripotent cells (10). Targeted gene modification is achieved through the DNA repair machinery acting upon the chromosomal DNA double-strand breaks (DSB) caused by the TALENs that are expressed in the target cells. A DSB is principally repaired by one of two pathways, nonhomologous end-joining repair (NHEJ) or homologous recombination (HR). Repair by NHEJ often results in deletions/insertions at the site of DNA breakage and eventually the alteration of gene function. Moreover, DSBs can stimulate HR between the endogenous target

James C. Samuelson (ed.), *Enzyme Engineering: Methods and Protocols*, Methods in Molecular Biology, vol. 978, DOI 10.1007/978-1-62703-293-3_5, © Springer Science+Business Media New York 2013

locus and an exogenously introduced homologous DNA fragment with desired genetic information, a process called gene replacement or genome editing (11–13).

The DNA recognition of a TAL effector is carried out by the central tandem repeats; each repeat comprising 33–35 (34 in most cases) amino acids. The multiple repeats are nearly identical except for two variable amino acids at positions 12 and 13, the so-called RVD (14). To recognize the specific DNA sequence, the superhelix repeat domain of the TAL effector (e.g., PthXo1) wraps around the major groove of double-stranded DNA, aligning the DNA sense strand in a manner of one repeat or one RVD to one nucleotide sequentially and in an orientation of N- to C-terminus of the TAL effector corresponding to the 5′–3′ end of the target DNA (15, 16). In the fusion protein, the chimeric TALENs retain the ability of TAL effectors to recognize the specific double-stranded DNA and the ability of the FokI cleavage domain to make nonspecific DNA DSB in vitro or in vivo (1–3). To efficiently make a DSB, the FokI domain needs to dimerize (17) and, therefore, two TALENs are needed (1–3, 18). As a functioning pair, one molecule of TALEN (monomer) binds to one target site (also called the effector binding element, EBE) in one strand, while the other TALEN (mono- or heterodimeric) binds to another adjacent EBE in the complementary strand of a double-stranded DNA; and the two EBEs are separated by a spacer sequence of an appropriate length (in a range of 10–31 bp dependent on the lengths of the C-termini of the TAL effectors fused to the FokI domain). This allows the two FokI domains to be in close vicinity in order to dimerize and consequentially execute double-strand cleavage (Fig. 1).

The specificity of DNA binding by the TAL effector repeats is dictated by a fairly simple code where one nucleotide of the DNA target site corresponds to one RVD (19, 20). At least 23 RVDs have been found among the naturally occurring TAL effectors from *Xanthomonas* bacteria, but the following four RVDs are the most prevalent: NI, NG, NN, and HD (in single letter amino acid code). Each of the four RVDs recognizes a particular nucleotide preferentially over others—that is, the RVDs NI, NG, NN, and HD recognize the nucleotides “A,” “T,” “G,” and “C,” respectively (14). The specificity of the four RVDs in native TAL effectors made it possible to use them exclusively to de novo synthesize or assemble artificial TAL effector repeat arrays to target preselected sequences of interest. Several methods to custom-engineer designer TAL effectors or TALENs using the prevalent RVD repeats including ours have been reported (4, 5, 18, 21–26).

We have developed a modular assembly method to custom-engineer designer TALENs (dTALENs) for targeted gene modification. The method has been successfully used to make dTALENs to target endogenous genes in yeast and rice (5, 27).

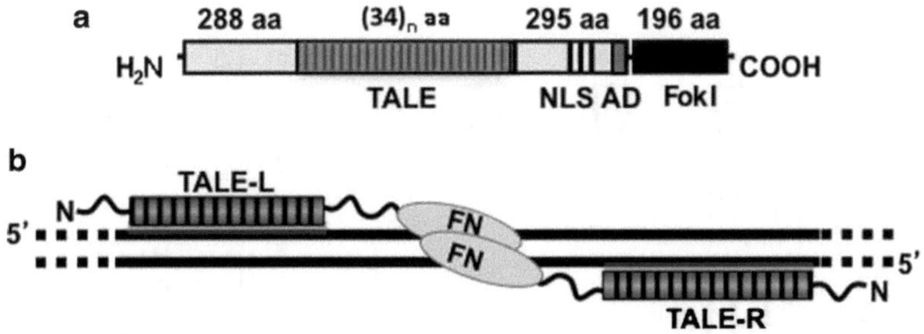

Fig. 1. Schematics of TALEN and the active TALEN/DNA complex for double stranded DNA cleavage. (**a**) TALEN protein is comprised of a TAL effector (TALE) and the FokI DNA cleavage domain (FokI). The architecture shown here is of a full-length TAL effector (e.g., AvrXa7) with the number of amino acids of each domain indicated above. The C-terminus of the TALE contains nuclear localization signals (NLS) directing the TALEN into the nuclei of target cells. AD is the activation domain. (**b**) Paired TALENs (homo- or heterodimeric).

Here we describe an improved version of that method. Instead of ligating eight predigested PCR-derived single-repeat units in one reaction, the present method involves digesting and ligating eight plasmid borne single-repeat units into one receptor plasmid in one tube by adapting the "Golden Gate" cloning strategy described by Engler et al. (28). The present method uses four basic repeats with RVDs of NI, NG, NN, and HD to generate 51 ready-to-use modules of single-TAL repeat units each in a plasmid. With one receptor vector for assembling the repeat array and one TALEN scaffold vector, the whole kit contains only 53 plasmids. These modules can be used to make TALENs through two rounds of DNA restriction and ligation. Assembly of the TALEN repeats can be in any predetermined order and number of repeat units can be in a range of 2–24. The modular assembly method is simple, fast, and inexpensive and can be performed in most academic or industrial molecular biology laboratories.

2. Materials

2.1. Plasmids

The 53 *Escherichia coli* strains, each containing a plasmid, are available from the Yang laboratory under a Material Transfer Agreement (MTA). The 51 modular repeats are summarized in Fig. 2. Sequence information for the modular repeats is available upon request. The receptor vector pTL-n contains the *ccdB* gene and confers tetracycline resistance to the *E. coli* strain DB3.1. Restriction of pTL-n with *Bsm*BI releases the *ccdB* gene and yields a vector fragment with one end compatible to the 5′ end of the modular repeats from Set 1 and the other end compatible to the 3′ end of the modular repeats from Set 8. The pTL-n vector can be used to

a

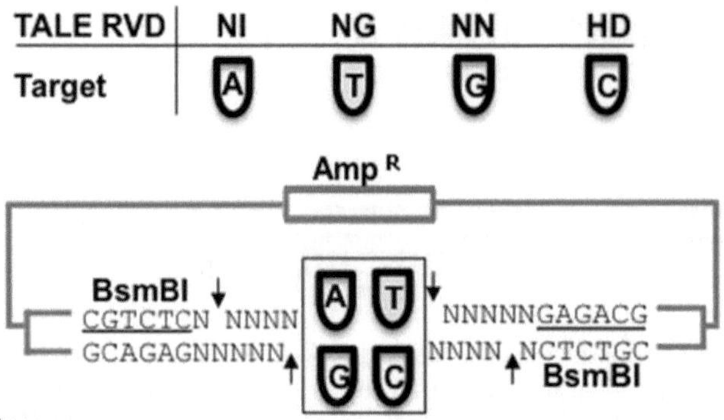

Set	5′ Overhang	3′ Overhang
1	CTGC	TGAT
2	ACTA	AGAA
3	TCTT	CAAC
4	GTTG	CGAG
5	GCTC	TCAA
6	AGTT	ACAT
7	TGTA	GCAC
8	CGTG	CAGA

b

Table of single-repeat units for assembly of TALEN repeat array

1st 8-mer:

Head Set (an internal SphI at 5′)	Set 2	Set 3	Set 4	Set 5	Set 6	Set 7	Pst-R Set (An internal PstI at 3′)
Head A, T, G, C	A2, T2, G2, C2	A3, T3, G3, C3	A4, T4, G4, C4	A5, T5, G5, C5	A6, T6, G6, C6	A7, T7, G7, C7	PstR-A, T, G, C
ctgc..../....tgat	acta..../...agaa	tctt..../....caac	gttg..../...cgag	gctc..../....tcaa	agtt..../....acat	tgta..../....gcac	cgtg..../....caga

2nd 8-mer:

Pst-F Set (An internal PstI at 5′)	Set 2	Set 3	Set 4	Set 5	Set 6	Set 7	BsrG-R Set (An internal BsrGI at 3′)
PstF-A, T, G, C							BsrR-A, T, G, C
ctgc..../....tgat	acta..../...agaa	tctt..../....caac	gttg..../...cgag	gctc..../....tcaa	agtt..../....acat	tgta..../....gcac	cgtg..../....caga

3rd 8-mer

BsrG-F Set (An internal BsrGI at 5′)	Set 2	Set 3	Set 4	Set 5	Set 6	Set 7	Tail-8 (Last common half repeat with an internal AatII at 3′)
BsrF-A, T, G, C							
ctgc..../....tgat	acta..../...agaa	tctt..../....caac	gttg..../....cgag	gctc..../...tcaa	agtt..../....acat	tgta..../....gcac	cgtg..../....caga

Additional single Tail modules

	Tail-2	Tail-3	Tail-4	Tail-5	Tail-6	Tail-7	
	acta..../...caga	tctt..../....caga	gttg..../....caga	gctc..../...caga	agtt..../....caga	tgta..../....caga	

Fig. 2. (**a**) Design of eight basic single-repeat sets each with four core RVD-coding modules (for A, T, G, and C, respectively) and bearing 5′ and 3′ terminal sequences as dictated by the table to the right. (**b**) Detailed repeat sets with variations for Set 1 (Head Set, Pst-F Set, and BsrG-F Set) and Set 8 (Pst-R Set, BsrG-R Set, and Tail-2 to -8 Set). Restriction of each single-repeat containing plasmid with the type IIS restriction enzyme *Bsm*BI generates the unique 5′ and 3′ overhangs with 2-bp polymorphisms (as described in the last rows of each panel). Tail-2 to -8 can be used to substitute any repeat at the respective position (e.g., Tail-2 for Set 2, Tail-3 for Set 3, and etc.) to put an end to the last repeat array, resulting in a varying number of the last repeat arrays.

clone any one of the repeat arrays. Another plasmid is pSK/TALEN-ΔR, a repeat-less fusion of full-length AvrXa7 and the FokI catalytic domain. The sequence information is also available upon request.

2.2. Other Reagents and Equipment

1. T4 DNA ligase or T7 DNA ligase.
2. 10 mM Adenosine-5′-triphosphate (ATP).
3. Restriction enzymes *Bsm*BI or *Esp*3I, *Sph*I, *Pst*I, *Bsr*GI, *Aat*II.
4. *E. coli* JM109 or DH5α competent cells.
5. SOC medium (0.5% yeast extract, 2% tryptone, 10 mM NaCl, 2.5 mM KCl, 10 mM $MgCl_2$, 10 mM $MgSO_4$, and 20 mM Glucose).

6. LB solid and liquid media with ampicillin (amp, 100 μg/ml) or tetracycline (tet, 12.5 μg/ml).
7. 2× YT medium (1.6% peptone, 1% yeast extract, and 0.5% NaCl).
8. DNA miniprep kit (Qiagen).
9. Low-EEO/Multipurpose agarose.
10. TAE electrophoresis buffer (40 mM Tris-acetate, 1 mM EDTA).
11. DNA GENECLEAN® III kit (BIO101 Systems).
12. Oligonucleotides:

    Seq-F, 5′- TGGCCCGTGTCTCAAAATCTCTG-3′.

    Seq-R, 5′- ATCTTTTCTACGGGGTCTGACG-3′.
13. PCR thermal cycler.

3. Methods

3.1. Choose the Target Site

Criteria:

(a) "T" precedes each EBE sequence (or at zero position of target site).

(b) EBE length ranges from 12 to 24 bp (see Note 1).

(c) Optimal spacer between the dual EBEs ranges from 16 to 20 bp.

Example:

```
TGAACTTCTTAATTTACCAAAGCAgatcat
   ttttgtcggcctATGATTGGGGTGCTTGTTTGGCCA

ACTTGAAGAATTAAATGGTTTCGTctagtaaaaacagccggaTA
   CTAACCCCACGAACAAACCGGT

Left EBE (23 bp) spacer (18 bp) Right EBE (23 bp)

Left EBE: 5'-GAACTTCTTAATTTACCAAAGCA-3'

Right EBE: 5'-GGCCAAACAAGCACCCCAATCAT-3'
```

3.2. Prepare Modular Repeat Plasmid DNA

1. Streak out the stock strains on LB plates with appropriate antibiotics (amp or tet).
2. Initiate bacterial cultures for all or specific individual plasmids in 5 ml 2× YT liquid medium with appropriate antibiotics (amp or tet) from single colonies, and grow the cell culture at 37°C for 12–16 h with vigorous shaking.
3. Prepare plasmid DNA using the QIAprep Spin Miniprep kit; measure and adjust the DNA concentration to 200 ng/μl.

3.3. Assemble 8-Mer Repeats

1. Arrange all the 51 modular repeats in an order as in Fig. 2.
2. Assemble the first 8-mer repeat array:
   Based on the first 8 nucleotides of EBE, arrange 8 plasmids in a predetermined order. For example, to assemble 8 repeats to recognize GAACTTCT, pick "Head G" from Head Set, "A2" from Set 2, "A3" from Set 3, etc., "PstR-T" from Pst-R Set; Add 1 μl of each plasmid into a PCR tube, and also add 1 μl of pTL-n.
3. Add up the reaction volume to 19 μl with 7 μl of H_2O, 2 μl of 10× ligation buffer, and 1 μl of *Bsm*BI.
4. Predigest the DNA at 37°C for 30 min.
5. Add 1 μl of T4 DNA ligase and 1 μl of ATP.
6. Set and run a thermocycler with the following program:
   16°C for 30 min.
   50 cycles of 37°C for 5 min and 16°C for 5 min.
   55°C for 15 min.
   80°C for 15 min.
   4°C for 1 h.
7. Transform JM109 (or DH5α) with 5 μl of ligation product by electroporation or chemical competent cell method.
8. Recover the cells by incubating in SOC medium for 1 h and plate all of the cells on a tetracycline plate.
9. Assemble the second or third 8-mer repeat array:
   Pick the plasmids from each set of modular repeats (Pst-F Set, Set 2–7, and BsrG-R Set) for the second 8-mer and modular sets (BsrG-F Set, Set 2–7, Tail-8) for the third 8-mer. The plasmid from Tail-2–7 can be used to replace the respective plasmid from Set 2–7 to assemble a varying number of repeats (e.g., 2-, 3-, to 7-mer repeats).
10. Perform the ligation and transformation procedure according to the assembly of the first 8-mer.
11. Confirm each 8-mer repeat array.
12. Pick seven colonies for cell culture and plasmid DNA isolation.
13. Digest the plasmid DNA with XbaI and XhoI to release the repeat array and enable analysis of insert size (see Note 2) (Fig. 4).
14. Sequence one clone of the correct size with either primer Seq-F or Seq-R (see Note 3).

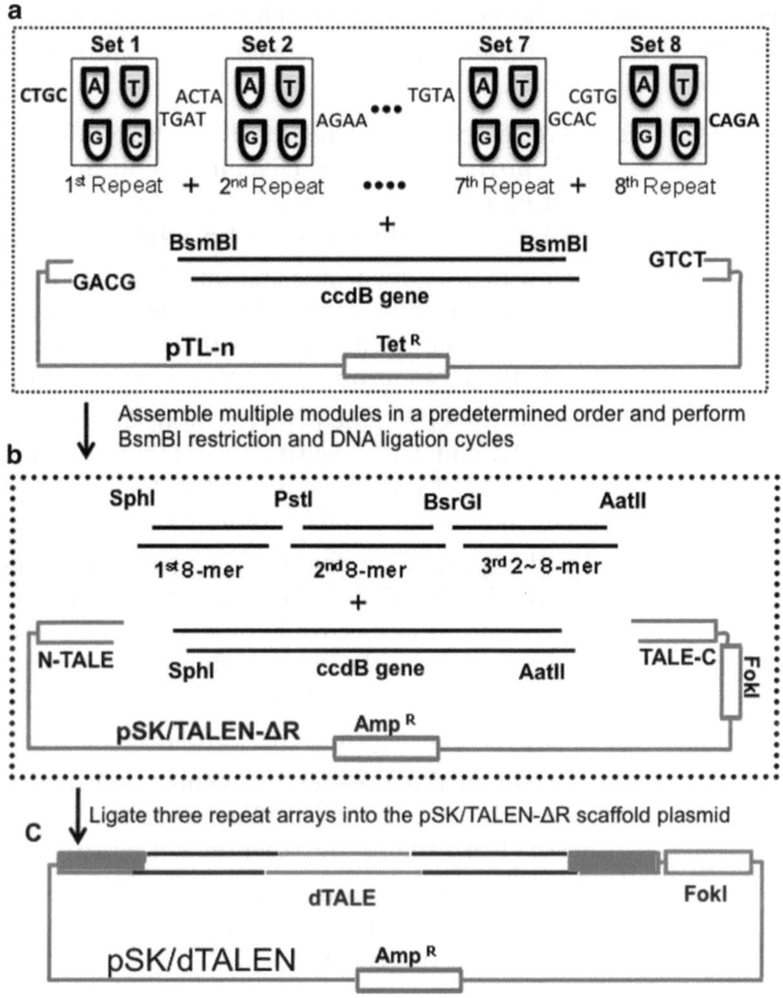

Fig. 3. Assembly of a typical 8-mer array and the final TALEN gene by restriction and ligation.

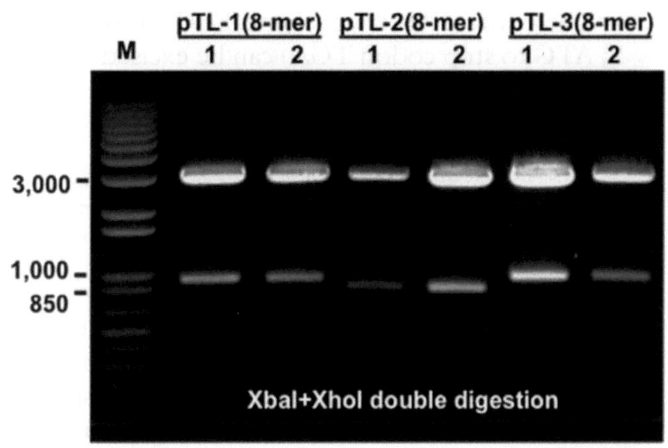

Fig. 4. Gel image for three 8-mer repeat arrays digested with XbaI and XhoI. Two clones (#1 and 2) from each 8-mer are shown in an ethidium bromide stained gel. M, 1 kb Plus DNA ladders from Invitrogen with the size (bp) indicated at the left side.

3.4. Assemble the Repeat Arrays into the TALEN Scaffold Vector

After isolating the three correct 8-mers, carry out the second round of digestion and ligation to create the final product TALEN gene (see Fig. 3).

1. Vector preparation: Digest pSK/TALEN-ΔR vector with *Sph*I and *Aat*II. Separate the DNA fragments in 1% agarose gel by electrophoresis. Collect the 5.2 kb (larger) fragment and purify the DNA using the GENECLEAN III kit.
2. First 8-mer fragment preparation: Digest pTL-1st 8-mer with *Sph*I and *Pst*I, separate the DNA fragments in 1% agarose gel by electrophoresis. Collect the 0.8 kb (smaller) fragment and purify the DNA using the GENECLEAN III kit.
3. Second 8-mer fragment preparation: Digest pTL-2nd 8-mer with *Pst*I and *Bsr*GI, separate the DNA fragments in 1% agarose gel by electrophoresis. Collect the 0.8 kb (smaller) fragment and purify the DNA using the GENECLEAN III kit.
4. Third repeat array fragment preparation: Digest pTL-3rd 8-mer with *Bsr*GI and *Aat*II, separate the DNA fragments in 1% agarose gel by electrophoresis. Collect the smaller fragment (size depends on the repeats number being used) and purify the DNA using the GENECLEAN III kit.
5. Ligate three 8-mers into the scaffold vector (see Note 4):

   Add all three 8-mer fragments (~60 ng/each) into one tube with vector DNA (~100 ng). Add 10× ligase buffer (1 μl), ligase (1 μl), and set the reaction at 4 or 16°C overnight.
6. Transform the ligation products into *E. coli*, plate for single colonies, grow cell culture from single colonies. Isolate plasmid DNA as described above.
7. Confirm the clone of interest by digestion with *Pst*I (1.6 + 6.1 kb), or with *Bsr*GI and *Sal*I (1.4 + 6.3 kb).

3.5. Construct Expression Vector

The entire TAL effector and FokI fusion gene (from start codon ATG to stop codon TGA) can be excised with *Kpn*I or *Bgl*II at the 5′ end and *Spe*I at the 3′ end for cloning into the desired expression vector.

4. Notes

1. Tail-2–8 each contains only half of one repeat (20 amino acids with asparagine at position 12 and the amino acid missing at position 13, thus conferring flexible specificity for any one of four nucleotides). For fewer than 24 repeats, for example, the 11.5-repeat array may be linked through one first 8-mer and a second 3.5-mer (ligation of one modular plasmid from Pst-F

Set, Set 2 and Set 3, respectively, and Tail-4) and inserted into the TALEN scaffold by ligating the two fragments (*Sph*I–*Pst*I and *Pst*I–*Aat*II) into pSK/TALEN-ΔR at *Sph*I and *Aat*II sites.

2. The size of different repeats: Head repeat, 160 bp; Set 2–7, 102 bp; Tail-2–8, 224 bp; so the first 8-mer fragment is ~890 bp (Fig. 4, lanes 2–3), the second 8-mer ~830 bp (Fig. 4, lanes 5–6), and the third 8-mer ~940 bp (Fig. 4, lanes 6–7). If <8 repeats, the fragment size of the third repeat array is $[940-102\times(8-n)]$ bp.
3. A good sequence read should cover the whole repeat region.
4. DNA may be concentrated by vacuum if the DNA concentration is too low.

Acknowledgment

This work was supported by US National Science Foundation Plant Genome Research Project (award no. 0820831 to BY).

References

1. Li T, Huang S, Jiang WZ, Wright D, Spalding MH, Weeks DP, Yang B (2011) TAL nucleases (TALNs): hybrid proteins composed of TAL effectors and FokI DNA-cleavage domain. Nucleic Acids Res 39:359–372
2. Christian M, Cermak T, Doyle EL, Schmidt C, Zhang F, Hummel A, Bogdanove AJ, Voytas DF (2010) Targeting DNA double-strand breaks with TAL effector nucleases. Genetics 186:757–761
3. Mahfouz MM, Li L, Shamimuzzaman M, Wibowo A, Fang X, Zhu JK (2011) De novo-engineered transcription activator-like effector (TALE) hybrid nuclease with novel DNA binding specificity creates double-strand breaks. Proc Natl Acad Sci USA 108:2623–2628
4. Cermak T, Doyle EL, Christian M, Wang L, Zhang Y, Schmidt C, Baller JA, Somia NV, Bogdanove AJ, Voytas DF (2011) Efficient design and assembly of custom TALEN and other TAL effector-based constructs for DNA targeting. Nucleic Acids Res 39:e82
5. Li T, Huang S, Zhao X, Wright DA, Carpenter S, Spalding MH, Weeks DP, Yang B (2011) Modularly assembled designer TAL effector nucleases for targeted gene knockout and gene replacement in eukaryotes. Nucleic Acids Res 39:6315–6325
6. Sander JD, Cade L, Khayter C, Reyon D, Peterson RT, Joung JK, Yeh JR (2011) Targeted gene disruption in somatic zebrafish cells using engineered TALENs. Nat Biotechnol 29:697–698
7. Huang P, Xiao A, Zhou M, Zhu Z, Lin S, Zhang B (2011) Heritable gene targeting in zebrafish using customized TALENs. Nat Biotechnol 29:699–700
8. Tesson L, Usal C, Menoret S, Leung E, Niles BJ, Remy S, Santiago Y, Vincent AI, Meng X, Zhang L et al (2011) Knockout rats generated by embryo microinjection of TALENs. Nat Biotechnol 29:695–696
9. Wood AJ, Lo TW, Zeitler B, Pickle CS, Ralston EJ, Lee AH, Amora R, Miller JC, Leung E, Meng X et al (2011) Targeted genome editing across species using ZFNs and TALENs. Science 333:307
10. Hockemeyer D, Wang H, Kiani S, Lai CS, Gao Q, Cassady JP, Cost GJ, Zhang L, Santiago Y, Miller JC et al (2011) Genetic engineering of human pluripotent cells using TALE nucleases. Nat Biotechnol 29:731–734
11. Jasin M (1996) Genetic manipulation of genomes with rare-cutting endonucleases. Trends Genet 12:224–228

12. Vasquez KM, Marburger K, Intody Z, Wilson JH (2001) Manipulating the mammalian genome by homologous recombination. Proc Natl Acad Sci U S A 98:8403–8410
13. Bibikova M, Beumer K, Trautman JK, Carroll D (2003) Enhancing gene targeting with designed zinc finger nucleases. Science 300:764
14. Boch J, Bonas U (2010) Xanthomonas AvrBs3 family-type III effectors: discovery and function. Annu Rev Phytopathol 48:419–436
15. Mak AN, Bradley P, Cernadas RA, Bogdanove AJ, Stoddard BL (2012) The crystal structure of TAL effector PthXo1 bound to its DNA target. Science 335:716–719
16. Deng D, Yan C, Pan X, Mahfouz M, Wang J, Zhu JK, Shi Y, Yan N (2012) Structural basis for sequence-specific recognition of DNA by TAL effectors. Science 335:720–723
17. Bitinaite J, Wah DA, Aggarwal AK, Schildkraut I (1998) FokI dimerization is required for DNA cleavage. Proc Natl Acad Sci USA 95:10570–10575
18. Miller JC, Tan S, Qiao G, Barlow KA, Wang J, Xia DF, Meng X, Paschon DE, Leung E, Hinkley SJ et al (2011) A TALE nuclease architecture for efficient genome editing. Nat Biotechnol 29:143–148
19. Boch J, Scholze H, Schornack S, Landgraf A, Hahn S, Kay S, Lahaye T, Nickstadt A, Bonas U (2009) Breaking the code of DNA binding specificity of TAL-type III effectors. Science 326:1509–1512
20. Moscou MJ, Bogdanove AJ (2009) A simple cipher governs DNA recognition by TAL effectors. Science 326:1501
21. Morbitzer R, Romer P, Boch J, Lahaye T (2010) Regulation of selected genome loci using de novo-engineered transcription activator-like effector (TALE)-type transcription factors. Proc Natl Acad Sci USA 107:21617–21622
22. Zhang F, Cong L, Lodato S, Kosuri S, Church GM, Arlotta P (2011) Efficient construction of sequence-specific TAL effectors for modulating mammalian transcription. Nat Biotechnol 29:149–153
23. Geissler R, Scholze H, Hahn S, Streubel J, Bonas U, Behrens SE, Boch J (2011) Transcriptional activators of human genes with programmable DNA-specificity. PLoS One 6:e19509
24. Morbitzer R, Elsaesser J, Hausner J, Lahaye T (2011) Assembly of custom TALE-type DNA binding domains by modular cloning. Nucleic Acids Res 39:5790–5799
25. Li L, Piatek MJ, Atef A, Piatek A, Wibowo A, Fang X, Sabir JS, Zhu JK, Mahfouz MM (2012) Rapid and highly efficient construction of TALE-based transcriptional regulators and nucleases for genome modification. Plant Mol Biol 78:407–416
26. Weber E, Gruetzner R, Werner S, Engler C, Marillonnet S (2011) Assembly of designer TAL effectors by golden gate cloning. PLoS One 6:e19722
27. Li T, Liu B, Spalding M, Weeks D, Yang B (2012) High efficiency TALEN-based gene editing for producing disease resistance in crop plants. Nat Biotechnol 30(5):390–392
28. Engler C, Kandzia R, Marillonnet S (2008) A one pot, one step, precision cloning method with high throughput capability. PLoS One 3:e3647

Chapter 6

In Vitro Evolution of Enzymes

Misha V. Golynskiy, John C. Haugner III, Aleardo Morelli, Dana Morrone, and Burckhard Seelig

Abstract

In the past decade, in vitro evolution techniques have been used to improve the performance or alter the activity of a number of different enzymes and have generated enzymes de novo. In this review, we provide an overview of the available in vitro methods, their application, and some general considerations for enzyme engineering in vitro. We discuss the advantages of in vitro over in vivo approaches and focus on ribosome display, mRNA display, DNA display technologies, and in vitro compartmentalization (IVC) methods. This review aims to help researchers determine which approach is best suited for their own experimental needs and to highlight that in vitro methods offer a promising route for enzyme engineering.

Key words: Enzyme, Directed evolution, In vitro selection, Ribosome display, mRNA display, DNA display, In vitro compartmentalization

1. Introduction

In vitro enzyme evolution offers a means to engineer enzymes by exploring enormous libraries of protein variants that exceed the capabilities of in vivo methods. The development of cell-free protein production systems made it possible to evolve enzymes outside of cells, in a test tube. In vitro evolution techniques have been used to improve existing enzymes and, in addition, have enabled the generation of biocatalysts de novo from a non-catalytic protein library.

All methods used for enzyme evolution require that each protein in a pool of mutants can be traced back to its encoding gene for identification, and potentially for the purpose of amplification, expression, and further evolution (1). A stable genotype–phenotype linkage allows for many enzyme variants to be mixed in a single reservoir while maintaining the ability to amplify genes of individual desired variants. Those variants are isolated from the reservoir

James C. Samuelson (ed.), *Enzyme Engineering: Methods and Protocols*, Methods in Molecular Biology, vol. 978,
DOI 10.1007/978-1-62703-293-3_6, © Springer Science+Business Media New York 2013

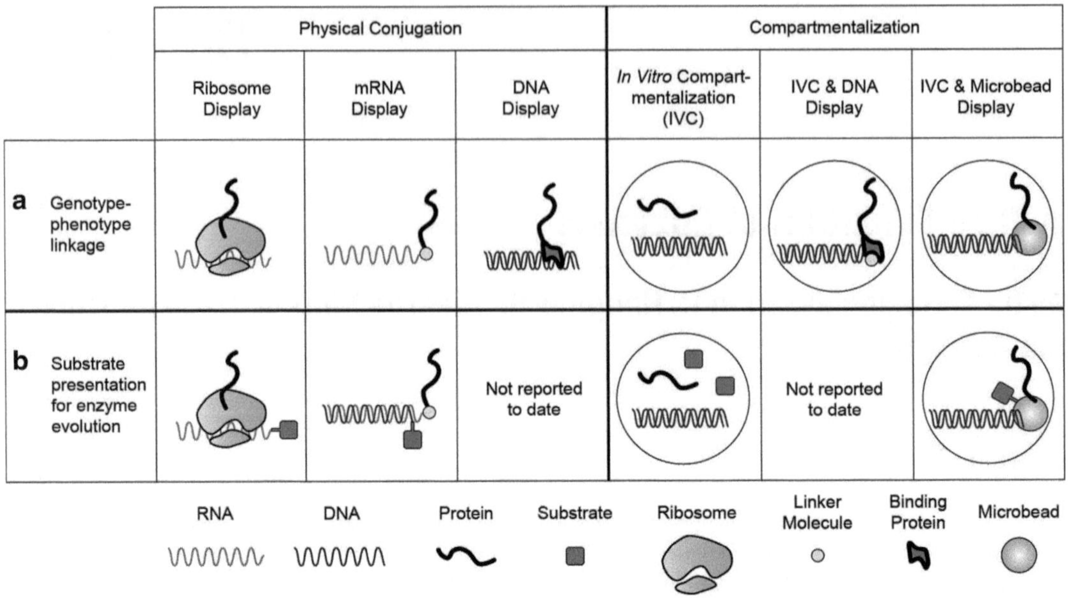

Fig. 1. Overview of methods for the in vitro selection or screening of proteins discussed in this review. (**a**) The *top row* shows different strategies to establish the crucial linkage between gene and protein. (**b**) The *bottom row* illustrates the introduction of substrate into the selection scheme to enable the evolution of enzymes.

using suitable screening or selection approaches. In the case of in vivo evolution methods, the genotype and phenotype are linked as the protein and its gene are contained in the same cell. With partial in vitro methods, proteins are translated by the host's cellular machinery and then displayed in an extracellular fashion, for example, on the surface of a phage in the phage display approach. In contrast, the methods described in this review are carried out entirely in vitro and do not require any step to be performed inside a host cell. The crucial genotype–phenotype link is maintained through either a direct physical link or through artificial compartmentalization. This review will discuss in vitro enzyme evolution technologies such as ribosome display, mRNA display, in vitro compartmentalization (IVC), and different versions of DNA display (Fig. 1). We will focus on protein enzymes only and therefore exclude the work on in vitro evolution of ribozymes and deoxyribozymes (2, 3). We will first describe general features common to all in vitro protein evolution strategies and then review the individual methods in more detail, highlighting examples of enzymes that have been evolved. We will also include several in vitro evolution strategies that have so far only been used for proof-of-principle model selections. We believe that this overview and the number of general considerations presented here will provide the reader with sufficient information to decide which of these broadly applicable in vitro techniques may be best suited for their own research. We hope to encourage scientists to harness the advantages of in vitro methods to evolve their specific enzyme of interest.

2. Benefits of In Vitro Evolution

In vitro methodologies have several advantages over in vivo and partial in vitro methods because they are not limited by cell survival, growth, or function. The three main advantages are (1) the ability to work with larger libraries of variants, (2) the tolerance to conditions that would be deleterious to cell survival, and (3) the ability to directly manipulate the DNA after each round of evolution.

As in vitro evolution is not dependent on library transformation into a host, the number of unique sequences that can be evaluated in a single experiment exceeds in vivo approaches. The largest reported in vitro libraries contain 10^{14} DNA sequences (4). By comparison, phage display libraries produce up to 10^{10} unique variants in a single transformation (5). Library sizes up to 10^{12} variants were reported for phage display by the pooling of dozens of separate transformations, but such scale-up may not be feasible for most laboratories (6). Most typical library sizes for in vivo selections are between 10^6 and 10^8 variants. Because in vitro evolution can search a larger sequence space, it is particularly well suited for isolating beneficial enzyme mutations that may be very rare.

The evolution of enzymes in vitro greatly expands the range of substrates and environmental conditions that can be investigated. The presentation of substrate to the enzymes is simplified as no cell walls have to be crossed, which are impermeable to many potential substrates. Most importantly, substrates and enzymes can be used that would be toxic to a cell (7). Furthermore, enzymes can be engineered with in vitro methods for increased stability under extreme conditions of pH, temperature, ion concentration, or in the presence of denaturants or organic solvents. In vitro evolution also allows for a more accurate representation of enzyme performance. Cellular evolution, in contrast, can generate complex phenotypes that falsely suggest increased activity through increased enzyme accumulation, rather than improved catalysis (8).

Finally, in vitro evolution allows for direct manipulation of the DNA library between each round of evolution. Unlike in vivo methods that require time-consuming purification of the target gene, DNA from in vitro evolution is amplified directly through PCR. This facilitates the introduction of diversity through methods like error-prone PCR or in vitro recombination. In comparison, in vivo methods may introduce genetic diversity by using microbial strains deficient in DNA repair pathways to eliminate the need for DNA purification. However, these mutations may occur anywhere in the genome, necessitating a low mutation rate for continued survival (9). Thus, by combining a selection or screen with methods to add genetic diversity, full Darwinian evolution can be carried out more conveniently in vitro.

While in vitro evolution greatly expands the tools available for the creation and engineering of new enzymes, in vivo approaches have certain advantages, too. As in vitro methods require purification of the genotype–phenotype components, in vivo evolution may involve fewer discrete steps. Furthermore, some enzymes are being developed for in vivo use, such as enzymes that need to function within a metabolic pathway. Those enzymes could initially be evolved in vitro, but ultimately need to be evolved in their native environment to optimize their intracellular compatibility. Thus, the two approaches can complement each other.

3. General Workflow for In Vitro Methods

All in vitro directed evolution methods follow a similar scheme. The initial DNA library encoding the protein variants is transcribed and translated, either sequentially or in a one-pot reaction. Next, the genotype–phenotype link and then genotype–phenotype-substrate link is established. This may be accomplished through a physical connection (ribosome display, mRNA display, and DNA display), or through compartmentalization (IVC, IVC-based DNA, and microbead displays) (Fig. 1). Active enzyme variants that convert substrate to product result in co-localization of genotype–phenotype with product and are then isolated by screening or selection. Finally, the genotype is recovered and either analyzed directly by sequencing or subjected to additional diversification for subsequent rounds of evolution. In the following sections, we will briefly discuss aspects of the construction of DNA libraries, present the different in vitro methods individually, and discuss their application for enzyme engineering.

4. Library Construction

In any directed evolution procedure, the size and quality of the starting DNA library are of great importance as they affect the probability of finding the desired mutant. Numerous reviews are available to guide researchers through library design and construction (27–31). Here, we will highlight only a few general considerations and point out aspects specific to in vitro evolution methods.

Although in vitro selection methods can sift through comparably large libraries of trillions of mutants, the sheer size of the protein sequence space prevents us from sampling more than an exceedingly small fraction of all possibilities. For example, the largest protein libraries used to date contain about 10^{13} variants (Table 1). This vast number of mutants will just be enough to include one molecule of all possible combinations for a sequence of ten amino acid positions. In comparison, most natural proteins are more than 100 amino acids in length. Therefore, libraries of mutants should be designed wisely to increase the chances of success in a directed evolution experiment. Accordingly, one should consider randomizing specific amino acid positions by using degenerate codons. Instead of randomizing positions with NNN codons (N = A,C,G,T), NNK codons (K = G,T), NNS codons (S = C,G) or even a reduced alphabet of NDT codons (D = A,G,T) can be used to reduce oversampling caused by codon degeneracy (32). The use of degenerate codons can also reduce the likelihood of introducing unintended stop codons. For example, the NNN codon includes three stop codons, whereas the NNK or NNS codons include only one. Alternatively, a given library can be assembled from fragments that have been preselected to decrease the occurrence of premature stop codons (4). More recently, DNA synthesis via phosphoramidite trinucleotides has become commercially available (33). Codon by codon synthesis using trinucleotides offers full control of the library composition by defining the set of desired amino acid mutations at any position while avoiding stop codons (see Note 1). This method is still costly but will likely become affordable in the near future (34).

In order to use a DNA library for a specific in vitro evolution technique, the sequences at both termini of the DNA have to be made compatible to the method of choice. The 5′-end includes promoter and enhancer sequences necessary to facilitate transcription and translation, respectively (see Note 2). The nature of these sequences depends on the type of transcription and translation system used (see Note 3). Other sequence elements might be included such as a terminator, stabilizing hairpins, affinity purification tags, or sequences that are specific to the particular in vitro evolution method (35) (see Note 4).

5. Methods for In Vitro Enzyme Evolution

Ribosome display. The ribosome display technology creates the genotype–phenotype linkage through a ternary complex of a stalled ribosome, the translated protein and its encoding mRNA (Fig. 1). The complex is stabilized by high magnesium concentrations and low temperatures. Ribosome display was initially described for the purification of specific mRNA sequences based on immunoprecipitation of the encoded protein (36). Subsequently, this method was developed further to select and evolve peptides and proteins (37, 38). Although ribosome display has mostly been used for selection

Table 1
Comparison of in vitro technologies discussed in this review

Method	Genotype–phenotype link	Reported variants in single experiment	Results
Ribosome display	Non-covalent complex of mRNA-ribosome-protein	~10^{13}	Proof-of-concept selections for sialyltransferase (10), beta-lactamase (11), dihydrofolate reductase (12), DNA ligase (13), and sortase (14)
mRNA display	Covalent fusions of mRNA-protein via puromycin	~10^{13}	Selection for de novo RNA ligase (15, 16)
DNA display	Covalent or non-covalent complex of DNA-protein	~10^{12}	Proof-of-concept selection of binders, but no enzymes (17, 18)
In vitro compartmentalization (IVC)	Spatial confinement	~10^{9} (selection) ~10^{6}–10^{8} (screening by FACS/microfluidics)	Selection for methyltransferase (19) and restriction nuclease (20); proof-of-concept screening for β-galactosidase (21, 22)
IVC and microbead display	Non-covalent complex of DNA-microbead-protein	~10^{9} (selection) ~10^{6}–10^{8} (screening by FACS/microfluidics)	Screening for phosphotriesterase (23); proof-of-concept screening for hydrogenase (24); proof of concept selection of biotin ligase (25)
IVC and DNA display	Covalent or non-covalent complex of DNA-protein	~10^{8}–10^{9}	Selection of antibodies as heterodimers, but no enzymes (26)

of binders, several model selections for enzymatic activity have been reported and will be reviewed here in more detail.

Several criteria must be met in order to generate ribosome-displayed proteins. Most importantly, the terminal stop codon of the gene of interest must be removed. This will prevent the ribosome from dissociating and releasing the nascent protein and will instead promote stalling of the ribosome and therefore maintain the ternary complex. Stem-loop structures are often added to flank the gene on both termini to increase RNA stability during translation and subsequent manipulations. Since the protein is not released from the ribosome, a C-terminal protein spacer (>100 amino acids) is added to ensure that the displayed protein has exited the protein-conducting channel of the ribosome and can fold properly. Typically, ribosome-displayed proteins are generated through sequential transcription and translation, as coupled transcription/translation systems can result in 100-fold reduced protein yield (38, 39). The translation is stopped by decreasing the temperature and increasing the Mg^{2+} concentration to stabilize the ternary complex. To maintain the genotype–phenotype linkage, the subsequent selection process also has to be performed at low temperatures and in presence of elevated Mg^{2+} concentrations. The ribosome-displayed proteins are mostly used in selections without any additional purification. The RNA is recovered after the selection by dissociating the ternary complex through chelation of Mg^{2+} with EDTA. A detailed protocol has been published elsewhere (40) (see Note 5).

Ribosome display has been utilized in a number of model selections for enzymatic activity. Most selections were performed by selecting for binding to an immobilized substrate, substrate analog, or inhibitor. These model selections demonstrated enrichment of the desired enzyme (10- to 100-fold per round of selection) compared to an inactive control (Table 1) (10–12, 14). While enzyme selection strategies based on binding can be successful in isolating enzymes with known properties (e.g., searching through metagenomic libraries for a desired activity), they are not well suited for changing substrate specificity or substantially improving activity (41, 42). In one example of a truly product-driven model selection, ribosome display has been employed for isolation of a T4 DNA ligase (13). Active enzymes able to ligate a DNA adaptor to the 3′-end of their encoding mRNA were selectively amplified via an adaptor-specific primer and were enriched 40-fold over known inactive mutants. Similar to this selection approach, the 3′-end of the mRNA could be used for the attachment of alternative substrates which would allow for a selection of other catalysts by ribosome display.

mRNA display. mRNA-displayed proteins are covalently attached to their encoding mRNA via the small linker molecule puromycin

(Fig. 1) (43, 44). This stable covalent link allows for the selection of proteins under a wide range of conditions. mRNA display has been used to select for a novel enzymatic activity from a non-catalytic library of randomized proteins (15, 16). This is the first example of a de novo enzyme generated by directed evolution from a naïve library. In addition, and similar to ribosome display, mRNA display has been widely employed for isolation of binders (45).

Central to the mRNA display method is the modification of the stop codon-free 3′-end of the messenger RNA with a puromycin-containing DNA linker prior to translation (46, 47). During the subsequent in vitro translation, the ribosome synthesizes the polypeptide until it reaches the DNA-puromycin-modified 3′-end of the mRNA where it stalls. Puromycin, which is an antibiotic that mimics the aminoacyl end of tRNA, enters the ribosome and becomes covalently attached to the C-terminus of the nascent polypeptide. The resulting mRNA-displayed proteins are typically purified from unfused proteins and mRNA using purification tags. The mRNA-displayed proteins are reverse transcribed to produce the cDNA. Reverse transcription also minimizes potential RNA secondary structure and increases RNA stability. Detailed protocols on mRNA display have been published recently (16, 48, 49). Through slight modifications of the mRNA display protocol, covalent fusions of protein and encoding cDNA can be generated (cDNA display) (50, 51).

mRNA display is the first directed evolution method that has produced an entirely artificial enzyme without a predecessor in nature. Starting from a non-catalytic protein scaffold containing two zinc fingers with each loop randomized (52), the authors isolated an RNA ligase enzyme that catalyzes the splinted ligation of a 5′-triphosphorylated RNA strand to the 3′-hydroxyl end of a second RNA (15, 16). This particular catalytic activity has not been reported in any natural enzyme. For this selection, product formation was the only selection criterion. The authors attached one of the substrates to the mRNA-displayed protein during the reverse transcription step forming a protein-mRNA-cDNA-substrate complex. The incubation with the second substrate, which was labeled with biotin, allowed any active enzymes to ligate the biotin moiety to their own cDNA enabling the selective immobilization on streptavidin beads. The isolated enzyme accelerates the reaction more than 10^6-fold. The ligase shows multiple turnover, although the selection scheme only requires a single catalytic event. While this example has been the only reported application of mRNA display for the isolation of enzymes to date, the general selection scheme is applicable for a wide range of bond-forming reactions. Furthermore, variations of this scheme have been proposed to apply mRNA display to the evolution of enzymes for bond-breaking and other modification reactions (53).

In vitro compartmentalization (IVC). Directed evolution by IVC mimics in vivo evolution inside a cell by using water-in-oil emulsions to enclose proteins and their encoding DNA within the same droplet compartment thereby creating the genotype–phenotype link through spatial confinement (54). IVC has been employed not only in several model enzyme selections, but also to improve the performance of existing enzymes through screening and selection methods.

Compartmentalization by droplet formation is achieved by stirring an aqueous solution of genes and a coupled transcription/translation (TS/TL) system into a mixture of mineral oil and surfactants (55). The DNA concentration is chosen such that the average droplet contains no more than a single gene. The low volume of the droplets (5–10 femtoliters) corresponds to a low nanomolar concentration of the single DNA molecule, which is efficiently transcribed and translated inside the droplet (22, 54, 56). Although droplet composition is similar across different IVC experiments, in some cases the oil/surfactant mixtures need to be optimized for compatibility with the specific TS/TL solution used and the enzymatic activity that is being evolved (54, 57). It has been shown that the droplets are stable up to 100 °C for many days and do not exchange DNA or protein between each other (54, 58). Detailed protocols for the IVC method have been published (55).

IVC-based selections have been used to evolve enzymes that process nucleic acid substrates. Here, the encoding DNA is also the substrate for the enzyme and the selection is dependent on successful DNA modification. In one approach, the activity of the methyltransferase (M.HaeIII) was improved toward a nonnative, although already recognized, DNA sequence (19). A library of variants of M.HaeIII was made by mutating the DNA contacting residues. The 3′-end of the DNA library was modified with a biotin moiety and connected to the remaining gene via the target methylation site that can be cleaved by endonuclease NheI unless the site is has been methylated by M.HaeIII. Therefore, only methylated genes were not cleaved by NheI and were captured on streptavidin beads. A similar approach was used for the model selection of a restriction endonuclease activity from a randomized library of the restriction enzyme FokI. Three specific residues were randomized in the catalytic domain, and cleavage sites for FokI were introduced in the 3′-UTR (20). Only the genes coding for an active FokI variant were cleaved and captured on beads after incorporation of biotinylated deoxyuracil triphosphate at the cohesive ends generated by the restriction enzyme.

The IVC methodology has also been used in combination with screening approaches. This allows for the evolution of enzymes for non-nucleic acid-related reactions, but also reduces the number of mutants that can be interrogated compared to selection strategies. In the screening approach, either fluorescence-activated cell sorting

(FACS) or microfluidics-based droplet sorting is used to separate active and inactive enzymes based on the conversion of non-fluorescent substrate into fluorescent product. For FACS mediated screening, water-in-oil-in-water emulsions (double emulsions) are generated since FACS instrumentation is incompatible with oil as the main medium (59). Exploiting this principle, the very low β-galactosidase activity of the Ebg enzyme from *Escherichia coli* was increased at least 300-fold by in vitro evolution using a commercially available fluorogenic substrate (22). Recently, the same researchers reported a model enrichment of β-galactosidase using a homemade microfluidic system (21). Although the throughput in the microfluidic system is about tenfold less than in FACS-based screening, this loss is offset by other advantages. First, the microfluidic system generates highly monodisperse droplets, enabling quantitative kinetic analysis (21, 60). Second, the authors utilized microfluidic components that allowed them to fuse droplets together and introduce new content into droplets. This conferred multiple benefits as the authors were able to perform emulsion PCR in droplets and then merge them with droplets containing the TS/TL mix. By generating about 30,000 gene copies per droplet prior to TS/TL, low enzymatic activity is more likely to be detected due to the elevated enzyme concentration (21). Furthermore, reagents can be readily added to the droplets after translation, in case the translation conditions are not compatible with enzymatic assay (61). The use of microfluidics is a promising route for IVC-based enzyme engineering due to the modularity and potential for customization of individual components. However, in contrast to commercially available FACS instruments, assembly of microfluidics devices still requires substantial expertise.

IVC has also been used in conjunction with in vivo enzyme evolution by generating compartments that contain cells. To keep the focus of this review we are not discussing this in vivo application.

DNA display. Strategies that either directly or indirectly establish a physical link between the DNA and the encoded protein are referred to as DNA display (Table 2). Although several different DNA display methods have been developed, only the IVC-mediated microbead display has been used to evolve enzymes. This method generates the genotype–phenotype link through the capture of DNA and its translated protein onto the same streptavidin-coated microbeads inside a droplet (Fig. 1) (23, 24). This approach requires multiple biotinylated reagents such as primers, antibody, and reaction substrate in order to capture the template DNA, the protein modified with an epitope tag and the substrate onto the microbead, respectively.

Using microbead display, Tawfik and Griffiths improved the catalytic performance of an already very efficient phosphotriesterase enzyme 63-fold ($k_{cat} > 10^5\ s^{-1}$) through FACS-based screening (23).

Table 2
DNA display methods. Only the microbead display has been used to evolve enzymes

Method	Principle of attachment	DNA—point of attachment	Protein fusion partner
Microbead display (23–25)	Non-covalent binding of DNA to streptavidin microbead and of HA-tagged protein via anti-HA antibody to same bead, IVC is needed	Biotinylated	HA tag
STABLE (26, 62)	Non-covalent attachment of protein to DNA, IVC is needed	Biotinylated	Streptavidin
CIS-display (17)	Non-covalent attachment of protein to DNA	RepA gene	DNA replication initiator (RepA)
Covalent DNA display (63, 64)	Covalent attachment of enzyme to suicide inhibitor that is linked to DNA, IVC is needed	Modified with 5-fluoro-deoxycytidine	HaeIII methyltransferase
Covalent antibody display (18)	Covalent attachment of enzyme to DNA	P2A gene	Endonuclease P2A
SNAP display (65, 66)	Covalent attachment of enzyme to suicide inhibitor that is linked to DNA, IVC is needed	Modified with benzyl guanine	SNAP tag

This work demonstrated the ability to generate, break, and regenerate the IVC droplets and purify the genotype–phenotype product attached to the microbeads. Furthermore, a substrate was used that carried a photo-caged biotin. Therefore, the substrate stays in solution until the biotin is uncaged, which causes the immobilization of substrate and resulting product on the beads. Incubation with a fluorescent product-specific antibody enabled the specific labeling and isolation by FACS of only those microbeads to which functional enzymes and their coding DNA were attached (23).

In a different proof-of-concept experiment, a modified microbead display protocol was performed as a selection instead of a screen, thereby potentially harnessing larger library sizes (25). In this experiment, an active biotin ligase was enriched from a mixture of inactive genes. Following product formation and immobilization, the purified microbeads were incubated with product-specific antibodies that were conjugated to a cleavable, gene-specific PCR primer instead of a fluorophore. Re-emulsification and droplet PCR with a solution lacking this primer resulted in a 20-fold enrichment of the desired genes.

Another microbead display model screen employing FACS used an indirect readout for activity to isolate (FeFe) hydrogenases (24).

Because the hydrogenase activity (H_2 breakdown) is difficult to measure directly, the authors employed a redox-sensitive dye that can generate a fluorescent signal. Purified microbeads carrying the immobilized DNA and enzymes were re-compartmentalized in the presence of the redox dye. This dye was modified with a C12-alkyl chain and therefore interacts nonspecifically with the hydrophobic polystyrene beads. Hydrogenase activity resulted in fluorescence of the dye and enabled flow cytometric sorting of the microbeads to recover the DNA of active enzymes, yielding a 20-fold enrichment over inactive genes. This proof-of-concept study used microfluidics to generate monodisperse droplets and microbeads with a larger diameter (5.6 μm rather than 1 μm) to increase the bead surface allowing more fluorescent substrate to bind, thereby improving the signal to noise ratio. The indirect readout as described here could be applied to other screening strategies if environmentally sensitive fluorophores are available (pH, redox potential).

Presently, only microbead display has been employed to evolve enzymes. Yet other DNA display methods could potentially be used for this purpose. In contrast to microbead display, all other DNA display methods directly attach the protein to its encoding gene via a fusion protein which binds to a specific DNA sequence within the parent gene or to a small molecule attached to the parent gene (Table 2). The IVC method is often used in conjunction with DNA display as the physical genotype–phenotype linkage allows for the microcompartments to be broken up and generated again in order to introduce new components into the system (e.g., substrates). However, two proof-of-concept studies conducted without IVC demonstrated the production of DNA-displayed proteins solely by incubating templates with the *E. coli* cell extract (17, 18).

6. General Principles and Comparison of Different Methods

In the previous section, we highlighted individual examples for the use of in vitro enzyme evolution. In this section, we will compare the different methods and discuss aspects that several methods have in common.

The types of reactions catalyzed by enzymes can be divided into transformation reactions, bond-forming reactions, and bond-breaking reactions (Fig. 2a). Depending on the reaction type, the strategy by which enzymes can be selected varies slightly. In general, affinity selections are used to isolate enzymes by methods that create a physical link between phenotype and genotype such as ribosome display, mRNA display, and DNA display (Figs. 1 and 2b). To enable an enzyme affinity selection, the substrate has to be linked to the gene-enzyme complex. Enzymes for a transformation reaction

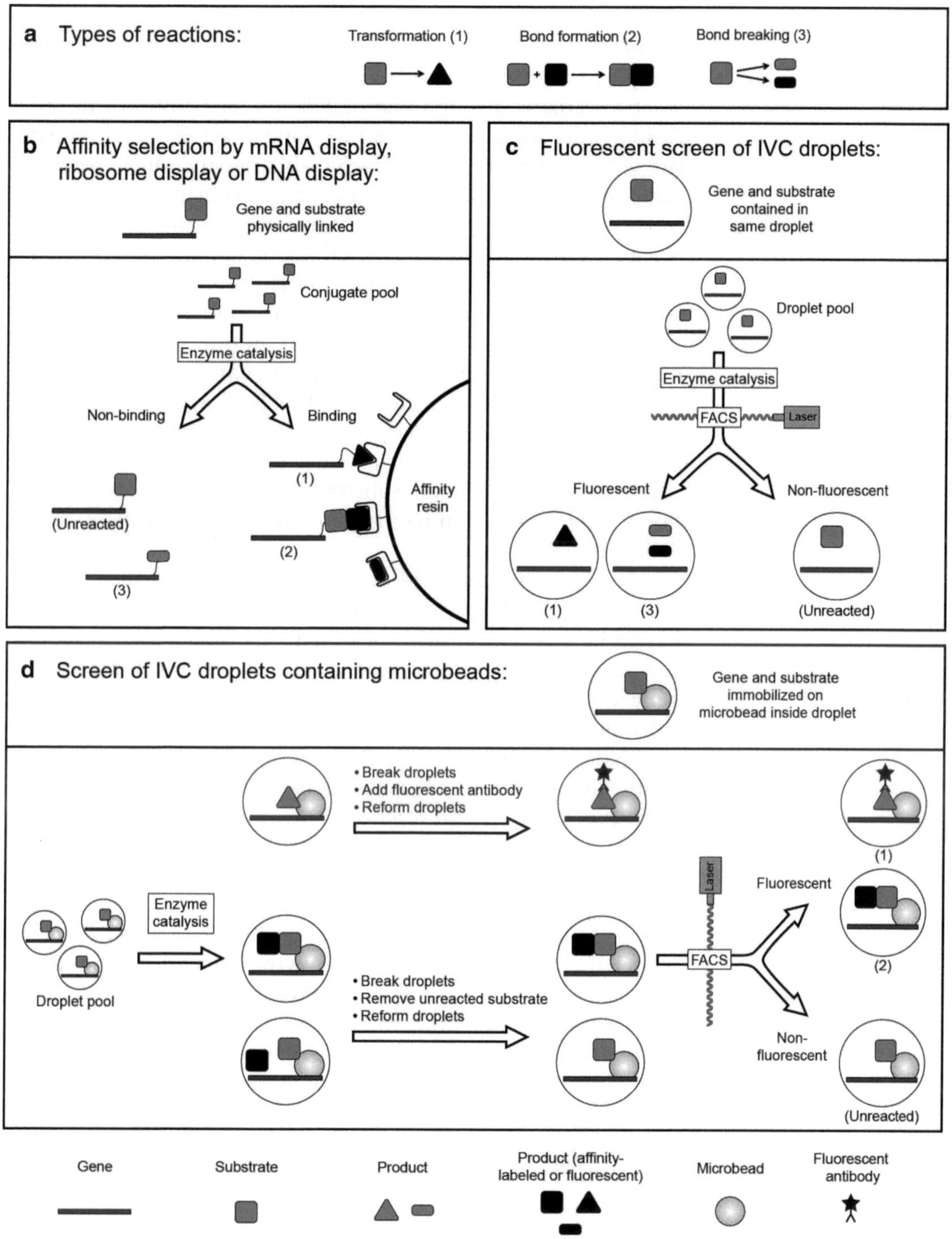

Fig. 2. Isolation of enzymatic activities using in vitro technologies. (**a**) Types of enzymatic activities that can be evolved using in vitro approaches. (**b**) Affinity selection of physically linked gene-substrate/product conjugates. The enzyme itself is also linked to the gene-substrate complex, but is omitted from the figure for improved clarity. (**c**) Screen of IVC droplets that become fluorescent as a result of catalysis by the enzyme (not shown) contained in same compartment. Separation is achieved through fluorescence-activated cell sorting (FACS) or microfluidics. (**d**) Screen for enzyme catalysis by FACS of IVC droplets containing microbeads. The enzyme contained in each compartment is not shown to improve clarity. Numbers in brackets refer to the type of activity as shown in (**a**).

can then be isolated if a product-specific affinity reagent, such as an antibody, is available (reaction type 1). Via the antibody, the ternary complex of product, active enzyme, and gene is separated from inactive variants through immobilization (see Note 6). In the case of an affinity selection for bond-forming enzymes (reaction type 2), the second substrate carries a selectable moiety. Only proteins that catalyze the bond formation between two substrates will attach this moiety to the gene-protein-substrate complex and can therefore be isolated. For bond-breaking reactions (reaction type 3), the whole complex of gene, protein, and substrate is immobilized via the substrate and only variants that cleave the bond will be released and selected (see Note 7). In contrast to affinity selections, the IVC methodology mostly employs fluorescent screening to isolate evolved enzyme variants either by FACS or microfluidics (Fig. 2c, d). This can be achieved if the product of the reaction becomes fluorescent or a fluorescent product-specific antibody is available. Alternatively, the second substrate, which will be attached in a bond-forming reaction to the gene-microbead-substrate complex, is fluorescent.

For any enzyme evolution experiment regardless of which methodology is used, the specific selection or screening strategy has to be customized with respect to the underlying reaction. In the case of affinity selections, the need to link the substrate to the gene complex without substantially changing the nature of the substrate can be challenging especially for small substrates (see Notes 8–10). On the other hand, suitable fluorophores that enable the screening of IVC droplets might not be compatible with some types of chemical reactions.

Two important questions have to be considered when deciding on which enzyme evolution strategy to use: Is the desired mutant potentially very rare such as a mutant exhibiting a novel activity? Or, alternatively, is the goal of the evolution experiment to generate a highly proficient enzyme? Selection strategies can search larger libraries and are therefore more likely to discover rare mutants, compared to screening approaches. At the same time, affinity selections only select for a single turnover event and cannot evolve an enzyme for high substrate affinity as the substrate is linked to the enzyme and therefore present at a high local concentration. In contrast, IVC-based screening methods can directly evolve an enzyme for high turnover and substrate affinity, yet, the library size of screening methods is several orders of magnitude smaller than those of selections. Therefore, it might be most beneficial to combine the two strategies and first use an affinity selection method to isolate potentially rare enzyme variants with altered activity or substrate specificity and then switch to an IVC-based screening method to optimize enzymatic proficiency.

7. Conclusions

Enzyme engineering by in vitro enzyme evolution has made tremendous progress in the past decade. We now have a range of powerful in vitro methods available that can efficiently evolve biocatalysts in a test tube by searching protein libraries orders of magnitude larger than those used by conventional in vivo evolution approaches. In vitro enzyme evolution is uniquely suited to address two of the greatest challenges in biocatalyst design: de novo generation of novel activity and activity within harsh environments. Applying the repertoire of in vitro evolution methods, exciting new examples of enzyme engineering are expected to emerge, thereby solving problems in biocatalysis that have previously been difficult to address.

8. Notes

1. *Codon usage and compatibility.* Depending on the translation system used, the codon usage can vary substantially. During library construction, it is important to try to avoid rare codons that would reduce the translation yield. Furthermore, enzymes evolved using eukaryotic systems (e.g., rabbit reticulocyte) might employ codons that cause difficulty with protein expression and evolution in prokaryotes, requiring the use of strains that supplement rare/eukaryotic tRNAs.
2. *Increasing in vitro translation yields.* Translation can be controlled or improved by enhancer sequences such as a ribosome binding site for *E. coli*-based cell-free extracts (19), by an AMV enhancer for eukaryotic systems, or a TMV translation enhancer for both eukaryotic systems and *E. coli*-based systems (67). Furthermore, the optimization of translation conditions (lysate, salt, and template concentrations) can also increase the protein yield.
3. *Translation systems.* Several commercial and homemade options are available for in vitro protein translation such as *E. coli*, rabbit reticulocyte, and wheat germ lysates. *E. coli* translation systems are attractive because of their low cost and high protein production but suffer from abundant nuclease contamination and simple folding machinery. Rabbit reticulocyte lysates conversely are expensive but robust in promoting proper folding and contain fewer nucleases. Wheat germ lysate provides an intermediate between rabbit reticulocyte and *E. coli* extracts in that it is both inexpensive and promotes folding, but it requires fine optimization of ionic concentrations for each gene.

Minimal reconstituted systems of purified individual components (e.g., commercially available PURE kits) have also been employed and have been shown to improve stability and efficiency of mRNA-based methods (68). Such systems are also more amenable to unnatural amino acid incorporation. Ultimately, the nature of the in vitro method and the enzyme of interest dictate the type of lysate that might be best suited for the desired application. For review articles see ref. 69–71.

4. *UTR reconstitution between rounds of evolution.* The 5′-UTR, and in some cases 3′-UTR (20), are lost during transcription and translation and need to be reconstituted before a subsequent round. This can be done by PCR (15), overlap extension PCR (20), or by a coupled uracil excision–ligation strategy (72).

5. *Ribosome display variations.* Several modifications to the ribosome display technology can increase the stability of the ternary complex and were recently highlighted in Methods in Molecular Biology (73). Ribosome-inactivation display (RID) utilizes a ricin toxin to inactivate and stall the ribosome, resulting in covalent attachment of protein to the ribosome. Although use of the ricin toxin substantially increases the total gene size by about 1 kb, the ribosomes no longer need to be kept at low temperature and high salt concentrations to maintain the genotype–phenotype link (74). Alternatively, the use of translation mixtures assembled from purified components (68) or depleted of transfer-messenger RNA (tmRNA) (38) have also been used to increase yield and stability of the ternary complex by eliminating stalled ribosome rescue mechanisms.

6. *Minimizing nonspecific interactions during immobilization.* Attachment of large biopolymers to the enzyme of interest (e.g., DNA-, RNA-, or ribosome display) may result in nonspecific interaction with the resins used during immobilization and purification, lowering the overall enrichment of the desired enzymes. With the exception of pure IVC and mRNA display, all in vitro methods require use of fusion proteins, further increasing size of the genotype–phenotype complex. Nonspecific interactions with resins can be counteracted by including an excess of salmon sperm DNA, tRNA, or BSA in the buffers. Recently, it was suggested to use of polylysine "wrappers" as coatings to mask the negatively charged RNA and minimize its impact on the outcome of selections (75); however, this strategy has not yet been applied to in vitro evolution of enzymes. Furthermore, nonspecific resin interactions can be selected against by first incubating the genotype–phenotype complex with the resin alone and then using only the flow through for the real selection resin that is modified with the appropriate capture agents.

7. *Counter selections.* Alternating cycles of selection and counter selection can be used to ensure that the bond-forming or bond-breaking activity occurs at the desired site within the substrate. For example, when isolating nucleases or proteases, counter selections should be employed to improve the specificity of the enzyme toward a desired sequence within the recognition site. Previous work has shown that selections for bond-breaking activity without counter selection can enrich for catalysts that break bonds outside of the expected region (76, 77). In one case, the selection for a peptidase resulted in the isolation of DNA nucleases instead (76).
8. For small molecule substrates, the site of modification may be close to the site recognized or acted upon by an enzyme. As with any directed evolution experiment, it is important to confirm that the isolated enzyme also processes the unmodified substrate.
9. *Spacers.* Attachment of the genotype, substrate, or additional protein domains to the enzyme of interest requires sufficient spacing to minimize impact of these fusions on the enzyme's performance. Depending on the location of the protein termini, simple flexible linkers composed of $GnSn$ (4) or rigid linkers using the (EAAAAK)n motif can be used to provide appropriate spacing (78). Similarly, polyethylene glycol spacers or alkyl chains of varied length can be used as connectors (3).
10. The decision of whether to modify the substrate or employ product-specific antibodies depends on the substrate size and antibody availability. For example, the use of product-specific antibodies would be preferred for small molecule substrates where the derivatization may affect the binding to the enzyme. However, if the substrate is a large molecule, it may be simpler to derivatize the substrate (without affecting the enzyme's performance) than to generate a product-specific antibody.

References

1. Cohen N, Abramov S, Dror Y, Freeman A (2001) *In vitro* enzyme evolution: the screening challenge of isolating the one in a million. Trends Biotechnol 19:507–510
2. Fiammengo R, Jäschke A (2005) Nucleic acid enzymes. Curr Opin Biotechnol 16:614–621
3. Joyce GF (2004) Directed evolution of nucleic acid enzymes. Annu Rev Biochem 73:791–836
4. Cho G, Keefe AD, Liu RH, Wilson DS, Szostak JW (2000) Constructing high complexity synthetic libraries of long ORFs using *in vitro* selection. J Mol Biol 297:309–319
5. Kehoe JW, Kay BK (2005) Filamentous phage display in the new millennium. Chem Rev 105:4056–4072
6. Sidhu SS, Lowman HB, Cunningham BC, Wells JA (2000) Phage display for selection of novel binding peptides. Methods Enzymol 328:333–363
7. Renesto P, Raoult D (2003) From genes to proteins—*in vitro* expression of rickettsial proteins. Ann N Y Acad Sci 990:642–652
8. Bulter T, Alcalde M, Sieber V, Meinhold P, Schlachtbauer C, Arnold FH (2003) Functional

expression of a fungal laccase in Saccharomyces cerevisiae by directed evolution. Appl Environ Microbiol 69:987–995

9. Chusacultanachai S, Yuthavong Y (1994) Random mutagenesis strategies for construction of large and diverse clone libraries of mutated DNA fragments. Methods Mol Biol 270:319–333
10. Bieberich E, Kapitonov D, Tencomnao T, Yu RK (2000) Protein-ribosome-mRNA display: affinity isolation of enzyme-ribosome-mRNA complexes and cDNA cloning in a single-tube reaction. Anal Biochem 287:294–298
11. Amstutz P, Pelletier JN, Guggisberg A, Jermutus L, Cesaro-Tadic S, Zahnd C, Plückthun A (2002) *In vitro* selection for catalytic activity with ribosome display. J Am Chem Soc 124:9396–9403
12. Takahashi F, Ebihara T, Mie M, Yanagida Y, Endo Y, Kobatake E, Aizawa M (2002) Ribosome display for selection of active dihydrofolate reductase mutants using immobilized methotrexate on agarose beads. FEBS Lett 514:106–110
13. Takahashi F, Funabashi H, Mie M, Endo Y, Sawasaki T, Aizawa M, Kobatake E (2005) Activity-based *in vitro* selection of T4 DNA ligase. Biochem Biophys Res Commun 336:987–993
14. Quinn DJ, Cunningham S, Walker B, Scott CJ (2008) Activity-based selection of a proteolytic species using ribosome display. Biochem Biophys Res Commun 370:77–81
15. Seelig B, Szostak JW (2007) Selection and evolution of enzymes from a partially randomized non-catalytic scaffold. Nature 448:828–831
16. Seelig B (2011) mRNA display for the selection and evolution of enzymes from *in vitro*-translated protein libraries. Nat Protoc 6:540–552
17. Odegrip R, Coomber D, Eldridge B, Hederer R, Kuhlman PA, Ullman C, FitzGerald K, McGregor D (2004) CIS display: *in vitro* selection of peptides from libraries of protein-DNA complexes. Proc Natl Acad Sci U S A 101:2806–2810
18. Reiersen H, Lobersli I, Loset GA, Hvattum E, Simonsen B, Stacy JE, McGregor D, FitzGerald K, Welschof M, Brekke OH, Marvik OJ (2005) Covalent antibody display—an *in vitro* antibody-DNA library selection system. Nucleic Acids Res 33:e10
19. Cohen HM, Tawfik DS, Griffiths AD (2004) Altering the sequence specificity of *HaeIII* methyltransferase by directed evolution using *in vitro* compartmentalization. Protein Eng Des Sel 17:3–11
20. Doi N, Kumadaki S, Oishi Y, Matsumura N, Yanagawa H (2004) *In vitro* selection of restriction endonucleases by *in vitro* compartmentalization. Nucleic Acids Res 32:e95
21. Fallah-Araghi A, Baret JC, Ryckelynck M, Griffiths AD (2012) A completely *in vitro* ultrahigh-throughput droplet-based microfluidic screening system for protein engineering and directed evolution. Lab Chip 12:882–891
22. Mastrobattista E, Taly V, Chanudet E, Treacy P, Kelly BT, Griffiths AD (2005) High-throughput screening of enzyme libraries: *in vitro* evolution of a beta-galactosidase by fluorescence-activated sorting of double emulsions. Chem Biol 12:1291–1300
23. Griffiths AD, Tawfik DS (2003) Directed evolution of an extremely fast phosphotriesterase by *in vitro* compartmentalization. EMBO J 22:24–35
24. Stapleton JA, Swartz JR (2010) Development of an *in vitro* compartmentalization screen for high-throughput directed evolution of (FeFe) hydrogenases. PLoS One 5:1–8
25. Kelly BT, Griffiths AD (2007) Selective gene amplification. Protein Eng Des Sel 20:577–581
26. Sumida T, Doi N, Yanagawa H (2009) Bicistronic DNA display for *in vitro* selection of Fab fragments. Nucleic Acids Res 37:e147
27. Arnold FH, Georgiou G (2003) Directed evolution library creation: methods and protocols, vol 231, Methods in molecular biology. Humana Press Inc., Toyota, NJ
28. Labrou NE (2010) Random mutagenesis methods for *in vitro* directed enzyme evolution. Curr Protein Pept Sci 11:91–100
29. Shivange AV, Marienhagen J, Mundhada H, Schenk A, Schwaneberg U (2009) Advances in generating functional diversity for directed protein evolution. Curr Opin Chem Biol 13:19–25
30. Denault M, Pelletier JN (2006) Protein library design and screening. Methods Mol Biol 352:127–154
31. Neylon C (2004) Chemical and biochemical strategies for the randomization of protein encoding DNA sequences: library construction methods for directed evolution. Nucleic Acids Res 32:1448–1459
32. Reetz MT, Kahakeaw D, Lohmer R (2008) Addressing the numbers problem in directed evolution. Chembiochem 9:1797–1804
33. Virnekas B, Ge LM, Pluckthun A, Schneider KC, Wellnhofer G, Moroney SE (1994) Trinucleotide phosphoramidites—ideal reagents for the synthesis of mixed oligonucleotides for random mutagenesis. Nucleic Acids Res 22:5600–5607

34. Janczyk M, Appel B, Springstubbe D, Fritz HJ, Muller S (2012) A new and convenient approach for the preparation of beta-cyanoethyl protected trinucleotide phosphoramidites. Org Biomol Chem 10:1510–1513
35. Ahn JH, Kang TJ, Kim DM (2008) Tuning the expression level of recombinant proteins by modulating mRNA stability in a cell-free protein synthesis system. Biotechnol Bioeng 101: 422–427
36. Schechter I (1973) Biologically and chemically pure mRNA coding for a mouse immunoglobulin L-chain prepared with the aid of antibodies and immobilized oligothymidine. Proc Natl Acad Sci U S A 70:2256–2260
37. Mattheakis LC, Bhatt RR, Dower WJ (1994) An *in vitro* polysome display system for identifying ligands from very large peptide libraries. Proc Natl Acad Sci U S A 91:9022–9026
38. Hanes J, Plückthun A (1997) *In vitro* selection and evolution of functional proteins by using ribosome display. Proc Natl Acad Sci U S A 94:4937–4942
39. Lipovsek D, Plückthun A (2004) In vitro protein evolution by ribosome display and mRNA display. J Immunol Methods 290:51–67
40. Zahnd C, Amstutz P, Plückthun A (2007) Ribosome display: selecting and evolving proteins *in vitro* that specifically bind to a target. Nat Methods 4:269–279
41. Jestin JL, Kaminski PA (2004) Directed enzyme evolution and selections for catalysis based on product formation. J Biotechnol 113:85–103
42. Turner NJ (2009) Directed evolution drives the next generation of biocatalysts. Nat Chem Biol 5:568–574
43. Roberts RW, Szostak JW (1997) RNA-peptide fusions for the *in vitro* selection of peptides and proteins. Proc Natl Acad Sci U S A 94:12297–12302
44. Nemoto N, Miyamoto-Sato E, Husimi Y, Yanagawa H (1997) *In vitro* virus: bonding of mRNA bearing puromycin at the 3′-terminal end to the C-terminal end of its encoded protein on the ribosome *in vitro*. FEBS Lett 414:405–408
45. Leemhuis H, Stein V, Griffiths AD, Hollfelder F (2005) New genotype–phenotype linkages for directed evolution of functional proteins. Curr Opin Struct Biol 15:472–478
46. Kurz M, Gu K, Lohse PA (2000) Psoralen photo-crosslinked mRNA–puromycin conjugates: a novel template for the rapid and facile preparation of mRNA–protein fusions. Nucleic Acids Res 28:e83
47. Liu RH, Barrick JE, Szostak JW, Roberts RW (2000) Optimized synthesis of RNA-protein fusions for *in vitro* protein selection. Methods Enzymol 318:268–293
48. Takahashi TT, Roberts RW (2009) *In vitro* selection of protein and peptide libraries using mRNA display. Methods Mol Biol 535:293–314
49. Cotten SW, Zou JW, Valencia CA, Liu RH (2011) Selection of proteins with desired properties from natural proteome libraries using mRNA display. Nat Protoc 6:1163–1182
50. Kurz M, Gu K, Al-Gawari A, Lohse PA (2001) cDNA—protein fusions: covalent protein-gene conjugates for the *in vitro* selection of peptides and proteins. Chembiochem 2:666–672
51. Ueno S, Nemoto N (2012) cDNA display: rapid stabilization of mRNA display. Methods Mol Biol 805:113–135
52. Cho GS, Szostak JW (2006) Directed evolution of ATP binding proteins from a zinc finger domain by using mRNA display. Chem Biol 13:139–147
53. Golynskiy MV, Seelig B (2010) *De novo* enzymes: from computational design to mRNA display. Trends Biotechnol 28:340–345
54. Tawfik DS, Griffiths AD (1998) Man-made cell-like compartments for molecular evolution. Nat Biotechnol 16:652–656
55. Miller OJ, Bernath K, Agresti JJ, Amitai G, Kelly BT, Mastrobattista E, Taly V, Magdassi S, Tawfik DS, Griffiths AD (2006) Directed evolution by *in vitro* compartmentalization. Nat Methods 3:561–570
56. Bernath K, Hai MT, Mastrobattista E, Griffiths AD, Magdassi S, Tawfik DS (2004) *In vitro* compartmentalization by double emulsions: sorting and gene enrichment by fluorescence activated cell sorting. Anal Biochem 325: 151–157
57. Ghadessy FJ, Holliger P (2004) A novel emulsion mixture for *in vitro* compartmentalization of transcription and translation in the rabbit reticulocyte system. Protein Eng Des Sel 17:201–204
58. Ghadessy FJ, Ong JL, Holliger P (2001) Directed evolution of polymerase function by compartmentalized self-replication. Proc Natl Acad Sci U S A 98:4552–4557
59. Eisenstein M (2006) Tiny droplets make a big splash. Nat Methods 3:71
60. Song H, Ismagilov RF (2003) Millisecond kinetics using nanoliters of reagents. J Am Chem Soc 125:14613–14619
61. Mazutis L, Baret JC, Treacy P, Skhiri Y, Fallah Araghi A, Ryckelynck M, Taly V, Griffiths AD (2009) Multi-step microfluidic droplet processing: kinetic analysis of an *in vitro* translated enzyme. Lab Chip 9: 2902–2908

62. Doi N, Yanagawa H (1999) STABLE: protein-DNA fusion system for screening of combinatorial protein libraries *in vitro*. FEBS Lett 457:227–230
63. Bertschinger J, Neri D (2004) Covalent DNA display as a novel tool for directed evolution of proteins *in vitro*. Protein Eng Des Sel 17: 699–707
64. Bertschinger J, Grabulovski D, Neri D (2007) Selection of single domain binding proteins by covalent DNA display. Protein Eng Des Sel 20:57–68
65. Stein V, Sielaff I, Johnsson K, Hollfelder F (2007) A covalent chemical genotype–phenotype linkage for *in vitro* protein evolution. Chembiochem 8:2191–2194
66. Kaltenbach M, Stein V, Hollfelder F (2011) SNAP dendrimers: multivalent protein display on dendrimer-like DNA for directed evolution. Chembiochem 12:2208–2216
67. Gallie DR, Kado CI (1989) A translational enhancer derived from Tobacco Mosaic-Virus is functionally equivalent to a Shine-Dalgarno sequence. Proc Natl Acad Sci U S A 86: 129–132
68. Villemagne D, Jackson R, Douthwaite JA (2006) Highly efficient ribosome display selection by use of purified components for *in vitro* translation. J Immunol Methods 313:140–148
69. Jagus R, Joshi B, Miyamoto S, Beckler GS (1998) *In vitro* translation. Curr Protoc Cell Biol. 2001 May; Chapter 11:Unit 11.2. doi: 10.1002/0471143030.cb1102s00
70. Hillebrecht JR, Chong SR (2008) A comparative study of protein synthesis in *in vitro* systems: from the prokaryotic reconstituted to the eukaryotic extract-based. BMC Biotechnol 8:58
71. He MY, He YZ, Luo Q, Wang MR (2011) From DNA to protein: no living cells required. Process Biochem 46:615–620
72. Stein V, Hollfelder F (2009) An efficient method to assemble linear DNA templates for *in vitro* screening and selection systems. Nucleic Acids Res 37:e122
73. Douthwaite JA, Jackson RH (2011) Ribosome display and related technologies: methods and protocols, vol 805, Methods in molecular biology. Humana Press, New York
74. Fujita S, Zhou JM, Taira K (2007) Ribosome-inactivation display system. Methods Mol Biol 352:221–236
75. Lamboy JA, Tam PY, Lee LS, Jackson PJ, Avrantinis SK, Lee HJ, Corn RM, Weiss GA (2008) Chemical and genetic wrappers for improved phage and RNA display. Chembiochem 9:2846–2852
76. Chandra M, Sachdeva A, Silverman SK (2009) DNA-catalyzed sequence-specific hydrolysis of DNA. Nat Chem Biol 5:718–720
77. Wang TP, Su YC, Chen Y, Liou YM, Lin KL, Wang EC, Hwang LC, Wang YM, Chen YH (2012) *In vitro* selection and characterization of a novel Zn(II)-dependent phosphorothiolate thiolesterase ribozyme. Biochemistry 51:496–510
78. Yonezawa M, Doi N, Higashinakagawa T, Yanagawa H (2004) DNA display of biologically active proteins for *in vitro* protein selection. J Biochem 135:285–288

Chapter 7

Residue-Specific Incorporation of Unnatural Amino Acids into Proteins In Vitro and In Vivo

Amrita Singh-Blom, Randall A. Hughes, and Andrew D. Ellington

Abstract

The incorporation of noncanonical (unnatural) amino acids into proteins offers researchers the ability to augment the biochemical functionality of proteins for a myriad of applications including bioorthogonal conjugation, biophysical and structural studies, and the enhancement or de novo creation of novel enzymatic activities. The augmentation of a protein throughout its coding sequence by global residue-specific incorporation of unnatural amino acid analogs is an attractive technique for studying both the utility of individual chemistries available through unnatural amino acids and the general effects of unnatural amino acid substitution on protein structure and function. Herein we describe protocols to introduce unnatural amino acids into proteins using the *Escherichia coli* translation system either in vivo or in vitro. Special attention is paid to obtaining high levels of incorporation while maintaining high yields of protein expression.

Key words: Unnatural amino acids, Noncanonical amino acids, Amino acid analogs, Genetic code evolution, Tryptophan analogs, Cell-free synthesis, In vitro transcription and translation

1. Introduction

The incorporation of unnatural amino acids into proteins has become a powerful tool for augmenting protein function (1–3), studying protein and cellular processes (4), creating novel functional proteins (5), and expanding the functionality of laboratory-evolved proteins and peptides (6, 7). Over the last few decades, three major methodologies have been developed to introduce unnatural amino acids into proteins (illustrated in Fig. 1). These methodologies include global amino acid replacement, site-specific incorporation, and semisynthetic incorporation. These methods differ primarily in how the unnatural amino acid is introduced into proteins. Global or residue-specific amino acid replacement takes

James C. Samuelson (ed.), *Enzyme Engineering: Methods and Protocols*, Methods in Molecular Biology, vol. 978,
DOI 10.1007/978-1-62703-293-3_7, © Springer Science+Business Media New York 2013

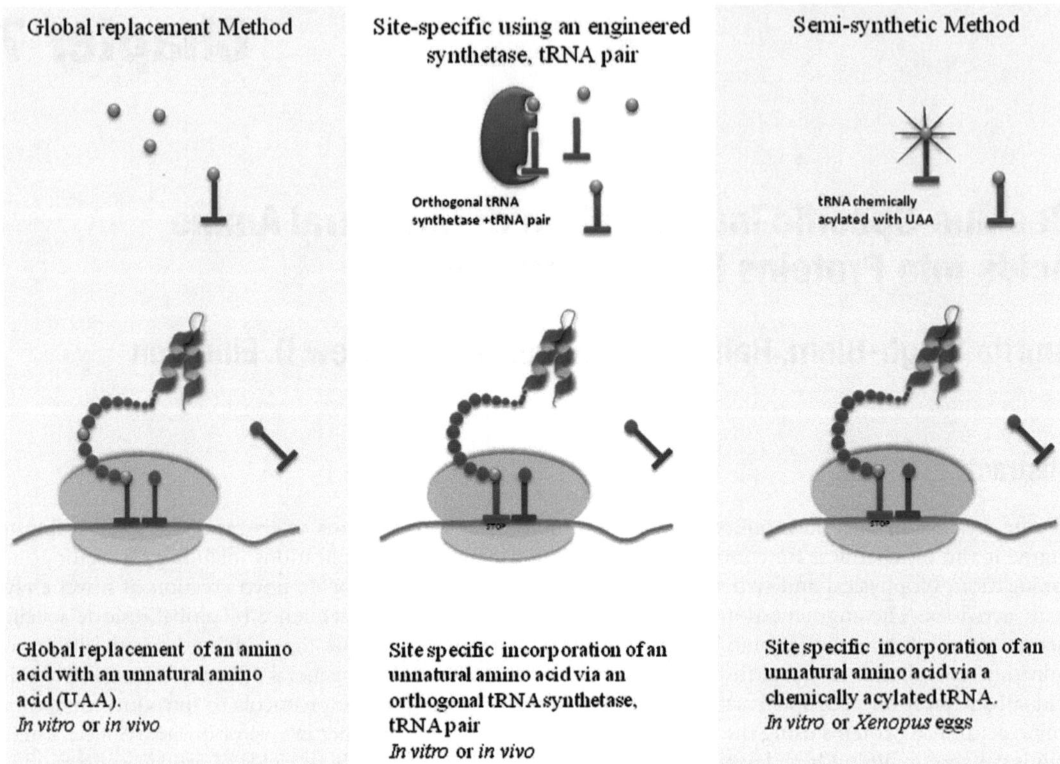

Fig. 1. Incorporation of unnatural amino acids into proteins. Global incorporation or residue-specific incorporation (*left panel*) requires only the addition of exogenous unnatural amino acid. The unnatural amino acid is charged onto the cellular tRNA by the endogenous tRNA synthetase and incorporated into proteins instead of the wild-type amino acid. Site-specific incorporation (*middle panel*) requires an exogenous tRNA and tRNA synthetase pair which function orthogonally to the cellular machinery. The exogenous tRNA synthetase charges the unnatural amino acid onto its cognate tRNA which recognizes a unique codon like a stop codon or a 4-base codon on the mRNA being translated. Semisynthetic incorporation (*right panel*) makes use of chemical or enzymatic acylation to charge a suppressor tRNA with the unnatural amino acid. This is added to the translation system and is inserted at a unique site like a stop codon.

advantage of the natural (or partially modified) substrate flexibility of aminoacyl-tRNA synthetases, the enzymes responsible for attaching amino acids to their appropriate tRNAs (8). These methodologies replace one of the twenty canonical amino acids with an amino acid analog via removing or limiting the availability of the natural amino acid such that the unnatural amino acid analog is incorporated into proteins instead of the natural amino acid. Residue-specific (global) amino acid replacement will incorporate the unnatural amino acid analog at every position within the protein in which the natural amino acid was encoded (e.g., tryptophan analogs would replace tryptophan at every UGG codon encoded in the messenger RNA). This would lead to the incorporation of *multiple* unnatural amino acids into a target protein of interest. This methodology has been used for years to incorporate over a hundred amino acid analogs into proteins (9) and has also been used to evolve proteome-wide incorporation and tolerance of unnatural amino acids by entire organisms (10–12). In contrast, site-specific incorporation methodologies insert an unnatural amino acid at a "unique" codon position within the protein sequence as defined by the researcher (usually a translation termination codon or "stop codon" such as UAG). This is usually done using a nonsense suppressor tRNA which has been mutated to recognize the natural translation termination codon sequence as a sense codon. Additionally, a mutant orthogonal aminoacyl-tRNA synthetase is often used in combination with its cognate suppressor tRNA to increase the productivity and specificity of protein expressed with an unnatural amino acid (13, 14). This methodology has led to the translational incorporation of over 50 amino acids with a variety of functionalities into proteins (15). The last group of methodologies used to incorporate unnatural amino acids into proteins are the semisynthetic methods which link the unnatural amino acid to a suppressor tRNA via a combination of chemical synthesis and enzymatic ligation (16, 17), rather than relying on the amino acylation activity of a tRNA synthetase. While semisynthetic methods can be used to incorporate diverse unnatural amino acids beyond the somewhat limited specificities of aminoacyl-tRNA synthetases, the amount of protein that can be produced is stoichiometrically limited to the amount of chemically acylated tRNA added to the translation reaction.

In this chapter, we will present protocols for the global, residue-specific incorporation of unnatural amino acids into proteins produced by *Escherichia coli* or in vitro translation systems derived from *E. coli*. In particular, we will detail the replacement of tryptophan in proteins with unnatural tryptophan analogs. While the protocols outlined herein are specific for the global incorporation of tryptophan analogs into proteins, in theory and in practice, one can adapt these protocols for any unnatural amino acid analog that is translationally compatible with the *E. coli* translation system (reviewed in ref. 9). This can be done by using a different amino acid auxotrophic strain and modifying the stated protocols to include tryptophan (where applicable) and exclude the natural amino acid you wish to replace. Both the in vitro and in vivo protocols in this chapter make use of a tryptophan auxotroph to control the availability of endogenous tryptophan and force the translational machinery to incorporate the analog that is supplied in molar excess (Fig. 2). While the *E. coli* tryptophanyl tRNA synthetase present in cells and lysates primarily recognizes tryptophan analogs that are similar in structure to tryptophan itself (Table 1, Fig. 3), engineered aminoacyl-tRNA synthetases could be introduced into cells to expand the list of compatible analogs.

An overview of the in vitro and in vivo methods of incorporating unnatural amino acid analogs into proteins that are described in this chapter is presented in Fig. 2. The choice of which protocol

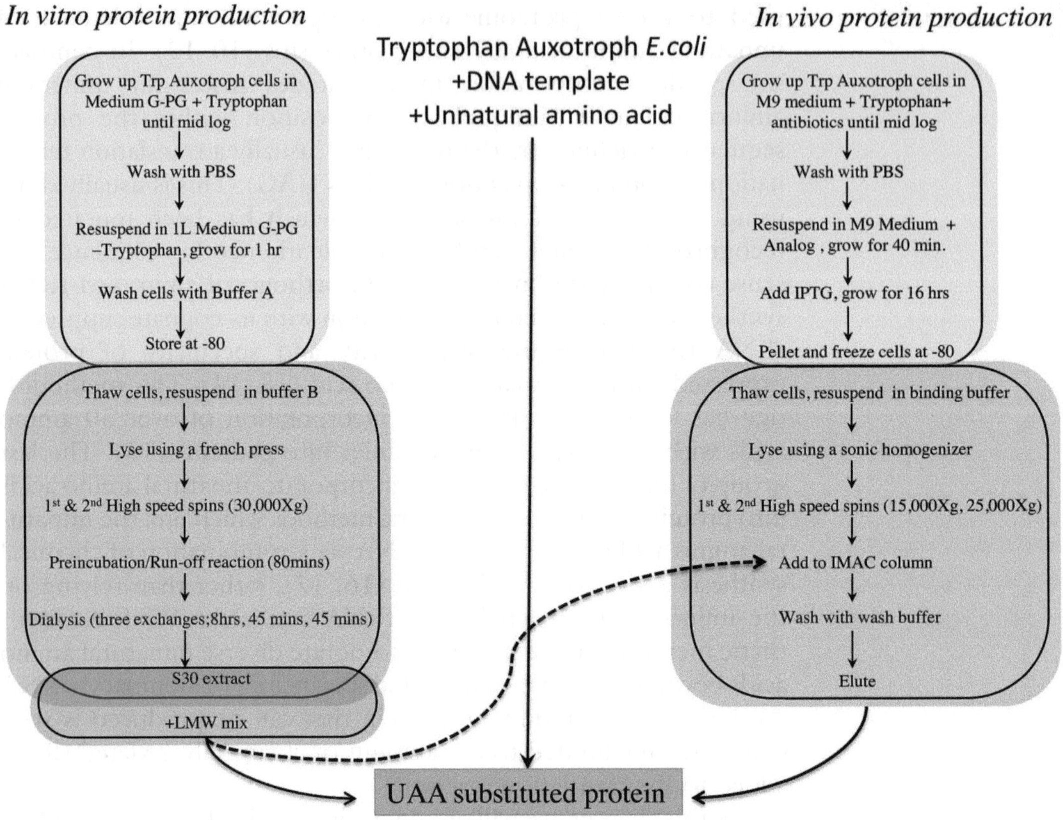

Fig. 2. Procedure for in vitro and in vivo incorporation of unnatural amino acid. The work flow for in vitro and in vivo protocols for production of unnatural amino acid-substituted protein presented in this chapter is contrasted above. Further details are presented within the text.

to employ depends on the desired outcome. The in vitro (cell-free) synthesis protocol will yield a system that can be used to test a variety of different amino acids in different target proteins. By setting up small batch reactions with template DNA coding for different proteins and various tryptophan analogs, a number of different combinations can be screened in short order. The template DNA used in cell-free synthesis can be a PCR product or plasmid. Depending on the protein produced, the function of the analog-substituted protein can even be assayed directly from the cell-free synthesis reaction. The in vitro format is also compatible with amino acid analogs that present difficulties in cellular uptake such as methylated tryptophans or analogs which are cytotoxic to the cells. Similarly, cytotoxic proteins can be expressed in vitro as long as it does not inhibit the transcription and translation processes in the reaction. The in vivo synthesis protocol requires the choice of target protein (and corresponding DNA template) and analog to be made in advance, but the protein yields are much higher which would make it the procedure of choice if a large amount of purified analog-substituted protein is required.

Table 1
Unnatural analogs of tryptophan

Structure in Fig. 3	Analog	In vivo or cell-free synthesis system (*Escherichia coli*)	Supplier/synthesis
1	L-tryptophan	Y	Commercially available
2	4-Methyl-DL-tryptophan	Y[a]	Commercially available
3	4-Fluoro-DL-tryptophan	Y[b]	Commercially available
4	5-Hydroxy-L-tryptophan	Y[c]	Commercially available
5	5-Methyl-DL-tryptophan	Y[d]/N[a]	Commercially available
6	5-Methoxy-DL-tryptophan	NA	Commercially available
7	5-Fluoro-L-tryptophan	Y[b]	Commercially available
8	5-Bromo-DL-tryptophan	N[e]	Commercially available
9	6-Methyl-DL-tryptophan	N[a]	Commercially available
10	6-Fluoro-DL-tryptophan	Y[b]	Commercially available
11	7-Methyl-DL-tryptophan	N[a]	Commercially available
12	7AzaW	Y[f]	Commercially available
13	3-(thianaphthen-3-yl)-L-alanine	N[g]	Commercially available
14	L-β(thieno[3,2 *b*]pyrrolyl)alanine	Y[h]	Synthesized[i]
15	L-β (thieno[2,3 *b*]pyrrolyl)alanine	Y[h]	Synthesized[i]

The structures of these analogs are presented in Fig. 3. This table lists some commonly available tryptophan analogs and whether they are known to be incorporated into proteins based on reports in the literature as well as how to source them

[a] Budisa et al. (20)
[b] Pratt and Ho (21)
[c] Hogue et al. (22)
[d] Lark (23)
[e] Kwon and Tirrell (24)
[f] Schlesinger (25)
[g] Hall et al. (28)
[h] Budisa et al. (26)
[i] Phillips et al. (27)

2. Materials

2.1. Materials for the In Vitro Incorporation of Unnatural Amino Acids into Proteins

All solutions and reagents were made using sterile, deionized water in glassware that has been triple rinsed with sterile, deionized water. The protocol below should yield between 7.5 and 8.5 mL of S30 lysate, which in turn can be used for 450–500 (50 μL) in vitro transcription and translation reactions.

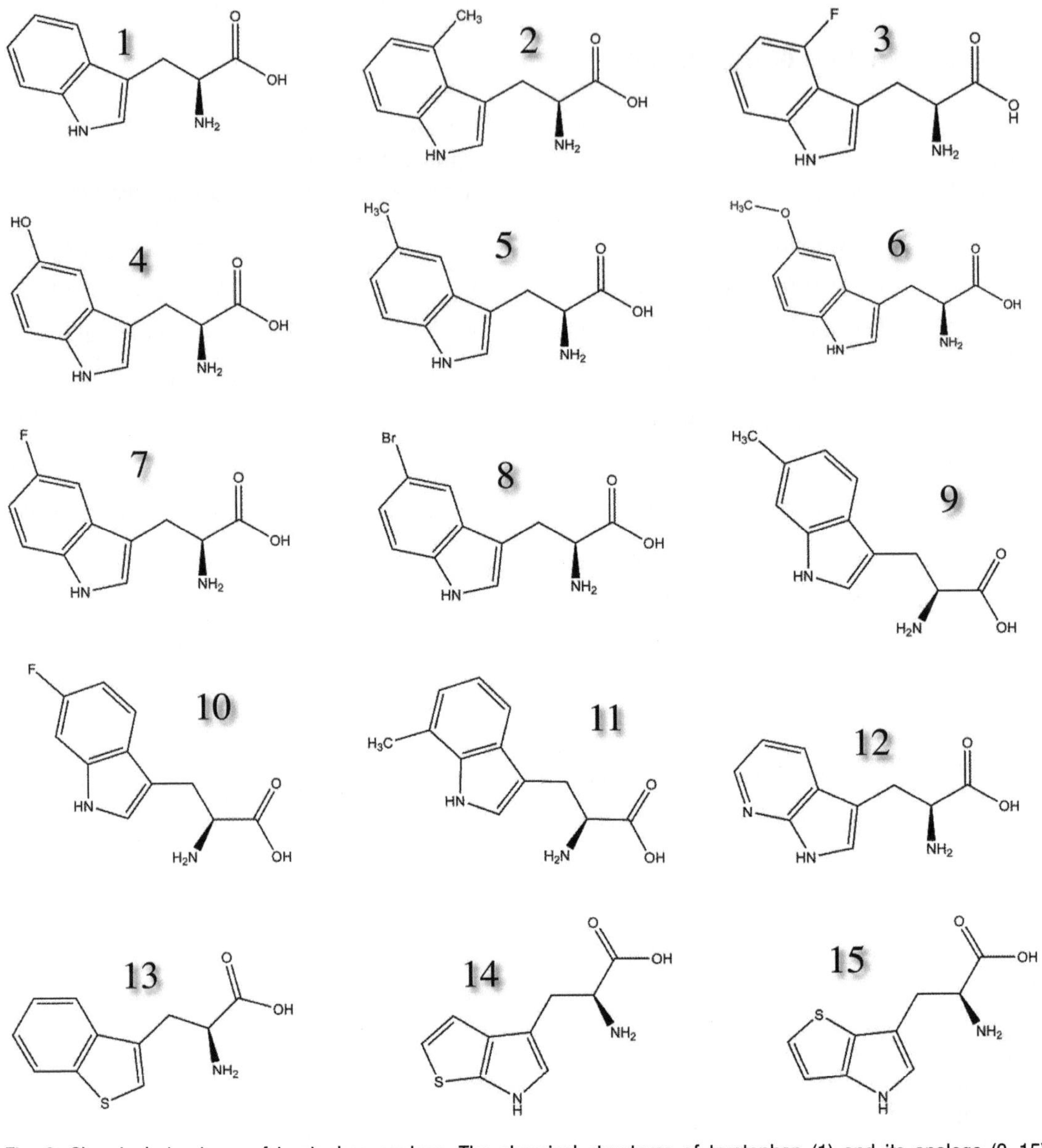

Fig. 3. Chemical structures of tryptophan analogs. The chemical structures of tryptophan (1) and its analogs (2–15) presented in Table 1 are listed above.

2.1.1. Medium G-PG

Lysates are made from cells grown in a rich, defined medium supplemented with phosphates and glucose (modified from ref. 10, 18).

1. Nucleoside solution (100×): to 50 mL of water, add 0.5 g each of guanosine, adenosine, cytidine and uridine, and 2.5 g of thymidine. Add KOH dropwise to solubilize (only a few drops should be necessary) and then make up the volume to 100 mL. Store at 4°C.

2. Vitamin amide solution (100×): to 50 mL of water add 0.1 g thiamine, 0.1 g niacinamide, 0.1 g pyridoxal, 0.2 g pantothenic acid, 0.01 g biotin, 0.01 g α-lipoic acid, 0.01 g p-aminobenzoic acid (PABA), 0.001 g folic acid, 0.001 g riboflavin, 1.5 g ribitol, 0.5 g glutamine, and 0.5 g asparagine. Add a few drops of KOH to solubilize and then make the volume up to 100 mL with water.
3. $CaCl_2$ solution (1,000×): add 0.2 g $CaCl_2$ to a final volume of 10 mL of water.
4. Alanine solution (400×): make a 40 mg/ml solution by adding 0.4 g DL-alanine to 100 mL of water.
5. Amino acid solution (400×): add 0.4 g of all L-amino acids except tryptophan and tyrosine to 30 mL water (see Note 1). Make the volume up to 50 mL with water.
6. $MnSO_4$ solution (1,000×): add 0.22 g $MnSO_4{\cdot}4H_2O$ to a final volume of 10 mL of water.
7. $FeCl_3$ solution (10,000×): add 0.06 g $FeCl_3{\cdot}6H_2O$ to a final volume of 10 mL of water.
8. Fill a 4 L beaker with 2.5 L of water. Add components according to Table 2 below.
9. Adjust the pH of the medium to 7 (with 10 M KOH) and bring up to 4 L with water.
10. Filter sterilize the medium G-PG through a 0.2 μm filter.
11. Dispense 1 L each into 4×4 L autoclaved, baffled flasks.
12. Store at 4°C.
13. Prepare 2 mL of 0.02 g/mL tryptophan solution by dissolving 40 mg of l-tryptophan in 1.5 mL of water. Add drops of KOH till the tryptophan goes into solution. Make the volume up to 2 mL with water. Store at −20°C.

2.1.2. Buffers and Consumables for Making the Lysate

1. *E. coli* strain that is auxotrophic for tryptophan.
2. 30 mL of Luria–Bertani broth (per liter: 10 g tryptone, 5 g yeast extract, 10 g NaCl).
3. Buffers A and B: make stocks of 3 M magnesium acetate, 6 M potassium acetate, and 1 M Tris–acetate, pH 8, 1 M DTT. For buffer B, add 20 mL of 3 M magnesium acetate, 40 mL of potassium acetate, 40 mL of Tris–acetate, pH 8, and 4 mL 1 M DTT a to a final volume of 4 L. Add 0.5 mL of β-mercaptoethanol to 1 L buffer B to make buffer A.
4. Slide-A-Lyzer dialysis cassette, with a 7000MWCO, syringe, and needle (18 G).

2.1.3. Low Molecular Weight Mix

Prepare stock solutions of the components and mix them according to Table 3 to obtain the low molecular weight (LMW) mix for

Table 2
Medium G-PG components

Component	Stock concentration	Final concentration	Amount to add to 4 L medium G-PG
Glutamic acid	N/A	0.15 g/L	0.6 g
Glycerol		0.15 g/L	0.6 g
KH_2PO_4	N/A	1.36 g/L	5.44 g
Na_2SO_4	N/A	0.11 g/L	0.44 g
NH_4Cl	N/A	0.54 g/L	2.16 g
NH_4NO_3	N/A	0.095 g/L	0.38 g
$MgSO_4 \cdot 7H_2O$	N/A	0.1 g/L	0.4 g
Glucose	N/A	18 g/L	72 g
NaH_2PO_4	N/A	3.03 g/L	12.12 g
Na_2HPO_4	N/A	10.72 g/L	42.88 g
L-tyrosine	N/A	0.02 g/L	0.08 g
$CaCl_2$	20 mg/mL (1,000×)	20 mg/L	4 mL
$MnSO_4$	22 mg/mL (1,000×)	22 mg/L	4 mL
$FeCl_3$	6 mg/mL (10,000×)	0.24 mg/L	400 μL
Nucleoside solution	100×	1×	40 mL
Vitamin amide solution	100×	1×	40 ml
DL-alanine	40 mg/mL (400×)	25 mg/L	2.5 mL
Amino acid solution	8 mg/mL (400×)	0.02 g/L	10 mL

800 × 50 μL IVTT reactions. Adding 23 μL of the LMW mix to a 50 μL reaction yields the appropriate final concentrations. Aliquot reactions into tubes and freeze at −80°C (see Note 2).

2.1.4. DNA Template

The DNA template used to express your protein of interest can be a plasmid (from a mini-prep) or a linear DNA such as a PCR product. The plasmid should be one suitable for expression in a cell-free synthesis system. For example, pIVEX plasmids contain a T7 promoter and ribosome binding site upstream of the protein-coding sequence and a T7 terminator downstream. Alternatively, many of the pET vector series are compatible with in vitro transcription and translation reactions (see Subheading 3.2.1 for more information on cloning your gene of interest into a pET vector). If you are using a linear template generated by PCR

Table 3
Low molecular weight (LMW) mix components

Component	Stock concentration	Final concentration	Volume to add to LMW mix (μL)
HEPES–KOH pH 7.5	1 M	55 mM	2,200
DTT	1 M	1.7 mM	68
ATP, pH 7	100 mM	1.2 mM	480
CTP, GTP, UTP	100 mM	0.8 mM	320
Creatine phosphate	0.8 M	80 mM	4,000
PEG-6000	40%	4.0%	4,000
cAMP	50 mM	0.64 mM	512
tRNA	175 mg/mL	175 μg/mL	400
Ammonium acetate	1.4 M	28 mM	800
Magnesium acetate	3 M	10.7 mM	142.4
Amino acids (–W,Y) (see Note 1)	50 mM	2 mM	1,600
Tyrosine	N/A	2 mM	0.012 g
Folinic acid	57 mM	68 μM	48
Potassium glutamate	3 M	210 mM	2,800

amplification, then the gene should again have a T7 promoter and ribosome binding site upstream of the protein-coding region and a T7 terminator downstream (see Note 3). These regulatory regions may need to be added during amplification of the linear template with appropriate primers that encode them.

2.1.5. Equipment

1. 37°C incubator shaker.
2. French pressure cell press.
3. SpeedVac.
4. SDS-PAGE gel running equipment.
5. Storage phosphor screen.
6. Storage phosphorimager imaging device.

2.2. Materials for the In Vivo Incorporation of Unnatural Amino Acids into Proteins

1. Tryptophan auxotrophic *E. coli* strain BL21(DE3)Δ*trp*C pLysS (genotype: F^- trpC (Ins: GrpII intron dfrA (trimethoprim)) ompT gal dcm lon hsdS_B($r_B^-m_B^-$) λ(DE3 (lacI lacUV5-T7 gene 1 ind1 sam7 nin5)) (pLysS: T7 LysS cat) is used for protein expression experiments (see Note 4).
2. DH10B strains are used for plasmid manipulation and storage.

3. Luria–Bertani media (per liter: 10 g tryptone, 5 g yeast extract, 10 g NaCl, and 1.5% agar for plates).
4. M9 minimal media base (5× stock solution, per liter: 64 g $Na_2HPO_4{\cdot}7H_2O$, 15 g KH_2PO_4, 2.5 g NaCl, 5 g NH_4Cl), supplemented with 2 mL/L 1 M $MgSO_4$, 20 mL/L 20% glucose, 100 μL 1 M $CaCl_2$ and 1 mL/L trace mineral supplement (ATCC, Manassas, VA) and 0.1–1.0 mM tryptophan analog or 0.1 mM tryptophan (see Note 5). The media was further supplemented with 100 μg/mL ampicillin.
5. Tryptophan analogs: 7-azatryptophan, 4-, 5-, 6-, and 7-fluorotryptophan (see Table 1 for translational compatibility).
6. Nickel–NTA resin.
7. Polypropylene protein purification columns (2 mL bed volume).
8. Benzonase nuclease.
9. 1 M $MgSO_4$.
10. Sonic homogenizer or French press (see Note 6).
11. Buffers for Ni–NTA (His-tag) purification:
    a. Binding buffer (1×): 20 mM Tris–HCl, pH 8.0, 300 mM NaCl, and 10 mM imidazole.
    b. Wash buffer (1×): 20 mM Tris–HCl, pH 8.0, 300 mM NaCl, and 50 mM imidazole.
    c. Elution buffer (1×): 20 mM Tris–HCl, pH 8.0, 300 mM NaCl, and 250 mM imidazole.
12. Materials and equipment for denaturing polyacrylamide gel electrophoresis (PAGE).

3. Methods

3.1. *In Vitro* Incorporation of Unnatural Amino Acids into Proteins

The procedure to make the tryptophan auxotroph S30 lysate takes ~3 days and will be described chronologically (see Note 7). Unless otherwise specified, all procedures are carried out at room temperature.

3.1.1. Making an S30 Lysate Depleted in Endogenous Tryptophan

1. Grow a starter culture of the tryptophan auxotroph in at least 20 mL of Luria–Bertani broth overnight at 37°C on a shaker.
2. Inoculate 1 L of medium G-PG + tryptophan solution to a final concentration of 20 mg/L with 10 mL (1:100 ratio) starter culture.
3. Prepare 1× PBS and keep at room temperature.
4. Prepare 4 L of 1× buffer B. Remove 1 L of this and make buffer A by adding 0.5 mL β-mercaptoethanol. Store buffer A and buffer B at 4°C.

5. Grow the cells at 37°C on a shaker at 200 rpm until the cells reach mid-log phase (see Note 8).
6. Spin down cells at 6,000 × *g* at 16°C and immediately wash with 500 mL 1× PBS per liter of culture. The cell pellet should be resuspended gently with minimal shear (see Note 9).
7. Spin down again at 6,000 × *g* and 16°C.
8. Resuspend the cells gently in 1 L medium G-PG, this time without tryptophan.
9. Grow cells at 37°C with shaking at 200 rpm for 50 min.
10. Spin down the cultures at 6,000 × *g* at 4°C. Wash the cells twice with cold S30 buffer A (1 L). After the second wash, flash freeze the cells (see Note 10).
11. Weigh the cell pellet and store at −80°C (see Note 11).
12. Thaw the cells on ice with 1 mL of buffer B for every gram of cell pellet. Resuspend the cells until they form a smooth cell paste.
13. Transfer the cell paste to a chilled French press cell and lyse the cells with a single pass through the French press at 1,100 psi. Collect the lysate in tubes on ice.
14. Centrifuge the lysate twice at 30,000 × *g* at 4°C. After the first spin, carefully transfer the supernatant to a new tube, taking care to avoid the pellet.
15. After the second spin, transfer the lysate to a tube wrapped in foil and incubate at 37°C on a rotary shaker at 150 rpm for 90 min. Take note of the volume of lysate.
16. Carefully pipette the lysate into a dialysis cassette with between MWCO of 7,000 and 8,000 Da, attach a floatation device, and place in 80× volume of cold buffer B in a beaker or other suitable container. Place a clean stir bar in the beaker and dialyze for 12 h with gentle stirring at 4°C.
17. Subsequent to the first 12 h dialysis, perform two buffer exchanges for 45 min each.
18. During the third dialysis exchange, set up a tray with ice. From the estimate of the volume of final lysate (step 12) set up tubes on ice to aliquot the lysate into (see Note 2).
19. Transfer the lysate to the chilled centrifuge tubes and spin at 4,000 × *g* for 10 min.
20. Aliquot the lysate into new tubes on ice and store at −80°C.

3.1.2. Setting Up a Cell-Free Translation Reaction

Set up each 50 μL reaction according to the recipe in Table 4. Control reactions should be set up and run in parallel with the sample reactions. The controls should include a positive control containing DNA template and tryptophan and two negative controls, viz., a reaction containing tryptophan but no DNA

Table 4
Each 50 μL cell-free synthesis reaction contains the following components

Component	Final concentration	Volume (μL)
S30 lysate		17
LMW mix		23
T7 RNA polymerase (Epicenter, Madison, WI)	~500 U	2
W/W analog (100 mM) (see Note 5)	2 mM	1
Methionine (8 mg/mL)	0.016 mg	2
S^{35}-labeled methionine (>1,000 Ci/m mol)		1–2
DNA template (see Note 14)	50–500 ng	
Water		Rest
Salmon sperm DNA (see Note 15)	1–2 μg	1–2
Total volume		50

template and a reaction containing DNA template but no exogenous tryptophan or analog (also, see Note 12).

1. Thaw the lysate and LMW mix on ice (see Note 15). Thaw the remaining components at room temperature.
2. Combine all the components (except the lysate) on ice. Take appropriate precautions and dispose of contaminated materials as necessary when using S^{35} methionine.
3. Add the lysate last.
4. Incubate at 30°C for 1–4 h and store at 4°C thereafter.
5. For purification of His-tagged protein products (see Note 16). If available, conduct a functional assay specific for the protein of interest. To analyze the production of radiolabeled protein, follow steps 6–13.
6. Remove 5 μL of the cell-free synthesis reaction and add it to 50 μL of ice cold acetone.
7. Incubate on ice for 10 min.
8. Spin at 8,000 × *g* for 10 min.
9. Remove the remaining acetone by evaporation at room temperature for 20 min or in a SpeedVac (Thermo Fisher) for 5 min.
10. Resuspend the pellet in 1× SDS-loading buffer and heat at 95°C for 15 min.
11. Run the samples on an SDS-PAGE gel with an appropriate protein ladder and/or size standards.

12. Transfer the gel to a nitrocellulose membrane and expose the transferred membrane to a storage phosphor screen overnight.
13. After reading the screen on the phosphorimager, use ImageQuant or other image quantitation software to quantitate the bands and analyze the results (see Note 12).

3.2. In Vivo Incorporation of Unnatural Amino Acids into Proteins

The methods described below detail:

1. Construction of an Expression Plasmid Encoding the Gene of Interest (see also Note 17).
2. The expression of proteins containing unnatural amino acids (tryptophan analogs) in place of one of the natural amino acids.
3. The purification and characterization of proteins containing unnatural amino acids.

3.2.1. Construction of an Expression Plasmid Encoding the Gene of Interest

The expression system described herein relies on a tryptophan auxotrophic mutant of the commonly used BL21 (DE3) strain, and hence can support the overexpression of genes driven by the T7 RNA polymerase promoter. This strain is compatible with any of a number of commercially available *E. coli* expression vectors, especially the pET series (Novagen-EMD) (see Note 17 for an example of how to clone into pET21 using *Xho*I and *Nde*I restriction sites).

3.2.2. Expression of Proteins Containing Unnatural Tryptophan Analogs In Vivo

This protocol relies on the forced incorporation of a tryptophan analog into a protein of interest using a tryptophan auxotroph under conditions where tryptophan is removed from the media and replaced with a tryptophan analog. To achieve high levels of incorporation of the unnatural amino acids into proteins, the bacterial strain should first be grown under permissive conditions (tryptophan in media) and then switched to restrictive conditions (analog but no tryptophan in the media) at the time that protein expression is induced. Careful monitoring of the bacterial growth cycle is required to obtain optimal results.

1. Transform the pET21 expression construct into the BL21(DE3)Δ*trp*C pLysS strain and plate onto LB + 100 μg/mL ampicillin and 34 μg/mL chloramphenicol plates. Incubate at 37°C overnight.
2. Inoculate a 3 mL LB + ampicillin/chloramphenicol culture by transferring a colony from the transformed plate into the liquid media. Incubate the culture overnight with shaking at 37°C.
3. Inoculate a 250 mL culture of M9 media supplemented with 0.1 M tryptophan, 100 μg/mL ampicillin, and 34 μg/mL chloramphenicol with a 1:100 (v/v) (2.5 mL) sample from the overnight starter culture.

4. Grow the bacteria at 37°C with shaking to mid-log phase (optical density at 600 nm ≈ 0.5–0.6).
5. Transfer the culture to an appropriate centrifuge tube/bucket and spin at 6,000 × *g* for 20 min to pellet the cells. Discard the media.
6. Wash the cells with 50 mL of phosphate buffer (room temperature) to remove any residual tryptophan-containing media. Centrifuge the cells at 6,000 × *g* for 20 min to repellet the cells.
7. Resuspend the cell pellet in 250 mL of M9 media (room temperature) with supplements and antibiotics plus 0.3 mM tryptophan analog (see Note 5). Transfer resuspended cells to a shake flask.
8. Incubate the resuspended cell culture with shaking at 37°C for 40 min.
9. To the shake culture add IPTG to 1 mM final concentration. Incubate the culture at 37°C with shaking for an additional 16 h (overnight) (see Note 18).
10. Pellet the cells by centrifugation and discard the media. Either immediately proceed to the next section, for isolation of the protein containing the unnatural amino acids, or store the bacterial cell pellet at −80°C until purification can be performed.

3.2.3. Purification of Proteins Containing Unnatural Amino Acids

Proteins bearing unnatural amino acids can be isolated in much the same way as other overexpressed proteins. In this example, we recommend purification of proteins with C-terminal poly-histidine tags (His-TAG; encoded on the pET21 expression vector) via immobilized metal affinity chromatography (IMAC). In principle, any affinity tag can be used for purifying the protein of interest, though one should avoid affinity tag sequences where there is a possibility of replacing one of the natural amino acids in the tag with an unnatural amino acid that could disrupt binding. As with many protein purification procedures, the exact conditions for isolating highly purified protein containing unnatural amino acids will rely in part on the properties of the protein itself. Thus, the protocol below is a guide that will likely require further optimization based on the properties of the protein of interest.

1. Resuspend the bacterial cell pellet in 30 mL of binding buffer and lyse the cells using a sonic homogenizer (see Note 6).
2. Centrifuge the lysed cells at 15,000 × *g* for 15 min to pellet the cell debris.
3. Decant the cleared lysate into a clean Oak Ridge-style centrifuge tube and add 30 μL of 1 M $MgSO_4$ to yield a final concentration of 1 mM Mg^{2+}. Mix gently by inversion and store on ice.

4. Add Benzonase to the lysate to a final concentration of 8 U/mL (240 U total for 30 mL). Incubate on ice for 10 min with occasional mixing by inversion.
5. Centrifuge the lysate at 25,000×g for 20 min to pellet any remaining cell debris.
6. While the lysate is being clarified set up an IMAC column by adding 2 mL of Ni–NTA (Qiagen) to an empty poly-prep chromatography column. Allow the resin to settle and then wash the column with 10 mL (5 column volumes) of sterile water to remove residual buffer and ethanol from the IMAC matrix.
7. Equilibrate the column by gravity flow with 4 column volumes (8 mL) of binding buffer.
8. Carefully add the cleared cell lysate to the column and allow it to flow through the IMAC resin.
9. After all of the lysate has flowed through the resin, wash the column with 4 column volumes (8 mL) of binding buffer followed by 8 column volumes (16 mL) of wash buffer.
10. Elute the protein from the column using 4 column volumes (8 mL) of elution buffer. Collect 0.5 mL fractions from the column and analyze 5–10 μL fractions by electrophoresis on an SDS-PAGE gel.
11. Pool all of the fractions that contain the protein of interest at 95% or greater purity and concentrate (if desired) using an Amicon ultrafiltration column. Quantitate the isolated protein via the BCA (bicinchoninic acid) protein assay or Bradford protein assay using known standards.

3.3. Analysis of Proteins Containing Unnatural Amino Acids

Following purification it is often desirable to determine the extent to which the natural amino acid in the protein has been replaced by the unnatural amino acid analog. This is commonly done by mass spectrometry to determine the mass difference between the proteins (or their substituent peptides). A detailed protocol for mass spectrometry will be highly dependent on what facilities are available to the user, and it is beyond the scope of this work. However, a general protocol for preparation of protein samples for MS would be as follows:

1. Lyophilize 5–10 μg samples of your protein of interest that either contain the unnatural or natural amino acid and resuspend the dried protein in 0.1 M NH_4HCO_3.
2. Digest the resuspended protein with l-tosylamido-2-phenylethyl chloromethyl ketone-treated trypsin at 37°C for 10 h.
3. Remove the trypsin via centrifugation in a spin column with a 0.45 μm filter. The immobilized trypsin will be retained on the membrane, while the digested peptides from your protein of interest will be contained in the eluate.

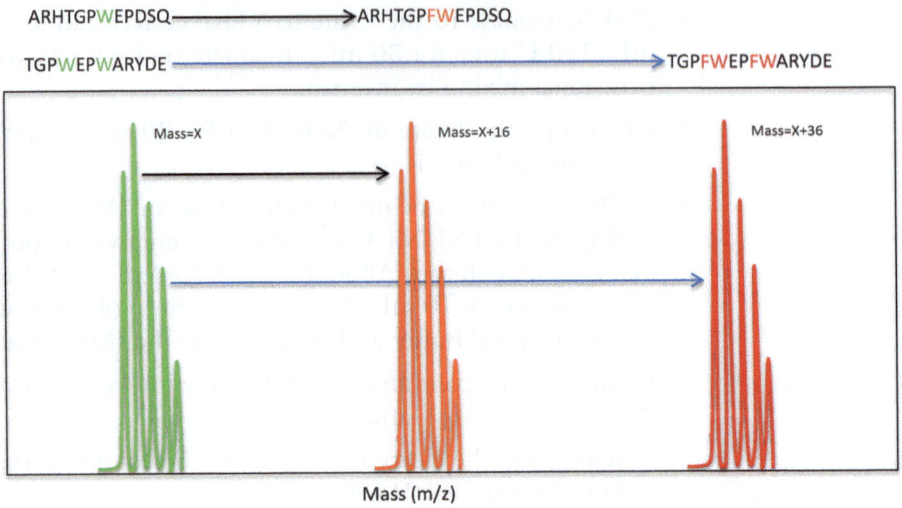

Fig. 4. Sample MALDI–TOF spectra of two hypothetical peptides. The first, ARHTGPWEPDSQ, contains a single tryptophan residue. Its spectrum (Mass = *X*) shifts to the right upon substitution by a tryptophan analog W*. For illustration, we have substituted it above with a fluorotryptophan (e.g., 4-, 5-, or 6-fluorotryptophan) so the mass of the substituted peptide is different by a single fluorine (Mass = *X* + 16). Similarly, the second peptide, TGPWEPWARYDE, containing two tryptophan residues (Mass = *X*) shifts to the right (Mass = *X* + 36) upon replacement with two fluorinated tryptophans.

4. Lyophilize the digested peptides and resuspend them in water to a final concentration of 210 μM.
5. Analyze the digested samples by LC–ESI–MS or MALDI–TOF–MS. Compare the peptide masses obtained with samples containing the unnatural amino acid to those lacking the unnatural amino acid and note any mass differences between the two. An example is shown in Fig. 4 (see Note 19).

Alternatively, a detailed protocol for characterization of proteins containing unnatural amino acids via protein hydrolysis and HPLC can be found in ref. 19.

4. Notes

1. Add the indicated amounts of valine, phenylalanine, and isoleucine to water and incubate at 37°C for 15 min with shaking. Then add leucine and cysteine and incubate at 37°C for 15 min with shaking. Add methionine, alanine, arginine, asparagine, aspartic acid, glutamic acid, glycine, and glutamine to the solution. Add drops of 10 N KOH until all amino acids go into solution. Finally, add histidine, lysine, proline, serine, and threonine and incubate at 37°C with shaking until all of the amino acids are in solution.

2. We usually set up 6× reaction and 12× reaction aliquots. However, your requirements may dictate different scales. Keep in mind that it is best to not subject the lysate or LMW mix to repeated freeze–thaw cycles.
3. The RTS 100 *E. coli* LinTempGenSet His-tag (5 Prime) is a commercially available kit containing DNA primers that can be used to append appropriate regulatory sequences to the coding sequence of your protein via PCR.
4. The plasmid pLysS contains an expression cassette for T7 lysozyme, which is a natural inhibitor of the T7 RNA polymerase expressed in a DE3 strain. This allows for tighter control of the expression of target proteins under the control of T7 RNA polymerase promoters. Induction of T7 RNA polymerase expression and concomitant production of the target protein only when unnatural amino acid is added to the media should result in greater incorporation of the unnatural amino acid.
5. Unnatural amino acid preparations are often supplied as racemic mixtures of D/L isomers. Since only the L-enantiomer is compatible with cellular translation the concentration of the unnatural amino acid used should be doubled to compensate for the decreased concentration in the mixture.
6. Bacterial cells can be lysed using any method of choice including using commercial lysis solutions such as BugBuster (Novagen-EMD) or BPER (Pierce).
7. If the growth characteristics of a bacterial strain are not known, a growth curve should be performed, either on a plate reader or manually to get an estimate of how quickly the cells might reach the mid- to late log phase of growth. The OD_{600} at stationary phase should also be determined.
8. We check the OD_{600} periodically during growth to ensure that the cells do not overgrow to stationary phase.
9. Gently pipetting with a 10 mL pipette seems to work best. Do not vortex.
10. The cells should be kept cold and the washes performed with a total of 1 L of buffer A. The washes can be performed in any kind of centrifuge bottle but the final wash should be in a bottle or 50 mL tube that can be flash frozen.
11. We have stored the cell pellet at −80°C for up to 3 days before proceeding to the next step with no measurable loss in cell-free protein synthesis activity in the final lysate.
12. Besides the usual controls (containing and omitting template DNA), we find it useful to gauge the levels of endogenous tryptophan remaining by setting up a control that contains template DNA, but does not contain tryptophan or an analog.

This provides a measure of how much full-length "background" expression occurs in the absence of exogenous tryptophan. The protein yield from a reaction containing template DNA and tryptophan is considered the positive control, while the protein yield from a reaction containing template DNA and no tryptophan or analog is considered the negative control. In our hands the background protein expression level is usually around 10% of the positive control. Reactions containing template and an analog will generally fall somewhere in between these two controls. The relative level of incorporation is a basal metric for how well the analog is taken up into cells and utilized by the translation machinery.

13. For a quick check of whether the cell-free synthesis system is working, a plasmid encoding GFP can be used as template. After the reaction, observe the tubes under UV transillumination (comparing tubes with and without template) and check for green fluorescence. For a more quantitative assay of lysate activity or protein yield, monitor S^{35} incorporation.
14. Salmon sperm DNA is only needed if the template is linear DNA, such as a PCR product.
15. Do not vortex or spin down the lysate. Pipette gently to mix the lysate. For the LMW mix, mix thoroughly before dispensing because tyrosine (which is not very soluble) may have settled to the bottom of the tube as a whitish precipitate.
16. The small amounts of protein produced in these reactions can still be purified using IMAC affinity resin, as described in Subheading 3.2.3, by scaling down the amount of resin and volumes of washes. Instead of a column, Ni–NTA resin can be added directly to the reactions and the purification performed in a batch as opposed to a column format. Set up 500 μL of cell-free synthesis reactions. Add 50 μL Ni–NTA resin (washed with binding buffer 1, without imidazole) to the reaction and incubate at room temperature with end-over-end rotation for 45 min. Pellet the resin by centrifugation at 2,000 × *g* for 1 min. Remove the supernatant (save a small aliquot for analysis) and wash sequentially with 750 μL binding buffer 1 without imidazole and then 750 μL binding buffer 1. Incubate the wash buffers with the resin for 5 min with end-over-end rotation before pelleting the resin. Again, save a small aliquot of each wash for analysis. Elute using 200 μL elution buffer. Analyze 5–10 μL of the washes and eluate on an SDS-PAGE gel.
17. The following protocol details a simple scheme to clone your gene of interest (GOI) into a pET21 vector or similar vector with *Xho*I and *Nde*I cloning sites.
    a. Amplify the GOI from the source DNA with a high-fidelity polymerase such as Phusion DNA polymerase (NEB).

b. Verify that the PCR product for the GOI is the correct size via agarose gel electrophoresis with an appropriate DNA ladder or size standards.

c. Purify the PCR product with a PCR cleanup kit.

d. Quantify the DNA by measuring absorbance at 260 nm using a spectrophotometer.

e. Digest the GOI DNA and the pET21 vector DNA with *Nde*I and *Xho*I restriction enzymes (NEB), employing the reaction conditions recommended by the supplier.

f. Run the *Nde*I/*Xho*I double-digested GOI and plasmid DNA on a 0.8% TAE low-melt agarose gel and purify the correct size bands from the agarose using a gel extraction kit.

g. Quantify the recovered DNA by measuring absorbance at 260 nm using a spectrophotometer.

h. Ligate the GOI into the digested pET21 vector using T4 DNA ligase following conditions recommended by the enzyme supplier.

i. Transform 2 μL of the ligation mixture into DH10B cells (Invitrogen) and plate on LB/Ampicillin selective media. Incubate the plates overnight at 37°C.

j. Inoculate 3 mL LB/Amp cultures from the transformation plates and grow overnight at 37°C.

k. Isolate the plasmid DNA from the overnight cultures using a Plasmid Mini Prep Kit following manufacturer's instructions.

l. Verify the sequence of your GOI within the pET21 expression plasmid.

18. Some optimization of the analog concentrations and the time of induction and length of expression may be needed to maximize the production of proteins containing unnatural amino acids. Typically tryptophan analog concentrations between 0.1 and 1 mM yield sufficient protein that largely contains the unnatural amino acid. Induction time and the length of the induction period are more critical. Typically the IPTG inducer should be added during mid-log phase. The optimum induction period (incubation time after the IPTG has been added to the cells) can be determined empirically. Generally an overnight induction period (~16 h) is a good place to start, although for some proteins a shorter induction period may be desirable. It is advisable to set up a series of smaller expression cultures to find the optimum induction time and analog concentration prior to preparing the protein in bulk. The solubility of the protein containing the unnatural amino acid may be an issue, as amino acid substitution can lead to protein misfolding and aggregation. The expressed state (soluble, insoluble, in

inclusion bodies, etc.) of your protein of interest should be determined via running fractions of the lysed cells on a gel and Western blot analysis to detect the C-terminal His-tag using an anti-polyhistidine antibody.

19. Analysis of the spectra obtained from a mass spectrometry experiment will depend on the target protein and the unnatural analog used. In general, in cases where complete substitution of the analog occurs, comparing the spectra of the peptide containing W with that of the peptide containing the unnatural analog will show a complete "shift" in mass of the analog, as shown in Fig. 4. However, in cases where incomplete substitution has occurred, some peptides will contain the analog while others will contain tryptophan. The spectrum obtained from proteins in which incomplete incorporation has occurred will correspondingly have at least two different sets of peaks (and if the peptide contains more than one tryptophan, there may be additional peaks reflecting mixtures of the natural and unnatural amino acid). However, depending on the analog used, additional peaks may not reflect incomplete incorporation, but instead species produced by neutral mass loss. Interpretation of such results is best done in consultation with someone who has extensive experience in mass spectrometry. The presence or absence of peaks of the predicted mass provides a qualitative indication of the success of unnatural amino acid incorporation. More quantitative assessments can only be made if it is known that both the analog-containing peptide and the W-containing peptide comparably ionize. Standards generated by synthesizing peptides containing W and the appropriate analog can help both with peak identification and quantitation.

Acknowledgments

This work was supported by the National Security Science and Engineering Faculty Fellowship (FA9550-10-1-0169), the Welch Foundation (F-1654), and the National Science Foundation (MCB-0943383). R.A.H. is supported by a postdoctoral fellowship from the Cancer Prevention and Research Institute of Texas (Project Nbr: RP101501) and is a postdoctoral fellow of the Applied Research Laboratories at The University of Texas at Austin. The content are solely the responsibility of the authors and do not necessarily represent the official views of the sponsors.

References

1. Chin JW, Martin AB, King DS, Wang L, Schultz PG (2002) Addition of a photocrosslinking amino acid to the genetic code of *Escherichia coli*. Proc Natl Acad Sci U S A 99: 11020–11024
2. Chin JW, Santoro SW, Martin AB, King DS, Wang L, Schultz PG (2002) Addition of p-azido-L-phenylalanine to the genetic code of *Escherichia coli*. J Am Chem Soc 124:9026–9027
3. Deiters A, Cropp TA, Summerer D, Mukherji M, Schultz PG (2004) Site-specific PEGylation of proteins containing unnatural amino acids. Bioorg Med Chem Lett 14:5743–5745
4. Lin S, Zhang Z, Xu H, Li L, Chen S, Li J, Hao Z, Chen PR (2011) Site-specific incorporation of photo-cross-linker and bioorthogonal amino acids into enteric bacterial pathogens. J Am Chem Soc 133:20581–20587
5. Bae JH, Rubini M, Jung G, Wiegand G, Seifert MH, Azim MK, Kim JS, Zumbusch A, Holak TA, Moroder L, Huber R, Budisa N (2003) Expansion of the genetic code enables design of a novel "gold" class of green fluorescent proteins. J Mol Biol 328:1071–1081
6. Tianero MD, Donia MS, Young TS, Schultz PG, Schmidt EW (2012) Ribosomal route to small-molecule diversity. J Am Chem Soc 134:418–425
7. Yoo TH, Link AJ, Tirrell DA (2007) Evolution of a fluorinated green fluorescent protein. Proc Natl Acad Sci U S A 104:13887–13890
8. Johnson JA, Lu YY, Van Deventer JA, Tirrell DA (2010) Residue-specific incorporation of non-canonical amino acids into proteins: recent developments and applications. Curr Opin Chem Biol 14:774–780
9. Budisa N (2006) Engineering the genetic code. Wiley-VCH, Weinheim
10. Wong JT (1983) Membership mutation of the genetic code: loss of fitness by tryptophan. Proc Natl Acad Sci U S A 80:6303–6306
11. Bacher JM, Bull JJ, Ellington AD (2003) Evolution of phage with chemically ambiguous proteomes. BMC Evol Biol 3:24
12. Bacher JM, Ellington AD (2001) Selection and characterization of *Escherichia coli* variants capable of growth on an otherwise toxic tryptophan analogue. J Bacteriol 183: 5414–5425
13. Wang L, Brock A, Herberich B, Schultz PG (2001) Expanding the genetic code of *Escherichia coli*. Science 292:498–500
14. Hughes RA, Ellington AD (2010) Rational design of an orthogonal tryptophanyl nonsense suppressor tRNA. Nucleic Acids Res 38:6813–6830
15. Young TS, Schultz PG (2010) Beyond the canonical 20 amino acids: expanding the genetic lexicon. J Biol Chem 285:11039–11044
16. Taki M, Hohsaka T, Murakami H, Taira K, Sisido M (2001) A non-natural amino acid for efficient incorporation into proteins as a sensitive fluorescent probe. FEBS Lett 507:35–38
17. Heckler TG, Chang LH, Zama Y, Naka T, Chorghade MS, Hecht SM (1984) T4 RNA ligase mediated preparation of novel "chemically misacylated" tRNAPheS. Biochemistry 23:1468–1473
18. Kim RG, Choi CY (2001) Expression-independent consumption of substrates in cell-free expression system from *Escherichia coli*. J Biotechnol 84:27–32
19. Bacher JM, Ellington AD (2007) Global incorporation of unnatural amino acids in *Escherichia coli*. Methods Mol Biol 352:23–34
20. Budisa N, Pal PP, Alefelder S, Birle P, Krywcun T, Rubini M, Wenger W, Bae JH, Steiner T (2004) Probing the role of tryptophans in Aequorea victoria green fluorescent proteins with an expanded genetic code. Biol Chem 385:191–202
21. Pratt EA, Ho C (1975) Incorporation of fluorotryptophans into proteins of *Escherichia coli*. Biochemistry 14:3035–3040
22. Hogue CW, Rasquinha I, Szabo AG, MacManus JP (1992) A new intrinsic fluorescent probe for proteins. Biosynthetic incorporation of 5-hydroxytryptophan into oncomodulin. FEBS Lett 310:269–272
23. Lark KG (1969) Incorporation of 5-methyltryptophan into the protein of *Escherichia coli* 15 T- (555-7). J Bacteriol 97:980–982
24. Kwon I, Tirrell DA (2007) Site-specific incorporation of tryptophan analogues into recombinant proteins in bacterial cells. J Am Chem Soc 129:10431–10437
25. Schlesinger S (1968) The effect of amino acid analogues on alkaline phosphatase. Formation in *Escherichia coli* K-12. II. Replacement of tryptophan by azatryptophan and by tryptazan. J Biol Chem 243:3877–3883
26. Budisa N, Alefelder S, Bae JH, Golbik R, Minks C, Huber R, Moroder L (2001) Proteins with beta-(thienopyrrolyl)alanines as alternative

chromophores and pharmaceutically active amino acids. Protein Sci 10:1281–1292

27. Phillips RS, Cohen LA, Annby U, Wensbo D, Gronowitz S (1995) Enzymatic synthesis of Thia-L-tryptophans. Bioorg Med Chem Lett 5:1133–1134

28. Hall LE, Hegeman GD, Bosin TR (1974) Incorporation of tryptophan and its benzo(b) thiophene, 1-methylindole, and indene analogs into protein of *Escherichia coli*. Res Commun Chem Pathol Pharmacol 9:145–153

Chapter 8

Reconstructing Evolutionary Adaptive Paths for Protein Engineering

Megan F. Cole, Vanessa E. Cox, Kelsey L. Gratton, and Eric A. Gaucher

Abstract

Reconstructing Evolutionary Adaptive Paths (REAP) is one of several methods to improve enzyme functionality. This approach incorporates computational and theoretical aspects of protein engineering to create a focused library of protein variety with a high degree of functionality. In contrast to other techniques like DNA shuffling, REAP allows a library to have diverse functionality among relatively few variants. REAP is a low-throughput method which takes advantage of natural selection and uses ancestral protein sequences to direct gene mutations, thereby creating a library with a high density of viable proteins. These proteins must then be assayed to characterize their functionality to identify which variants have the desired traits such as acid stability or thermostability.

Key words: Protein engineering, Ancestral sequence reconstruction, Phylogenetic analysis, Functional divergence, Molecular evolution

1. Introduction

Protein engineering uses a variety of techniques to modify or improve protein function for applications in industry or biomedicine. The engineering process consists of synthesizing a library of sequence variants, testing for the best ones, and using the knowledge gleaned from those assays to make improved variants. This process can be repeated until the desired activity or properties have been achieved. Desired protein traits could include altered kinetic activity, increased thermodynamic stability, enhanced lipophilicity, or increased pH stability. Sampling all sequence space, even for short peptides, generally results in a library too large to effectively sample for the desired activity. However, since the number of protein sequences available in databases has increased over the past decade, more focused methods that exploit such sequence information for protein engineering have become feasible.

James C. Samuelson (ed.), *Enzyme Engineering: Methods and Protocols*, Methods in Molecular Biology, vol. 978,
DOI 10.1007/978-1-62703-293-3_8, © Springer Science+Business Media New York 2013

These techniques include DNA shuffling, the consensus method, ancestral sequence reconstruction (ASR), ancestral mutation method (AMM), and REAP.

REAP is one method for identifying modern amino acid residues that can be replaced with ancient residues in an extant sequence. This technique uses sequence divergence to identify which amino acids may impart alternate activity to a protein (1, 2). When amino acids differ between subfamilies on a phylogenetic tree, this suggests that the branch separating the two subfamilies of the phylogeny has undergone functional divergence. One advantage of REAP is that it focuses on sites responsible for divergent evolution among homologous proteins. This focuses the number of sites that will be mutated and therefore results in a smaller library.

While DNA shuffling can be a powerful tool due to its unbiased approach, it results in large libraries that are expensive and time-consuming to assay for functionality (3). By providing more focused guidance for the amino acid replacements, REAP results in a much smaller library with a higher ratio of functional proteins to nonfunctional proteins. A consensus sequence can also be used to create small libraries; however, it ignores the context-dependent nature of amino acids and the resulting sequences have never been subjected to natural selection unlike ancient proteins (2). REAP takes advantage of natural selection by using ancestral sequences as the guide for gene mutations and, unlike say AMM, infers exactly which mutations are most likely to create functional diversity. Ancestral sequence reconstruction takes advantage of natural selection since such protein sequences have, theoretically, existed and been found fit for survival. However, this method does not involve sampling extant sequences and is dependent upon the accuracy of the computational models as well as the quantity of extant sequences used in the model. REAP is useful because it takes advantage of natural selection to modify extant sequences. This method is dependent upon the assumption that the divergence of an amino acid in an ancestral protein sequence may be the consequence of altered selective constraints acting on a protein.

Methods that create large libraries are expensive and costly to assay and frequently have too few enzymes that retain any activity or too few enzymes that display novel activity (the goldilocks problem). The concept behind REAP, using protein sequences that once (theoretically) existed, is that a greater percentage of protein variants will retain activity. By increasing the number of active variants, the probability of identifying proteins with the desired functionality increases as well. One experiment used a REAP library of only 93 variants to identify a DNA polymerase with increased promiscuity capable of faithfully incorporating nonstandard nucleosides (4). Advantages of a low-throughput technique for protein engineering include reduced cost of library synthesis and reduced dependence on high-throughput screens/selections that often serve as poor proxies for enzyme function. We anticipate that REAP will save laboratories' time and money by increasing the ratio of functional to nonfunctional protein variants.

2. Materials

The databases and software programs mentioned in Subheading 3 are listed here for reference.

NCBI	http://www.ncbi.nlm.nih.gov/
PFAM	http://pfam.sanger.ac.uk/
EBI	http://www.ebi.ac.uk/
GenBank	http://www.ncbi.nlm.nih.gov/genbank/
RefSeq	http://www.ncbi.nlm.nih.gov/RefSeq/
TPA	http://www.ncbi.nlm.nih.gov/genbank/tpa/
Swiss-Prot	http://www.uniprot.org/
PIR	http://pir.georgetown.edu/
PRF	http://www.prf.or.jp/index-e.html
BRENDA	http://www.brenda-enzymes.info/
PDB	http://www.wwpdb.org/
BLAST	http://blast.ncbi.nlm.nih.gov/Blast.cgi
ClustalW	http://www.ebi.ac.uk/Tools/msa/clustalw2/
T-Coffee	http://tcoffee.crg.cat/
MrBayes	http://mrbayes.sourceforge.net/
DIVERGE	http://xungulab.com/software/diverge/document/
Rate shift analysis server	http://www.daimi.au.dk/~compbio/rateshift/protein.html
PAML	http://abacus.gene.ucl.ac.uk/software/paml.html

3. Methods

An overview of the REAP methodology can be seen in Fig. 1.

3.1. Collect Homologous Sequences

The success of the REAP approach fundamentally relies on the natural functional diversity of homologous sequences included in an analysis. It is thus critical to collect sequence data from homologs

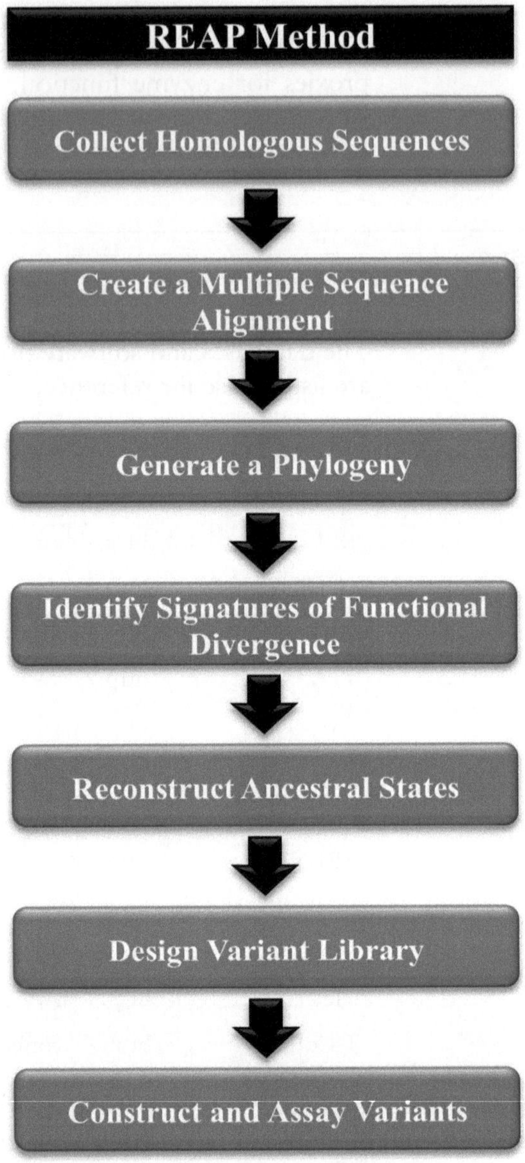

Fig. 1. Overview of the REAP method. First, homologous sequences are collected and used to create a multiple sequence alignment. The sequences and MSA are then analyzed to generate a phylogeny. Signatures of functional divergence along branches of the phylogeny are then identified. Ancestral sequence reconstruction is then used to reconstruct the ancestral states at the functionally divergent sites. These functionally important ancestral states then guide the design of the variant library. The resulting targeted library should capture meaningful functional diversity observed within extant and ancestral homologs in a relatively small number of variants. These variants can then be experimentally constructed and queried to identify desirable mutations for further analyses and optimization.

with diverse phenotypic properties. If possible, it is best to collect orthologous sequences from diverse families or domains of life. Paralogs can also be considered when duplicated sequences have enough similarity/identity to allow for reliable alignments and when it would offer relevant functional diversity.

It is important to collect a sufficient number of phylogenetically diverse homologous sequences in order to generate a well-articulated tree. However, too many sequences can be computationally challenging without offering much additional functional diversity. Generally, REAP works well with between 50 and 200 sequences that are approximately evenly spread throughout the phylogenetic space covered. In order to create a rooted phylogenetic tree in subsequent steps in the process, one must also include an out-group. The out-group is usually one or two sequences that while still homologous are from a more distant phylogenetic space. Often the out-group comes from a domain of life or a paralog that was otherwise not sampled.

Amino acid sequences rather than nucleotide sequences are usually used for REAP analyses as they allow for identification and alignment of more distantly related homologs with greater accuracy while still capturing phenotypic diversity. However, REAP can certainly be applied to nucleotide sequences when there is reason to believe that the nucleotide sequence itself is important for function. Many databases can be used to identify and collect homologous sequences, including NCBI, PFAM, EBI, GenBank, RefSeq, TPA, Swiss-Prot, PIR, PRF, BRENDA, and PDB. Homologs can be identified based on annotation, such as name or EC number, or by sequence similarity via a BLAST (basic local alignment search tool) search (see Note 1). When using annotation to identify homologs, it is important to verify that the sequence retrieved is a legitimate homolog since annotations can be incorrect or misleading (5). It may be useful to collect a few more sequences than your target number as some sequences may have to be dropped at the next step of sequence alignment.

3.2. Create a Multiple Sequence Alignment

The collected sequences must be assembled into a multiple sequence alignment (MSA) in order to analyze the variation between extant sequences and to reconstruct the ancestral states. Software programs such as ClustalW (6) or T-Coffee (7) can be used to create an initial alignment, but this must be manually inspected and adjusted. This is often a time-consuming step that should be repeated until the MSA is suitable for the next analyses in REAP. Refinement of the MSA is somewhat of an art form and care and time must be taken at this step because the phylogenetic analyses are based entirely on the quality of the MSA used as input and the assumption that aligned residues share a common ancestor.

Upon the initial automated creation of an MSA, one should first consider whether any of the gathered sequences should be dropped from the analysis. If sequences were identified by annotation, it is possible that some may have been misannotated and are not true homologs. These sequences will not fit well within the MSA and should be dropped. It is also common to find some sequences that should be dropped because they break up the MSA due to insertion/deletion events. While some insertion/deletion events can simply be trimmed from individual sequences, when a single sequence has an inordinate number of them, it is often simpler to just drop the sequence entirely. After dropping any sequences, the remaining data should be rerun through the MSA software to generate an updated alignment.

In addition to removing difficult sequences entirely, it is usually necessary to manually refine sequences so as to remove gaps in the MSA. Sequences can be trimmed to normalize the overall lengths of the sequences. This is often necessary at the N- and C-termini of proteins as these regions are especially prone to sequence diversity (see Note 2). Effort should be made to eliminate as many gaps as possible. When a single sequence creates a gap in the MSA due to an insertion event, that portion of that sequence can simply be deleted. However, when the case is less straightforward, care needs to be taken to determine whether the gap should be removed or not. This often involves utilizing knowledge of the evolutionary relationships between sequences to determine what is the correct course of action. Any gaps left in the MSA will be included in the reconstructed ancestral sequences so insight into whether the gaps might provide important function to the protein should be used to aid these decisions. As sequences are trimmed it is often useful to rerun the MSA software to generate an updated alignment.

Once the general assembly of the MSA is established, another round of manual inspection should be done to fine-tune any remaining issues. For example, in sequences that have gaps remaining, the gap borders should be carefully examined to make sure that the software has put residues on the correct side of the gap. This involves looking at an individual residue in a homolog and examining by eye whether it would align better with the MSA residues on the front or back end of the gap.

3.3. Generate a Phylogeny

A phylogenetic tree is necessary for REAP analysis and should be created de novo (rather than relying on published trees) for each REAP analysis as the evolutionary distances and relationships for a particular protein family will not match published trees (due to different mutation rates, horizontal gene transfer, and different sets of species data). Several evolutionary algorithms can be used to infer a phylogenetic tree. These include maximum parsimony, maximum likelihood, distance-based approaches such as neighbor joining, and Bayesian inference approaches (8). Basically, these methods search the set of potential trees for the one that best recapitulates the observed sequence data, each using its own metrics to determine the best fit. A popular Bayesian-based program that can be used to generate the phylogenetic tree is MrBayes (9).

After generation, the tree should be manually inspected to identify any inconsistencies with published evolutionary relationships. If inconsistencies are observed, one must decide whether the differences seem true (i.e., if a horizontal gene transfer event occurred) or seem to be in error (perhaps caused by an insufficient number of sequence or disparate representation across phyla). If corrections seem appropriate, then sequences should be added or removed to the MSA and the phylogenetic tree reconstructed until a consistent tree structure is observed.

3.4. Identify Signatures of Functional Divergence

The essence of the REAP approach lies in the identification of evolutionary signatures of functional divergence in order to focus the variant design (10). The underlying concept is that functionally significant sites tend to be conserved over evolutionary time. This higher degree of conservation can be observed in the sequence diversity of homologs and exploited for variant design.

One metric that can be used to identify sites undergoing functional divergence across branches within the phylogenetic tree is to model site-specific rate shifts. Under a heterotachous model, sites can have different mutation rates at different points in their evolutionary history, for example, across different phylogenetic branches (Fig. 2). When a site has a low mutation rate within one branch and a high mutation rate in another branch, it indicates that the site is associated with a function specific to the branch with a low mutation rate. This type of signal is termed type I functional divergence, heterotachy, or covarion-like.

Site-specific mutation rates can also be used to identify another type of functional divergence, type II. In type II functional divergence, a site has a low mutation rate across phylogenetic branches but the identity of the residue differs between the branches. This indicates that the site is functionally significant but that the function is different between the two branches. Software such as DIVERGE and Rate Shift Analysis Server can identify statistically significant signatures of functional divergence given an MSA and phylogenetic tree as input.

3.5. Reconstruct Ancestral States

Once the above analyses have identified sites and phylogenetic branches that are associated with functional divergence, ancestral sequence reconstruction can be performed to identify how these specific residues have evolved along the branches. This information can then be used to guide variant design by incorporating those mutations already sampled by nature and associated with functional divergence across the extant homologs. Thus, the variant library created can capture the range of phenotypes observed in modern proteins in a very targeted manner.

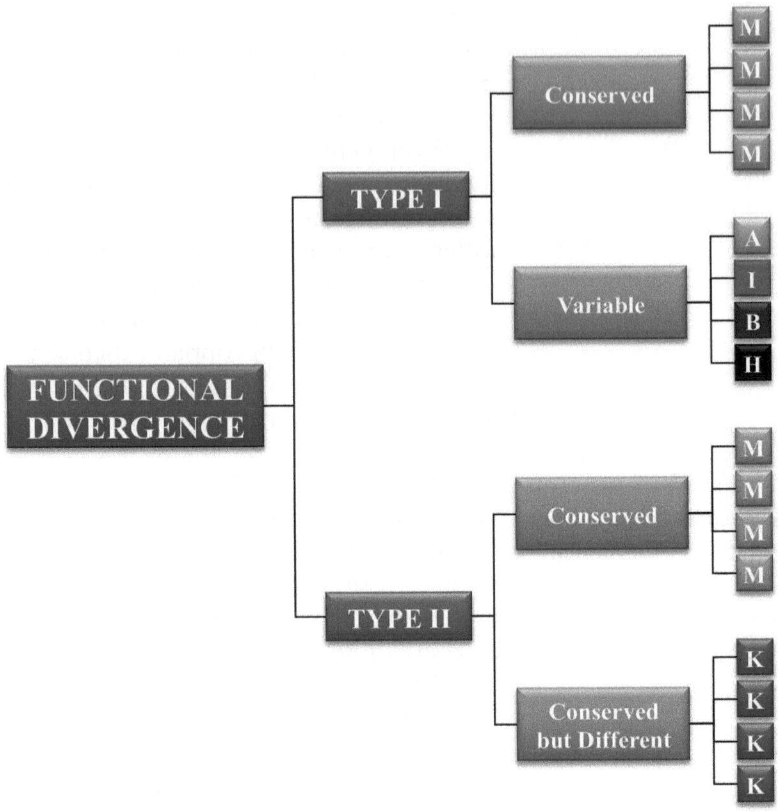

Fig. 2. Illustration of type I and type II functional divergence. Colored squares represent the amino acid identity of a particular site within the multiple sequence alignment for extant homologs. In type I functional divergence, a particular site is conserved (here as a conserved methionine) along one phylogenetic branch but variable along another branch. In type II functional divergence, the site is conserved but with a unique identity along both branches (methionine versus lysine). Both type I and type II functional divergence can be used to identify sites and identities associated with phenotype.

Software such as PAML (11) can be used to reconstruct the ancestral sequence states of interest given the MSA and phylogenetic tree. The program uses maximum likelihood to resurrect ancestral states and calculates the posterior odds of predictions. Thus, one has a metric to assess the uncertainty of a particular residue reconstruction and can observe the likelihood of the full set of potential residues. It is often useful to sample from the set of most likely candidates when trying to capture the ancestral state of a protein.

3.6. Design Variant Library

From these analyses, one is left with both a list of sites that are likely to be tied to phenotypic differences across the set of modern sequences and a reconstruction of their evolutionary history. The set of sites combined with their observed identities thus provide an extremely rich set of mutations that can be sampled to capture the functional diversity of the extant proteins. These potential mutations can be incorporated individually or in combinations with the parent protein sequence (see Note 3).

One can tightly control the sequence space covered by the library, depending on the synthesis method for variants (see below), to optimize the number of variants to be screened versus labor and expense. If inserting multiple mutations in a single variant, it is best to aim for each mutation to be covered by several variants so as to minimize the risk of epistasis impeding the discovery of beneficial mutations. It is also often useful to consider other factors such as knowledge of the protein's structure, amino acid chemistries, and computational models (see Note 4). These additional sources of knowledge can aid in further reducing the set or combinations of mutations to include or focus on in the library of variants.

3.7. Construct and Assay Variants

The method used to resurrect and query the variant library is highly specific to the library size and design, nature of the parent protein, and activity assay. In general, if single mutations will be introduced into the parent protein, then simple site-directed mutagenesis can be used. If multiple mutations are to be introduced into each variant or if the number of variants is extremely large, then it is usually better to use a commercial company to generate the template library. Companies can create a number of specific variants although the cost can become prohibitive if many variant designs are needed. Alternatively, companies can generate a template library where the frequency of each mutation as well as the average number of mutations per variant can be specified but the combination of mutations within variants will be random. This can be a very cost-effective method to test a large number of mutations without covering the entire sequence space represented by those mutations. However, it also requires additional analyses and experiments in order to parse out which mutations/combinations result in the optimal variant.

4. Notes

1. When identifying homologs via a BLAST search, one must first choose a sequence to use as the reference. This sequence is often the currently best-performing or most studied extant protein. It is sometimes necessary, however, to use different reference sequences when the homologs are quite divergent. For example, if searching for homologs in different domains, it may be necessary to use one reference for bacteria and another for eukaryotes.

   In performing the search it is often best to BLAST an individual genome and to select from among the best hits for that genome. If one were to BLAST against multiple genomes at once and then choose from the best hits, it is quite easy to introduce bias into the sampling of phylogenies by favoring the most closely related sequences. It is also critical to evaluate the validity of each hit as the homolog may not be present in each genome or several annotations for the same gene may be present. Performing the reciprocal search, where the homolog is used as the reference sequence against the original reference's genome to determine if the sequences are reciprocal best hits, can be useful to establish whether the sequences are true orthologs. If a true ortholog is present but has multiple annotations or copies, then the most suitable sequence should be selected based on alignment, gene size, and annotation.

2. The quality of the MSA commonly drops drastically at the N- and C-termini of the homologs as there may simply be too much divergence to accurately align the sequences. In such cases it is best to simply remove the N- and C-termini from where the alignment drops off. This will avoid these regions from contributing to the phylogenetic tree construction and will simplify the sequence reconstruction. It also avoids introducing mutations into the variant library in these regions, which often have species-specific signals important for protein expression. However, when necessary, it is acceptable to use the ends from an extant sequence.

3. Selection of the parent or backbone sequence in which to introduce mutations to create the variant library is critical as the surrounding sequence environment can affect the phenotypic effects of mutations. In general, it is common to use the currently best-performing sequence as the variant library backbone. This allows one to search for mutations that specifically improve the activity above the current standard. This is most useful when the desired characteristic is simply an improvement in a current phenotype or there is reason to believe that a small number of independent mutations will be needed to produce the desired protein.

   However, there may be instances when a different choice of parent sequence is more appropriate. One practical consideration is the expression and purification system. If, for example, experimental methods have been optimized for a particular homolog, then it may be wise to use that as the backbone sequence even when another more desired protein has been identified, but its expression has not been studied in detail or has been shown to be troublesome.

   Another key consideration is the stability of the parent protein. As many mutations can be destabilizing, it is important to start with a stable protein so it will still function when slightly destabilized. Often good choices in terms of stability are the ancestral or consensus proteins (12). The use of an ancestral homolog has the added benefit of placing some of the ancestral mutations into a more natural environment, thus minimizing epistatic barriers.

4. When other information such as experimental determination or computational prediction of protein structure is available, it should be utilized in variant design. For example, a subset of mutations identified by REAP could be selected for use based on their proximity to the substrate binding site. Similarly, computational models could be used to predict which REAP mutations would effect the desired phenotype. Literature knowledge can also be used to focus the set of REAP mutations to sites previously shown to effect phenotype. Chemical insights can also be used to direct variant design. For example, one may observe that a salt bridge is conserved among homologs and therefore introduce paired mutations to maintain this structure in variants. All of these outside sources of information can be used to guide expansion of the set of mutations to explore. For example, REAP may indicate that a particular site is associated with substrate specificity but may only identify a subset of amino acids as having been tested at these sites. With outside knowledge, such as insight that the structure will tolerate any amino acid at that site, one may very well find it worthwhile to include a wider range of mutations at these sites.

References

1. Cole MF, Gaucher EA (2011) Exploiting models of molecular evolution to efficiently direct protein engineering. J Mol Evol 72:193–203
2. Cole MF, Gaucher EA (2011) Utilizing natural diversity to evolve protein function: applications towards thermostability. Curr Opin Chem Biol 15:399–406
3. Fan Y, Fang W, Xiao Y et al (2007) Directed evolution for increased chitinase activity. Appl Microbiol Biotechnol 76:135–139
4. Chen F, Gaucher EA, Leal NA et al (2010) Reconstructed evolutionary adaptive paths give polymerases accepting reversible terminators for sequencing and SNP detection. Proc Natl Acad Sci U S A 107:1948–1953
5. Benner SA, Gaucher EA (2001) Evolution, language and analogy in functional genomics. Trends Genet 17:414–418
6. Larkin MA, Blackshields G, Brown NP et al (2007) Clustal W and Clustal X version 2.0. Bioinformatics 23:2947–2948
7. Notredame C, Higgins DG, Heringa J (2000) T-Coffee: a novel method for fast and accurate multiple sequence alignment. J Mol Biol 302:205–217
8. Yang Z, Rannala B (2012) Molecular phylogenetics: principles and practice. Nat Rev Genet 13:303–314
9. Huelsenbeck JP, Ronquist F, Nielsen R et al (2001) Bayesian inference of phylogeny and its impact on evolutionary biology. Science 294:2310–2314
10. Gaucher EA, Gu X, Miyamoto MM et al (2002) Predicting functional divergence in protein evolution by site-specific rate shifts. Trends Biochem Sci 27:315–321
11. Yang Z (2007) PAML 4: phylogenetic analysis by maximum likelihood. Mol Biol Evol 24:1586–1591
12. Tawfik DS, Bershtein S, Goldin K (2008) Intense neutral drifts yield robust and evolvable consensus proteins. J Mol Biol 379:1029–1044

Chapter 9

Oligonucleotide Recombination Enabled Site-Specific Mutagenesis in Bacteria

Bryan M. Swingle

Abstract

Recombineering refers to a strategy for engineering DNA sequences using a specialized mode of homologous recombination. This technology can be used for rapidly constructing precise changes in bacterial genome sequences in vivo. Oligonucleotide recombination is one type of recombineering that uses ssDNA oligonucleotides to direct chromosomal mutations. Oligo recombination occurs without addition of any exogenous functions, making this approach potentially useful in many different bacteria. Here we describe the basic technique for constructing a site-specific genomic mutation in *Pseudomonas syringae*.

Key words: Recombineering, Oligonucleotide recombination, Oligo-induced mutagenesis, Homologous recombination, Gene conversion

1. Introduction

Recombineering is based on the principal that bacterial genomic loci can be mutated to a specified sequence by homologous recombination with small linear DNA molecules. All recombineering strategies use genetic transformation to introduce the substrate DNA into bacterial cells, which introduces the desired mutation to a targeted DNA sequence in the genome. Generally, the substrate DNA is designed with target-specific sequences flanking the desired change, which can be a point mutation, insertion, or deletion. Until recently, it was assumed that recombineering reactions required the expression of exogenous functions derived from phage, such as lambda Red (1, 2), RecET (3), or mycobacterial phage Che9c proteins gp60 and gp61 (4). These recombinases catalyze very high

James C. Samuelson (ed.), *Enzyme Engineering: Methods and Protocols*, Methods in Molecular Biology, vol. 978, DOI 10.1007/978-1-62703-293-3_9, © Springer Science+Business Media New York 2013

levels of recombination in a narrow range of hosts (5), which limits their general utility for genetic engineering.

A new mode of recombination has recently been found in bacteria that occurs without addition of phage-encoded recombinases (6–8). These reactions are referred to as oligo recombination or oligo-induced mutagenesis. As it is currently understood, oligo recombination is distinct from other forms of genetic recombination because it does not require RecA or any other known recombinase (i.e., lambda Red). Because oligo recombination does not require addition of any specialized functions, it is likely that this strategy can be used to facilitate mutagenesis in a wide range of bacteria, including those where phage-derived recombineering systems have not been developed or when it is impractical to use an established phage protein to assist recombineering. Additionally, oligo recombination may provide a much-needed foothold in organisms where basic genetic systems have not been developed, such as new model organisms or organisms that posses qualities that make them attractive for biotechnology applications but are otherwise genetically intractable. Additionally, oligo recombination can be used to identify new species-specific phage-encoded recombinases to assist more sophisticated recombineering strategies (9).

As the name oligo recombination implies, the substrate in these reactions are ssDNA oligonucleotides that can be purchased from any commercial supplier. The efficiency of oligo recombination is influenced by the oligo concentration and sequence as well as the state of the transformed cells. Understanding the factors that influence oligo recombination frequency is useful because it provides information regarding the mechanism of recombination and can help when designing oligos for particular experiments.

First, the substrate oligo must enter the cell in sufficiently high numbers for recombination to take place. When linear ssDNA enters the cell, it is subject to degradation by endogenous DNA nucleases (10). This degradation led to the largely erroneous belief that bacteria are incapable of recombination with transformed linear DNA (10, 11). However, recent data suggests that the influence of these nucleases can be temporarily nullified by introduction of a large excess of DNA (6, 12). Transforming excess DNA is thought to saturate the nucleases while leaving enough additional substrate oligo to take part in recombination. Producing this state requires optimization of the transformation efficiency and oligo concentration.

Second, the oligos must stably anneal to the target location. Under physiological conditions, DNA annealing is primarily governed by oligo concentration, sequence, and length. It is most expedient to adjust the length of the oligo to meet the criteria for stable annealing. We have empirically determined that oligos with a $Tm \geq 62°C$ anneal to the target tightly enough to facilitate maximal recombination (6).

Third, mispairing of mutagenic nucleotides with the target sequence can reduce the recombination frequency if this distortion is recognized by the methyl-directed mismatch repair (mmr) system of the cell. The negative influence of mmr can be minimized in wild-type cells by designing oligos to install either a deletion or changes to more than three base pairs (6, 12).

Fourth, the recombination frequency is 10- to 100-fold higher if the oligo matches the lagging strand rather than the leading strand. This is thought to indicate that oligo recombination is intimately associated with DNA replication. A possible explanation for this observation is that oligos anneal to ssDNA near the replication fork effectively simulating an Okazaki fragment (6, 13).

Finally, the cells must be actively growing when they are made competent. This also is thought to indicate that replication is involved at some point in the process (8).

Oligo recombination is capable of producing mutant bacteria in a minimum of simple steps. Currently, finding mutants generated by oligo recombination requires a rigorous selection (either antibiotic resistance or prototrophy). Alternative strategies can also be envisaged that use differential growth to enrich for mutants that outperform the parental strain in a particular environment or bioprocess. Additionally, there are practical limitations on the types of mutations that can be introduced using this method because the mutation must be able to be encoded by an oligo. The types of mutations that oligo recombination is capable of include point mutations, small insertions, and deletions.

To date, oligo recombination has been demonstrated in *Pseudomonas syringae*, *Escherichia coli*, *Shigella flexneri*, *Salmonella typhimurium*, and *Legionella pneumophila* (6, 8). As an example of the steps required in an oligo recombination experiment, we describe the method for constructing a point mutation in the genomic *rpsL* gene that confers resistance to streptomycin in the plant pathogen *P. syringae*. This type of experiment is straightforward, demonstrates the important principles of this method, and is the type of experiment that would be a logical first step for anyone wanting to develop recombineering in new organisms.

2. Materials

1. *P. syringae* pv. *tomato* strain DC3000.
2. 100 mg/ml streptomycin stock solution.
3. 1 μg/μl *rpsL*K43R mutagenic oligo (5′CGCAATGCCGA GTTAGGTTTCCGAGGCGTAGTGGTATACACACG). Resuspend lyophilized oligo in sterile water to a final concentration 1 μg/μl (see Note 1).

4. Phosphate stock solution (100×): 0.86 M K_2HPO_4 (filter sterilize).
5. KB broth (14): 2% proteose peptone #3, 1.6 mM $MgSO_4 \cdot 7H_2O$, 1% Glycerol. Autoclave. After media has cooled, adjust to 1× phosphate stock (final concentration).
6. KB agar (14): 2% proteose peptone #3, 1.6 mM $MgSO_4 \cdot 7H_2O$, 1% Glycerol, 8.6 mM K_2HPO_4, 1.8% agarose.
7. 300 mM sucrose solution (filter sterilize).
8. Sterile dH_2O.
9. Gene Pulser (Bio-Rad Laboratories).
10. 0.2 cm electroporation cuvette.
11. Glucose: 20% solution in H_2O (filter sterilize).
12. Mg^{2+} stock solution (100×): 1 M $MgCl_2 \cdot 6H_2O$, 1 M $MgSO_4 \cdot 7H_2O$ (filter sterilize).
13. SOC broth (15): 2% Bacto tryptone, 0.5% yeast extract, 9.92 mM NaCl. Autoclave. After the media has cooled, adjust to 0.2% glucose and 1× Mg^{2+} using 100× stock solution.

3. Methods

All steps carried out at room temperature unless noted otherwise.

3.1. Preparation of Electrocompetent Cells (16)

1. Inoculate KB broth (14) with a single well-isolated *P. syringae* pv. *tomato* DC3000 colony. Grow overnight cultures at 30°C.
2. Dilute overnight culture 1:25 in 125 ml of fresh KB broth (see Note 2).
3. Grow subculture at 30°C until it reaches an OD_{600} of 0.8–1.0 and harvest cells by centrifugation at 20°C.
4. Wash cell pellet twice with equal volume of room temperature 300 mM sucrose.
5. Finally, resuspend cells in 1/60th the original culture volume using 300 mM sucrose.

3.2. Electroporation

1. Aliquot 100 μl of electrocompetent *P. syringae* cells into a 1.5 ml test tube.
2. Add 5 μl of 1 μg/μl *rpsL*K43R mutagenic oligo to the cell suspension and mix by pipetting several times.
3. Transfer oligo cell mixture to a 0.2 cm electroporation cuvette.
4. Electroporate cells with oligos using a 2.5 kV, 25 μF, 200 Ω pulse of electricity.

5. Immediately add electroporated cells to 5 ml of SOC broth and incubate at 30°C overnight with vigorous shaking (see Note 3).

3.3. Selection of Recombinants and Determination of Recombination Frequency

1. Aliquots of the outgrowth culture (step 3.1.5) are spread onto KB agar and KB agar supplemented with 100 μg/ml streptomycin. To determine the number of recombinants, spread 100 μl of undiluted and 10^{-1} dilution outgrowth culture on KB agar supplemented with 100 μg/ml streptomycin. To determine the number of viable cells, spread 100 μl of 10^{-5} and 10^{-6} dilutions on KB agar.
2. Incubate at 30°C for 3–4 days or until colonies are well formed.
3. Determine the number of recombinants and viable cells per milliliter by enumerating the number of colonies that grow on selective and nonselective growth media and accounting for dilution factors. Calculate the number of recombinants per 10^8 viable cells (see Note 4).

4. Notes

1. In this experiment, oligonucleotides are 44 nt long and are designed to have two 20 nt flanks that match the *P. syringae* DC3000 *rpsL* gene on either side of the four nucleotide change. The change encoded by the oligo alters the (AAA) lysine codon at position 43 to (CGG) arginine and a silent change in the wobble position of codon 42. The *rpsL* K43R mutation confers resistance to 100 μg/ml streptomycin in *P. syringae*. The sequence of the oligo, with nucleotides that are different from the genome underlined, is as follows: 5′CGCAATGCCGA GTTAGGTTT<u>CCGA</u>GGCGTAGTGGTATACACACG. Oligos were synthesized by IDT, Coralville IA with standard desalting. Further purification of oligos does not affect the recombination frequency.
2. The overnight culture of *P. syringae* pv. *tomato* DC3000 usually grows to an OD_{600} of 4.0. The desired density of the subinoculation is an OD_{600} of 0.25, which is typically achieved with a 1:25 dilution.
3. The cells are incubated overnight without selection. This extended recovery incubation is absolutely necessary for genes like *rpsL* where the wild-type (streptomycin-sensitive) allele is dominant to the K43R mutant allele. Because recombination only occurs on one strand, cells containing chromosomes

replicated immediately after recombination will contain one copy with the mutation and the other with the wild-type allele. No cells will survive streptomycin selection in this state. Extending the recovery outgrowth allows the sister chromosomes to segregate into daughter cells which then renders the mutant cells resistant to streptomycin.

4. The recombination frequency is determined by enumerating the number of recombinants per 10^8 viable cells. Normalizing the recombination frequency allows relevant comparisons to be made between cells subjected to different conditions during the process of generating recombinants. The decision to normalize to 10^8 viable cells was made based on the observations that this is the approximate number of cells in an electroporation reaction and has become the standard for expressing recombination frequencies in recombineering reactions.

References

1. Murphy KC (1998) Use of bacteriophage lambda recombination functions to promote gene replacement in *Escherichia coli*. J Bacteriol 180:2063–2071
2. Yu D, Ellis HM, Lee EC, Jenkins NA, Copeland NG, Court DL (2000) An efficient recombination system for chromosome engineering in *Escherichia coli*. Proc Natl Acad Sci U S A 97:5978–5983
3. Zhang Y, Buchholz F, Muyrers JP, Stewart AF (1998) A new logic for DNA engineering using recombination in *Escherichia coli*. Nat Genet 20:123–128
4. van Kessel JC, Marinelli LJ, Hatfull GF (2008) Recombineering mycobacteria and their phages. Nat Rev Microbiol 6:851–857
5. Datta S, Costantino N, Zhou X, Court DL (2008) Identification and analysis of recombineering functions from Gram-negative and Gram-positive bacteria and their phages. Proc Natl Acad Sci U S A 105:1626–1631
6. Swingle B, Markel E, Costantino N, Bubunenko MG, Cartinhour S, Court DL (2010) Oligonucleotide recombination in Gram-negative bacteria. Mol Microbiol 75:138–148
7. Swingle B, Markel E, Cartinhour S (2010) Oligonucleotide recombination: a hidden treasure. Bioeng Bugs 1:263–266. doi:10.4161/bbug.1.4.12098
8. Bryan A, Swanson MS (2011) Oligonucleotides stimulate genomic alterations of Legionella pneumophila. Mol Microbiol 80:231–247. doi:10.1111/j.1365-2958.2011.07573.x
9. Swingle B, Bao Z, Markel E, Chambers A, Cartinhour S (2010) Recombineering using RecTE from Pseudomonas syringae. Appl Environ Microbiol 76:4960–4968
10. Dutra BE, Sutera VA Jr, Lovett ST (2007) RecA-independent recombination is efficient but limited by exonucleases. Proc Natl Acad Sci U S A 104:216–221
11. Winans SC, Elledge SJ, Krueger JH, Walker GC (1985) Site-directed insertion and deletion mutagenesis with cloned fragments in *Escherichia coli*. J Bacteriol 161:1219–1221
12. Sawitzke JA, Costantino N, Li XT, Thomason LC, Bubunenko M, Court C, Court DL (2011) Probing cellular processes with oligo-mediated recombination and using the knowledge gained to optimize recombineering. J Mol Biol 407:45–59. doi:S0022-2836(11)00059-3 [pii] 10.1016/j.jmb.2011.01.030
13. Ellis HM, Yu D, DiTizio T, Court DL (2001) High efficiency mutagenesis, repair, and engineering of chromosomal DNA using single-stranded oligonucleotides. Proc Natl Acad Sci U S A 98:6742–6746
14. King EO, Ward MK, Raney DE (1954) Two simple media for the demonstration of pyocyanin and fluorescein. J Lab Clin Med 44:301–307
15. Hanahan D (1983) Studies on transformation of *Escherichia coli* with plasmids. J Mol Biol 166:557–580
16. Choi KH, Kumar A, Schweizer HP (2006) A 10-min method for preparation of highly electrocompetent *Pseudomonas aeruginosa* cells: application for DNA fragment transfer between chromosomes and plasmid transformation. J Microbiol Methods 64:391–397

Chapter 10

FX Cloning: A Versatile High-Throughput Cloning System for Characterization of Enzyme Variants

Eric R. Geertsma

Abstract

Methods for the cloning of large numbers of open reading frames (ORFs) into expression vectors are of critical importance for diverse disciplines in biology. Here I describe a system termed FX cloning that facilitates the high-throughput generation of expression constructs. FX cloning combines attractive features of established recombination- and single-strand-annealing-based cloning methods that were thus far not unified in one single method. FX cloning allows the straightforward transfer of a sequence-verified ORF to a variety of expression vectors, and it avoids the common but undesirable feature of significantly extending target ORFs with cloning-related sequences. It leaves a minimal seam of only a single amino acid to either side of the protein. Furthermore, FX cloning is highly efficient and economic in its use. The method is based on a class IIS restriction enzyme and negative selection markers. The full procedure takes place in one pot and does not require intermediate purifications. The method has proven to be very robust and suitable for all common pro- and eukaryotic expression systems.

Key words: High-throughput cloning, Type IIS restriction enzymes, Subcloning, *ccdB*, *sacB*, Counterselection marker

1. Introduction

Diverse disciplines in biology, ranging from biotechnology to structural biology, rely upon the rapid selection of enzyme variants or homologs with outstanding biochemical properties. Whether proteins with the desired phenotype can be identified critically depends on the size and quality of the initial population of genotypes collected. For the assembly of such large and high quality libraries, DNA cloning methods play a key role.

Preferably such a cloning system is robust, reliable, and inexpensive. In addition, it should require only minimal handling and be straightforward in use. Upon expression, the protein should not be extended with additional amino acids resulting from DNA

James C. Samuelson (ed.), *Enzyme Engineering: Methods and Protocols*, Methods in Molecular Biology, vol. 978,
DOI 10.1007/978-1-62703-293-3_10, © Springer Science+Business Media New York 2013

cloning sites. The primers for amplification of each open reading frame (ORF) should be comparably short and a single primer pair should suffice to generate all required expression constructs. In addition, the cloning system should be indiscriminate to the size and sequence of the ORFs to be cloned. Finally, straightforward transfer of a sequence-verified ORF to different expression vectors should be possible.

Though such an ideal cloning system might never exist, several cloning methods are currently available that comply with small subsets of these demands (Table 1). Traditional restriction and ligation-based cloning does not fall in this category. In general the efficiency of this method varies considerably; it is labor intensive due to the requirement for intermediate purification steps, and the frequent occurrence of restriction sites within genes prevents a uniform cloning strategy. Though some of these limitations are overcome by the Flexi Vector system (1), most popular high-throughput cloning approaches are either based on DNA recombination or on annealing of complementary single-stranded overhangs (Table 1). Recombination-based methods, like Gateway (2), rely on site-specific recombination using phage recombinases to insert an ORF into a plasmid vector. Methods based on the annealing of complementary single-stranded overhangs, such as LIC (3), SLIC (4), PIPE (5), Gibson assembly (6), and In-Fusion (7), use the exonuclease activity of DNA polymerases to generate long single-stranded overhangs on the insert that are complementary to the overhangs generated for the vector. Annealing of these overhangs generates a stable duplex that can be repaired *in vivo* (LIC, SLIC, PIPE, In-Fusion) or *in vitro* (Gibson assembly).

Both recombination- and annealing-based methods allow facile cloning independent of the DNA sequence of the target gene. A drawback of some systems, such as Gateway and LIC, is that they require the addition of long cloning sequences to the ORF that add up to nine amino acids to each protein terminus upon translation. The addition of such undesired and potentially interfering amino acid tails can be avoided using annealing-based systems that permit seamless integration of the insert into the plasmid. However, a limitation of these approaches is their requirement that the insert is provided as a dedicated PCR product. This implies that a separate primer pair is needed for each ORF/vector combination and, importantly, that the ORF needs to be sequence-validated repeatedly. At the price of extending the ORF with cloning sequences, Gateway and other recombination-based systems avoid dedicated primer pairs and resequencing by facilitating the subcloning of a sequence-verified ORF.

Fragment exchange (FX) cloning was developed as an alternative to these cloning approaches (8). FX cloning combines attractive features of recombination- and annealing-based approaches that were thus far not unified in one single cloning method (Table 1).

Table 1
Overview of popular methods for DNA (sub) cloning of PCR products based on recombination, single-strand annealing, or restriction–ligation

Cloning method[a]	Labor intensity	Size cloning sites		Sequence restrictions	Subcloning of insert	Preparation[c]		Reference
		Basepairs	Amino acids[b]			Insert	Vector	
Recombination-based								
Gateway	Low	25	8–9	None[e]	Yes	P/X[f]	X	(2)
Annealing-based								
EFC	Low	12–18	0	None	No	P	P	(16)
Gibson assembly	Low	40	0	None	No	P	L	(6)
In-Fusion[g]	Low	15	0	None	No	P	L	(17, 18)
LIC	Medium	12–15	4–5	None	No	P, T	L, T	(3)
PIPE	Low	15	0	None	No	P	P	(5)
SLIC	Medium	20–40	0	None	No	P, T	L, T	(4)
Restriction–ligation-based								
Traditional cloning	High	6–8	2–3	Many	Yes	P/X[f], D, G	D, G	(19)
Flexi Vector system	Medium[h]	8	3	Few[i]	Yes	P/X[f], D, G	D	(1)
FX cloning	Low	3	1	Few[j]	Yes	P/X[f]	X	(8)

[a]The list does not intend to provide a complete overview of all available cloning systems for each category

[b]Amino acids added at each protein terminus if the cloning sites are translated

[c]Required treatments of insert and vector prior to mixing, abbreviations: D (digestion by restriction enzymes), G (gel or column purification), L (linearization by PCR or digestion with a suitable restriction enzyme), P (PCR), T (T4 exonuclease treatment), X (no treatment required)

[d]Additional recombination-based cloning methods that allow subcloning are the Creator(20), Univector plasmid-fusion system(21), and MAGIC(22) systems. However, these systems do not facilitate initial assembly of a PCR product into a sequencing vector

[e]In rare occasions a cryptic *attB* site (effective size: 12–16 bp) might be encountered (2)

[f]Once the insert is cloned into a subcloning vector, plasmid material of this vector can replace the PCR product for subcloning into expression vectors

[g]In-Fusion cloning is considered single-strand-annealing-based as the proprietary In-Fusion enzyme is likely to be a vaccinia virus DNA polymerase (17, 18)

[h]Throughput of the Flexi Vector system is decreased by intermediate purifications

[i]Sequence to be cloned should not contain any internal SgfI and PmeI sites. As these enzymes have long 8 bp recognition sites internal sites occur infrequently

[j]Sequence to be cloned should not contain certain internal SapI sites that result in overhangs compatible with overhangs in the vector or with other internal overhangs. Sites fulfilling these criteria occur only sporadically

Like most annealing-based methods, FX cloning only minimally extends the ORF and is economic in its use. Similar to most recombination-based cloning systems, FX cloning allows the subcloning of a sequence-verified ORF into a variety of expression vectors and requires only one primer pair per target sequence. At the same time FX cloning remains compatible with direct cloning of PCR products into expression vectors. In addition, FX cloning does not require PCR amplification of vectors or pre-treatments of vector and insert. The method is straightforward and suitable for scientists new to molecular cloning. The insert, being a PCR product or a sequenced ORF located in a subcloning vector, is simply mixed with the target vector and subsequently the reaction proceeds in one cup. Transformation efficiencies are high and only about one order of magnitude less than for intact plasmids (8). This makes FX cloning highly suitable for applications where many transformants are required, such as the construction of gene libraries.

FX cloning relies on the use of type IIS restriction enzymes (RE) that cleave DNA outside their non-palindromic recognition sequence (9) (Fig. 1a). The resulting overhang is only defined by its distance away from the recognition site and not by its sequence. As a result, the sequence for a three nucleotide overhang can be any of the 64 different combinations of nucleotides. Upon digestion, the recognition site is physically separated from the cleavage site, thus minimizing the cloning-related sequences extending the ORF. The Achilles' heel of traditional restriction and ligation-based cloning, the frequent occurrence of detrimental restriction sites within genes, does not apply for FX cloning. This results from the use of the enzyme SapI that cuts infrequently in DNA sequences due to its comparably long recognition site of seven nucleotides. Equally important is that as a type IIS restriction enzyme, SapI generates cohesive ends that are defined by size but not by sequence. An occasional SapI site within a gene is likely to yield a different incompatible overhang and, consequently, internal sites are only rarely detrimental.

The overall cloning strategy is outlined in Fig. 1b, c. Initially, the target gene is amplified by PCR using a pair of comparably short primers each containing a SapI site. The SapI recognition sites are oriented so that they are removed from the ORF upon cleavage and the resulting overhangs differ in sequence to allow directional cloning (Fig. 1d). The cleaved ORF can be received by an intermediate sequencing vector (pINITIAL) or cloned immediately into one or more expression vectors. Both classes of vectors contain the same pair of SapI sites flanking the counterselection gene *ccdB* (10). Importantly in pINITIAL the sites have the same direction as in the PCR product: oriented towards the counterselection marker. In this case the respective recognition sites remain on the plasmid after cleavage to be used for subcloning. In the expression vectors, in contrast, the direction of the cleavage sites is reversed (Fig. 1e).

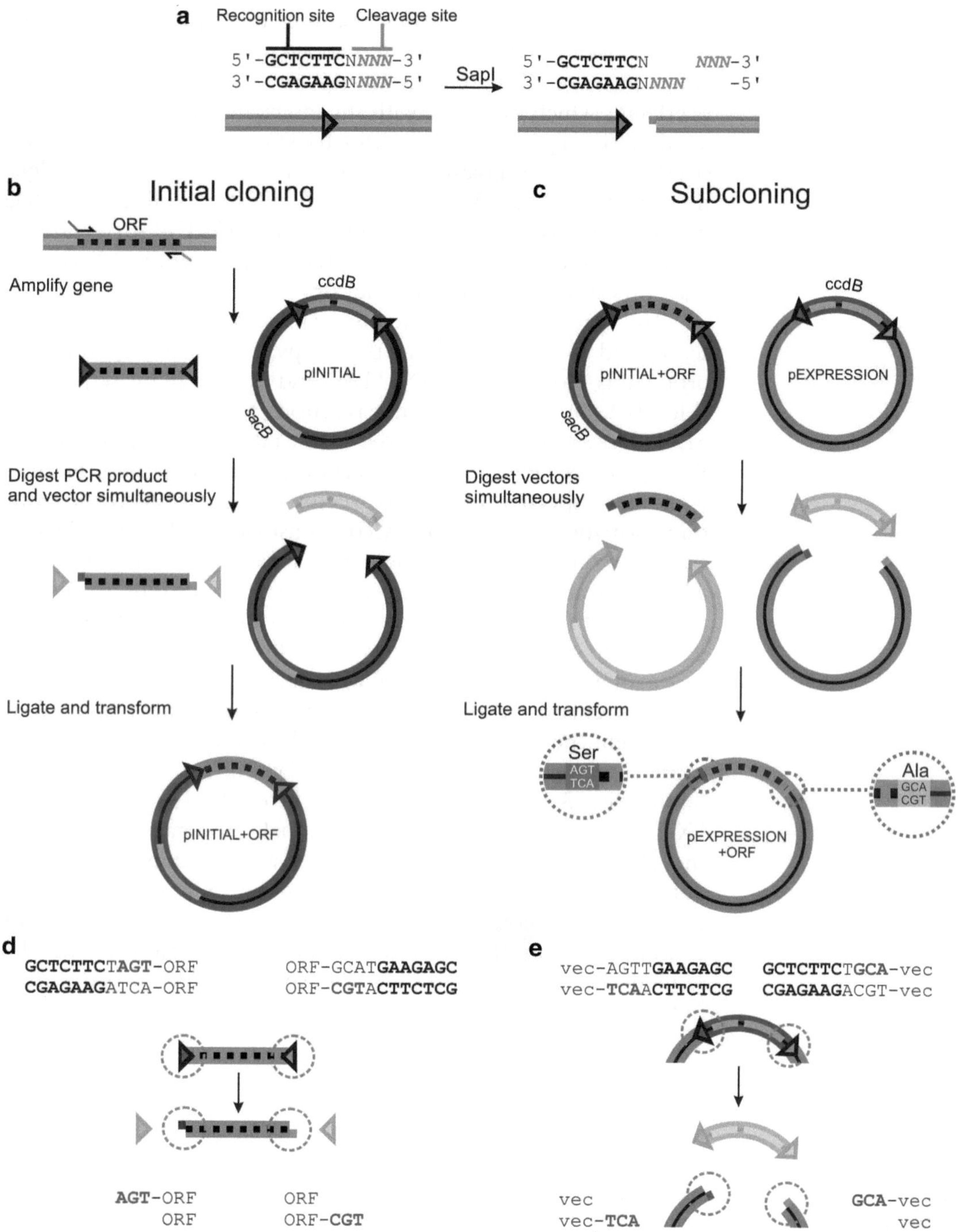

Fig. 1. Schematic overview of the FX cloning method. (**a**) SapI restriction site. The recognition site is shown in *bold letters*; nucleotides constituting the three base-pair single-stranded overhangs are in *italics*. *N* describes any of the four nucleotides. A schematic view of the cleavage is shown below. *Arrows* indicate the direction of the restriction site. (**b**) Cloning of a PCR product into pINITIAL. The amplified open reading frame (ORF) is shown as a *black dotted line*. The direction of the SapI restriction sites is indicated by *arrows*, which are colored corresponding to their respective overhangs generated after cleavage. The genes coding for the counterselection markers *ccdB* and *sacB* on pINITIAL are indicated. (**c**) Subcloning of an ORF into an expression vector (pEXPRESSION). The three nucleotides added to either terminus of the ORF are shown as insets (*circle*). (**d**) Orientation of the SapI cleavage sites in the PCR product and pINITIAL and (**e**) in expression vectors. The single-stranded overhangs generated upon cleavage are shown in *dark* and *light gray*, respectively. Adapted with permission from (8).

The recognition sites are thus lost after digestion, which ensures that the cloning-related sequences added to the ORFs remain small. After cleavage, both classes of vectors contain complementary overhangs which hybridize with the insert and are finally joined by ligation. Self-ligation of the vector backbone is not possible as the overhangs are incompatible. Subsequent transformation of the ligation mix to common CcdB-sensitive *E. coli* strains allows only cells with daughter plasmids containing the target ORF to survive. As no intermediate purification steps are required the entire FX cloning reaction can be performed in one cup.

The generation of different expression constructs from a sequenced ORF proceeds in two steps. In a first step the PCR product is cloned into pINITIAL and validated by sequencing (Fig. 1b). In a second similar reaction, the intermediate vector pINITIAL, now containing the ORF, takes over the role of the PCR product (Fig. 1c). The reaction proceeds in the same way, with the respective expression vector serving as acceptor. After transformation to a common CcdB-sensitive *E. coli* strain, the cells are plated on sucrose-containing media. Cells containing the pINITIAL-derivative are not viable under these conditions as this vector carries the counterselection marker *sacB* which renders *E. coli* sensitive to sucrose (11). Exclusively cells carrying expression vectors containing the target ORF will survive and form colonies.

This chapter contains all relevant information required to implement the FX cloning procedure. It first describes how specific expression vectors can be easily adapted for FX cloning in Subheading 3.1, though this step may be omitted should published FX cloning vectors (8) suffice. Additionally, it details protocols for the cloning of a PCR product into an FX cloning vector in Subheading 3.2. The vector receiving the PCR product can be either an expression vector or a sequencing vector. Finally, the subcloning reaction that allows the straightforward transfer of a sequence-verified ORF to different expression vectors is described in Subheading 3.3.

2. Materials

2.1. Adapting Expression Vectors to FX Cloning

1. Phusion high-fidelity DNA polymerase, dedicated buffer, and dNTPs.
2. Mutagenesis primers to remove SapI site(s) in the expression vector.
3. Autoclaved ultra-high-purity (UHP) water.
4. DpnI.
5. Dedicated primers to introduce SapI sites into the expression vector. The required 5′ overhangs are indicated in Table 2.

Table 2
PCR primer extensions required for FX cloning

Primer type	5′ Extension[a]	Potential target sequence of annealing part primer
Vector backbone reverse	5′ tatataGCTCTTCaACT nnn	Start codon, sequence coding for N-terminal protease site
Vector backbone forward	5′ atatatGCTCTTCtGCA nnn	Stop codon, sequence coding for C-terminal protease site
Insert forward	5′ atatatGCTCTTCtAGT nnn	Second codon of ORF
Insert reverse	5′ tatataGCTCTTCaTGC nnn	Penultimate codon of ORF

[a]Nucleotides in lowercase can be substituted for other bases if needed for primer optimization. The triplet nnn indicates the first ORF- or vector-specific codon of the primer

Table 3
Primers for sequencing of inserts and the construction of new FX cloning expression vectors

Primer name	Sequence	Purpose
ccdBfor	5′ TATATAAGTTGAAGAGCGACCTGCAGACTG GCTGTGTATAA	Amplify *ccdB-cat* cassette
catrev	5′ ATATATTGCAGAAGAGCTGAACTAGTGGAT CCCCAAAAAAG	Amplify *ccdB-cat* cassette
ccdBrevSQ	5′ GAAAATGACATCAAAAACGCCATTAACC	Sequence 5′ region flanking *ccdB-cat*
catforSQ	5′ CATTTTACGTTTCTCGTTCAGCTTTTTTG	Sequence 3′ region flanking ccdB-cat
INITforSQ	5′ ATCTGTTGTTTGTCGGTGAACGC	Sequence 5′ side of insert in pINITIAL
INITrevSQ	5′ TGGCAGTTTATGGCGGGCGT	Sequence 3′ side of insert in pINITIAL
ccdBrev	5′ TATATATGCAGAAGAGCAAAGCCAGATAAC AGTATGCGTATTTGCG	Replaces catrev for amplification of *ccdB* only

6. Primer pair ccdBfor/catrev (Table 3).
7. 50× TAE buffer: 249 g Tris-base, 57.1 ml acetic acid, 100 ml 0.5 M EDTA pH 8.0. Adjust volume to 1 l with water. Store at room temperature. Use a 1× TAE solution for gel electrophoresis.
8. DNA gel extraction kit.

9. FX cloning vectors: Available from the author upon request or constructed as described previously (8).
10. SapI and dedicated buffer.
11. 10 mM ATP pH 7: 10 mM Na_2-ATP, 10 mM $MgSO_4$. Dissolve in 50 mM KPi, pH 7.0 and adjust to pH 6.5–7.0 with NaOH (see Note 1). Store in small aliquots at –20°C.
12. T4 DNA ligase.
13. Chemical competent cells of a CcdB-resistant *E. coli* strain (such as *E. coli* DB3.1).
14. LB agar plates: 10 g tryptone, 5 g yeast extract, 10 g NaCl, and 15 g agar. Adjust volume to 1 l with water. Sterilize by autoclaving. Once the medium is cooled to ~60°C, add the appropriate antibiotics, mix, and pour the plates. Store plates at 4°C.
15. LB medium: 10 g tryptone, 5 g yeast extract, and 10 g NaCl. Adjust volume to 1 l with water. Sterilize by autoclaving. Supplement with the appropriate antibiotics immediately prior to use.
16. Restriction enzymes to perform analytical digests.
17. Autoclaved 87% w/v glycerol.
18. Plasmid miniprep kit.
19. Sequencing primers ccdBrevSQ and catforSQ (Table 3).

2.2. FX Cloning of PCR Products

1. Dedicated primer pair to amplify the ORF by PCR and introduce the required SapI sites. The required 5′ overhangs are indicated in Table 2. When ordering oligonucleotides, select the lowest synthesis scale offered by the supplier and mere desalting as the purification step. For large sets (>48 primers), order the primers in a 96-well plate. Upon delivery, spin down the primers and dissolve them to a concentration of 100 μM in UHP water. Make a substock at a concentration of 5 μM. The 100 μM stock can be stored for several months at –20°C. Store the 5 μM stock at 4°C. Spin down the material before use.
2. Chemical competent cells of an *E. coli* strain sensitive to CcdB (virtually all *E. coli* strains lacking the F plasmid, e.g., *E. coli* MC1061). Electro-competent cells may be substituted if very high transformation efficiencies are required.
3. Sequencing primers INITforSQ and INITrevSQ (Table 3) to verify sequences cloned into pINITIAL-derivatives or dedicated sequencing primers for sequences cloned into expression vectors.

2.3. Subcloning of Sequence-Verified ORFs

1. 70% w/v sucrose: 350 g sucrose. Adjust volume to 500 ml with hot water. Mix well and dissolve and sterilize the sucrose by autoclaving. Immediately after autoclaving determine whether the sucrose has completely dissolved and if needed carefully mix the solution until all the material is dissolved.

2. Low-salt LB agar supplemented with 7% w/v sucrose: 10 g tryptone, 5 g yeast extract, 5 g NaCl, and 15 g agar. Adjust volume to 0.9 l with water. Sterilize by autoclaving. Once the medium is cooled to ~60°C, add 100 ml sterile 70% w/v sucrose and the appropriate antibiotics, mix, and pour the plates. Store plates at 4°C.

3. Methods

3.1. Adapting Expression Vectors to FX Cloning

1. Determine whether the expression vector contains an internal SapI site. If this is not the case, proceed to step 2. Otherwise, remove the SapI site by making a single nucleotide substitution in any of the seven nucleotides constituting the recognition site using Quikchange site-directed mutagenesis (12) (see Note 2). Verify affected regions of the plasmid by sequencing.
2. Design primers to introduce FX cloning-compatible SapI sites in the expression vector by PCR (see Note 3). The SapI sites should be placed immediately adjacent to relevant expression elements such as start or stop codons or sequences coding for affinity tags or protease sites. The 5′ extensions required for the forward and reverse primers are indicated in Table 2 (see Note 4). Note that the SapI sites are oriented so that upon cleavage, only the overhangs remain associated with the vector backbone.
3. Introduce the SapI sites into the expression vector by PCR using the Phusion DNA polymerase. Prepare a 50 μl reaction mix according to the manufacturer's protocol and add the polymerase just prior to the start of the reaction. Place the sample in a PCR machine preheated to 98°C and start the reaction. A good starting point for a PCR program is: (a) 30 s at 98°C; (b) 10 s at 98°C; (c) 15 s at 61°C (decrease 0.5°C per cycle; see Note 5); (d) *X*s at 72°C (15–30 s/kb); Repeat (b)–(d) 14 times; (e) 10 s at 98°C; (f) 15 s at 53°C; (g) *X*s at 72°C; Repeat (e)–(g) 14 times; (h) 120 s at 72°C; (i) unlimited at 10°C.
4. Add 0.5 μl DpnI (5 U) and incubate 30 min at 37°C to digest the template DNA. Subsequently analyze all the material by TAE gel electrophoresis and gel purify the target band using a DNA gel extraction kit (see Note 6). Determine the DNA concentration spectrophotometrically. Store the material at 4°C until use.
5. Amplify the *ccdB-cat* cassette by PCR using the Phusion DNA polymerase. Use pINITIAL as template and primer pair ccdB-for/catrev (Table 3; see Note 7). The cassette can be amplified using the following PCR program: (a) 30 s at 98°C; (b) 10 s at 98°C; (c) 15 s at 61°C (decrease 0.5°C per cycle); (d) 30 s at 72°C; Repeat (b)–(d) 14 times; (e) 10 s at 98°C; (f) 15 s at

53°C; (g) 30 s at 72°C; Repeat (e)–(g) 14 times; (h) 120 s at 72°C; (i) unlimited at 10°C.

6. Add 0.5 μl DpnI (5 U) and incubate 30 min at 37°C to digest the template DNA. Subsequently analyze all the material by TAE gel electrophoresis and gel purify the 1.7 kb target band using a DNA gel extraction kit (see Note 6). Determine the DNA concentration spectrophotometrically. Store the material at 4°C until use.
7. Mix 100 ng of the PCR product of the expression vector with the *ccdB-cat* cassette PCR to have a vector:insert molar ratio of approximately 1:5 (see Note 8). Adjust the volume to 8 μl with UHP water and add 1 μl 10× SapI-buffer and 1 μl SapI (2 U; see Note 9). Incubate 1 h at 37°C (see Note 10) and heat inactivate SapI by incubating 20 min at 65°C. Cool the sample and add 1.25 μl 10 mM ATP and 1.25 μl T4 DNA ligase (1.25 U). Incubate 1 h at room temperature and subsequently heat inactivate T4 DNA ligase for 20 min at 65°C. Transform 5 μl sample to 100 μl chemical competent *E. coli* DB3.1 (see Note 11). Plate 0.9, 9, and 90% aliquots of the transformed cells on LB agar plates supplemented with the appropriate antibiotics. Incubate overnight at 37°C.
8. Pick a few single colonies from the plate to inoculate 5 ml LB supplemented with the appropriate antibiotics. Incubate overnight at 37°C. Subsequently, prepare glycerol stocks of the cultures and isolate the plasmids. Determine the DNA concentration and analyze the plasmids first by restriction analysis. Subsequently verify the regions flanking the *ccdB-cat* cassette by sequencing using primers ccdBrevSQ and catforSQ (Table 3) for two positive clones. Additionally confirm the correct sequence of other important regions of the plasmid, such as the promoter and sequences coding for transcriptional regulators. Discard glycerol stocks of negative clones. Store the plasmid DNA of the new FX cloning expression vector at −20°C.

3.2. FX Cloning of PCR Products

1. Design a forward and reverse primer for each gene to be cloned (see Note 12). The required 5′ extensions are indicated in Table 2. The gene-specific part of the primer should be designed so that it: (a) does not contain the start or stop codon of the target gene; (b) is sufficiently long to anneal in a stable and selective way with the target DNA; and (c) does not contain strong secondary structure elements that could interfere with PCR (see Note 13).
2. Amplify the desired ORFs using the Phusion DNA polymerase (see Note 14). Prepare a 50 μl reaction mix according to the manufacturer's protocol and add the polymerase just prior to the start of the reaction. Place the sample in a PCR machine preheated to 98°C and start the reaction. A good starting

point for a PCR program is: (a) 30 s at 98°C; (b) 10 s at 98°C; (c) 15 s at 61°C (decrease 0.5°C per cycle); (d) *X*s at 72°C (15–30 s/kb); Repeat (b)–(d) 14 times; (e) 10 s at 98°C; (f) 15 s at 53°C; (g) *X*s at 72°C; Repeat (e)–(g) 14 times; (h) 120 s at 72°C; (i) unlimited at 10°C.

3. If a plasmid was used as the template, add 0.5 μl DpnI (5 U) and incubate 30 min at 37°C. Analyze all the material by TAE gel electrophoresis and gel purify the target band using a DNA gel extraction kit (see Notes 15 and 16). Determine the DNA concentration spectrophotometrically. Store the material at −20°C until use.
4. Prepare 5 ml LB supplemented with the appropriate antibiotics and cultivate *E. coli* DB3.1 cells containing the desired FX cloning vectors overnight at 37°C. Isolate the plasmids using a miniprep kit and determine the DNA concentration. Store the material at −20°C until use.
5. Mix 50 ng of an FX cloning vector (see Note 17) with sufficient PCR product to have a vector:insert molar ratio of approximately 1:5. Add 1 μl 10× SapI-buffer and adjust the volume to 9 μl with UHP water. Subsequently add 1 μl SapI (2 U) and incubate 1 h at 37°C.
6. Heat inactivate SapI for 20 min at 65°C and allow the sample to cool to room temperature. Add 1.25 μl 10 mM ATP and 1.25 μl T4 DNA ligase (1.25 U) and incubate 1 h at room temperature.
7. Heat inactivate the T4 DNA ligase for 20 min at 65°C. Transform 5 μl of the ligation mix to 100 μl chemically competent cells of an *E. coli* strain that is CcdB-sensitive (virtually all *E. coli* strains lacking the F plasmid). Plate 1 and 10% aliquots on LB agar supplemented with the appropriate antibiotic (see Note 18). Incubate the plate overnight at 37°C.
8. Pick a few single colonies from the plate to inoculate 5 ml LB supplemented with the appropriate antibiotics (see Note 19). Incubate overnight at 37°C. Isolate the plasmids with a miniprep kit and determine the DNA concentration. Verify the insert by DNA sequencing. For inserts cloned into a pINITIAL-derivative, primers INITforSQ and INITrevSQ can be used (Table 3). Store the vector at −20°C until use. For inserts cloned into an expression vector one can proceed as required for the desired expression system.

3.3. Subcloning of Sequence-Verified ORFs

1. Prepare 5 ml LB supplemented with the appropriate antibiotics and cultivate *E. coli* DB3.1 cells containing the desired FX cloning expression vectors overnight at 37°C. Isolate the plasmids using a miniprep kit and determine the DNA concentration. Store the material at −20°C until use.

2. Mix 50 ng of an FX cloning expression vector with a sequenced derivative of pINITIAL holding the insert of interest to have an expression vector:sequencing vector molar ratio between 1:3 and 1:5. Add 1 μl 10× SapI-buffer and adjust the volume to 9 μl with UHP water. Subsequently add 1 μl SapI (2 U) and incubate 1 h at 37°C.
3. Heat inactivate SapI for 20 min at 65°C and allow the sample to cool to room temperature. Add 1.25 μl 10 mM ATP and 1.25 μl T4 DNA ligase (1.25 U) and incubate 1 h at room temperature.
4. Heat inactivate the T4 DNA ligase for 20 min at 65°C. Transform 5 μl of the ligation mix to 100 μl chemically competent cells of an *E. coli* strain that is CcdB-sensitive (virtually all *E. coli* strains lacking the F plasmid; see Note 20). Plate 1 and 10% aliquots on low-salt LB agar supplemented with 7% w/v sucrose and the appropriate antibiotic (see Note 21). Incubate the plate overnight at 37°C.
5. Pick a colony from the plate to inoculate 5 ml LB supplemented with the appropriate antibiotics. Incubate overnight at 37°C. Isolate the plasmid with a miniprep kit and determine the DNA concentration. The insert does not require additional verification by DNA sequencing. Proceed with the plasmid as required for the desired expression system.

4. Notes

1. For small volumes, an approximation of the pH is conveniently obtained by pipetting ~1 μl drops on pH strips.
2. The point mutation in the recognition site of an undesired SapI site should be silent if it is located in an ORF coding for an essential protein, e.g., a transcriptional repressor.
3. If primers exceeding 50 nucleotides are required, it is strongly recommended to order PAGE- or HPLC-purified material. This greatly increases the fraction of full-length primers thereby reducing the chance for internal truncations in the primer.
4. The 5′ extensions of the primers result in the extension of the ORF with triplets coding for an N-terminal Ser and C-terminal Ala residue. These overhangs were selected because they code for small neutral amino acids and are completely incompatible, thus excluding self-ligation of the vector. Other overhang sequences might be used as well but overhang pairs that have two complementary bases should be avoided. Due to the relaxed specificity of T4 DNA ligase and relatively high efficiency of intra-molecular ligation such

overhangs can lead to a high percentage of clones containing self-ligated vectors.

5. Touchdown PCR (13) is recommended as it favors the production of the desired product over products resulting from spurious priming. In addition, as a range of annealing temperatures is used the program does not require much fine-tuning for different targets.

6. Should the PCR not yield the desired product in sufficient amounts, consider the use of alternative buffers supplied by the manufacturer, the addition of dimethyl sulfoxide (DMSO), or the use of a freshly purified template. More detailed suggestions can be found in the *Molecular Cloning* protocol series (14).

7. Do not amplify the *cat* gene if the expression vector already contains a chloramphenicol resistance marker. To PCR the *ccdB* gene only (~0.7 kb), replace the reverse primer with ccdB reverse (Table 3).

8. A facile approximation for a 1:5 molar ratio of vector:insert is calculated using: (amount vector (ng) × size insert (bp) × 5)/ size vector (bp) = amount insert (ng) needed.

9. FX cloning was established using the enzyme SapI. However, SapI can in principle be replaced with its isoschizomers LguI, PciSI, and BspQI. Note that the latter requires 50°C for optimal activity.

10. Incubations of small volumes at temperatures above room temperature should be performed in a PCR machine with heated lid or in a water bath placed inside an incubator. This prevents suboptimal reaction conditions due to excessive condensation at the lid.

11. Plasmids containing the *ccdB* gene can only be maintained in ccdB-resistant strains such as *E. coli* DB3.1 (2). Once the *ccdB* gene has been exchanged with a DNA insert by FX cloning, the sample should be transformed to a CcdB-sensitive *E. coli* strain. The vast majority of *E. coli* strains is CcdB-sensitive.

12. ORFs containing an internal SapI site are cloned with approximately tenfold lower efficiency than fragments devoid of SapI sites (8). Due to the high efficiency of FX cloning, this will still result in many colonies. However, for some applications that require very high transformation efficiencies, such as the preparation of a library of variants of one gene, it will be beneficial to remove an internal SapI beforehand.

13. Design and optimization of primers for ORFs is most conveniently done using an automated Python script available from the author. In addition, this script analyzes the DNA sequence and determines whether detrimental internal SapI sites are present. To remove strong secondary structure elements in the

primers, mutations can be made in the first six 5′ nucleotides and the nucleotide between the recognition and cleavage site (both indicated in lowercase in Table 2). Additionally, silent mutations can be made in the gene-specific part of the primer.

14. If genomic DNA is used as a template, delicate handling is required and care should be taken not to compromise the material by repetitive freeze/thawing, vortexing, or vigorous pipetting.
15. Purification of the target PCR product from gel might be omitted if the quality of the PCR was superb. However, regularly small by-products are present and even trace amounts of these are cloned with high efficiency. As a result, more clones need to be analyzed in later steps thus decreasing the throughput. This is not a unique feature of FX cloning but applies to all (high-throughput) cloning procedures.
16. Due to the high efficiency of FX cloning and the use of primers with uniform restriction sites, care should be taken to avoid cross-contamination with PCR products run in adjacent wells. In addition, the gel container should be cleaned and the buffer should be replaced between runs to prevent contamination with PCR products analyzed in previous runs.
17. The choice of FX cloning vector depends on the purpose. To allow subcloning of sequence-verified ORFs, clone the insert first into a pINITIAL-derivative. Preferably the antibiotic markers of the pINITIAL-derivative is different from the expression vectors used later for subcloning as this provides another selection criterium next to *sacB* counterselection. Derivatives of pINITIAL holding antibiotic markers against kanamycin, chloramphenicol, or tetracycline are available. As most common expression vectors provide resistance against ampicillin or kanamycin, in general, pINITIAL-cat (providing chloramphenicol resistance) is recommended. If there is no need for subcloning, the ORFs can be cloned immediately into a FX cloning-compatible expression vector.
18. Especially for large numbers of samples, spreading the transformed cells over the plate is most easily done using glass beads (5 mm diameter). In addition, this prevents cross-contamination.
19. Due to the highly efficient *ccdB* counterselection marker the fraction of clones without insert will be extremely low. The reason that it is suggested to analyze more than one colony lies in the fact that mutations resulting from the primer synthesis or PCR amplification cannot be excluded.
20. If subsequent protein expression will take place in *E. coli* and if the expression system is sufficiently tight, such as the AraC/P_{BAD}-system (15), the ligation mixture can be transformed

immediately to the desired expression strain to avoid plasmid isolations and re-transformation.

21. Supplementing the medium with sucrose counterselects colonies containing an intact pINITIAL-derivative. Counterselection is even more efficient if the pINITIAL-derivative and the expression vector hold different antibiotic markers (see Note 17).

Acknowledgements

E.R.G. acknowledges a long-term fellowship from the Human Frontier Science Program and thanks Prof. Raimund Dutzler for critical reading of the manuscript.

References

1. Blommel PG, Martin PA, Wrobel RL, Steffen E, Fox BG (2006) High efficiency single step production of expression plasmids from cDNA clones using the Flexi Vector cloning system. Protein Expr Purif 47:562–570
2. Hartley JL, Temple GF, Brasch MA (2000) DNA cloning using *in vitro* site-specific recombination. Genome Res 10:1788–1795
3. Aslanidis C, de Jong PJ (1991) Coincidence cloning of Alu PCR products. Proc Natl Acad Sci U S A 88:6765–6769
4. Li MZ, Elledge SJ (2007) Harnessing homologous recombination in vitro to generate recombinant DNA via SLIC. Nat Methods 4:251–256
5. Klock HE, Koesema EJ, Knuth MW, Lesley SA (2008) Combining the polymerase incomplete primer extension method for cloning and mutagenesis with microscreening to accelerate structural genomics efforts. Proteins 71:982–994
6. Gibson DG, Young L, Chuang RY, Venter JC, Hutchison CA 3rd, Smith HO (2009) Enzymatic assembly of DNA molecules up to several hundred kilobases. Nat Methods 6:343–345
7. Berrow NS, Alderton D, Owens RJ (2009) The precise engineering of expression vectors using high-throughput In-Fusion PCR cloning. Methods Mol Biol 498:75–90
8. Geertsma ER, Dutzler R (2011) A versatile and efficient high-throughput cloning tool for structural biology. Biochemistry 50:3272–3278
9. Szybalski W, Kim SC, Hasan N, Podhajska AJ (1991) Class-IIS restriction enzymes—a review. Gene 100:13–26
10. Bernard P, Gabant P, Bahassi EM, Couturier M (1994) Positive-selection vectors using the F plasmid *ccdB* killer gene. Gene 148:71–74
11. Recorbet G, Robert C, Givaudan A, Kudla B, Normand P, Faurie G (1993) Conditional suicide system of Escherichia coli released into soil that uses the Bacillus subtilis *sacB* gene. Appl Environ Microbiol 59:1361–1366
12. Zheng L, Baumann U, Reymond JL (2004) An efficient one-step site-directed and site-saturation mutagenesis protocol. Nucleic Acids Res 32:e115
13. Don RH, Cox PT, Wainwright BJ, Baker K, Mattick JS (1991) 'Touchdown' PCR to circumvent spurious priming during gene amplification. Nucleic Acids Res 19:4008
14. Sambrook J, Russell DW (2001) Molecular cloning: a laboratory manual. Cold Spring Harbor, New York
15. Guzman LM, Belin D, Carson MJ, Beckwith J (1995) Tight regulation, modulation, and high-level expression by vectors containing the arabinose P_{BAD} promoter. J Bacteriol 177: 4121–4130
16. Tillett D, Neilan BA (1999) Enzyme-free cloning: a rapid method to clone PCR products independent of vector restriction enzyme sites. Nucleic Acids Res 27:e26
17. Marsischky G, LaBaer J (2004) Many paths to many clones: a comparative look at high-throughput cloning methods. Genome Res 14:2020–2028
18. Hamilton MD, Nuara AA, Gammon DB, Buller RM, Evans DH (2007) Duplex strand

joining reactions catalyzed by vaccinia virus DNA polymerase. Nucleic Acids Res 35: 143–151
19. Cohen SN, Chang AC, Boyer HW, Helling RB (1973) Construction of biologically functional bacterial plasmids *in vitro*. Proc Natl Acad Sci U S A 70:3240–3244
20. Colwill K, Wells CD, Elder K, Goudreault M, Hersi K, Kulkarni S, Hardy WR, Pawson T, Morin GB (2006) Modification of the creator recombination system for proteomics applications—improved expression by addition of splice sites. BMC Biotechnol 6:13
21. Liu Q, Li MZ, Leibham D, Cortez D, Elledge SJ (1998) The univector plasmid-fusion system, a method for rapid construction of recombinant DNA without restriction enzymes. Curr Biol 8:1300–1309
22. Li MZ, Elledge SJ (2005) MAGIC, an *in vivo* genetic method for the rapid construction of recombinant DNA molecules. Nat Genet 37:311–319

Chapter 11

Use of *Sulfolobus solfataricus* PCNA Subunit Proteins to Direct the Assembly of Multimeric Enzyme Complexes

Hidehiko Hirakawa and Teruyuki Nagamune

Abstract

In nature, enzymes often form multienzyme complexes to enhance their catalytic efficiencies and, moreover, evolve into genetically fused multidomain enzymes. Inspired by a natural fusion cytochrome P450 (P450) containing a monooxygenase domain and a reductase domain, we have developed a heterotrimeric protein-utilized method to form a multienzyme complex composed of a bacterial P450 and its catalytically essential two redox proteins. Three distinct proliferating cell nuclear antigens (PCNAs) from *Sulfolobus solfataricus*, each of which can be separately expressed, spontaneously form a heterotrimer. Fusion to the PCNAs enables complex formation of a bacterial P450 and two redox proteins through the self-assembling of the PCNAs and enhances the activity due to efficient electron transfer in the complex. This PCNA-mediated multienzyme complex formation will be available for other multienzyme reactions.

Key words: Cytochrome P450, Multienzyme, P450cam, Putidaredoxin, Putidaredoxin reductase, Heterotrimer, PCNA, Protein scaffold, Sulfolobus solfataricus, Self-assembly

1. Introduction

In nature, many enzymes construct multienzyme systems and form multienzyme complexes. For example, enzymes involved in cascade reactions such as tryptophan synthesis, Calvin cycle, and dhurrin synthesis form multienzyme complexes "metabolons" that directly transfer intermediates from one enzyme to another enzyme sequentially to enhance overall reaction rates, to protect unstable intermediates and to prevent diffusion of toxic intermediates into cells (1). Bacterial cellulosome accumulates cell wall-degrading enzymes, cellulases and hemicellulases, on a scaffold protein for the synergy effects of enzyme proximity and efficient substrate targeting (2). Thus, inspired by natural multienzyme complexes,

James C. Samuelson (ed.), *Enzyme Engineering: Methods and Protocols*, Methods in Molecular Biology, vol. 978,
DOI 10.1007/978-1-62703-293-3_11, © Springer Science+Business Media New York 2013

"artificial" enzyme assemblings have been developed to enhance the efficiency of multienzyme reaction (3).

Cytochrome P450s (P450s) are heme-containing monooxygenases that catalyze a wide variety of oxidation reactions including hydroxylation, epoxidation, and dealkylation with extraneous electrons that drive their catalytic cycles (4). Microsomal P450s accept electrons through an FAD- and FMN-containing NADP-dependent cytochrome P450 reductase (CPR), while many bacterial P450s accept electrons through ferredoxin and ferredoxin reductase. In addition to the above multicomponent P450 systems, natural fusion of P450s and redox proteins has been discovered. A cytochrome P450 from *Bacillus megaterium*, P450 BM3, which is a natural fusion protein composed of a heme-containing monooxygenase domain and a CPR-like reductase domain, shows a high catalytic activity due to efficient interdomain electron transfer. This suggests that artificial complex formation of proteins composing multicomponent P450 systems can enhance P450s' activities.

Recently, self-assembling proteins have attracted an interest as scaffolds to accumulate multiple enzymes (5–9). We have also demonstrated that a heteromeric self-assembling protein can be employed as a scaffold for complex formation of three proteins composing a bacterial P450 system (10). Proliferating cell nuclear antigen is a ring-shaped protein to encircle DNA. Although eukaryotes and many archaea have single homotrimeric PCNAs, *Sulfolobus solfataricus* has three distinct PCNA proteins, PCNA1, PCNA2, and PCNA3, that form a heterotrimer (11). The three PCNAs can be separately expressed in *Escherichia coli* as soluble proteins and self-assemble to form a heterotrimeric complex in a stepwise fashion. Thus, PCNA1 and PCNA2 form a stable heterodimer and subsequently the heterodimer and PCNA3 form a heterotrimer. Furthermore, fusion of proteins to the C-termini and the N-termini of the PCNAs does not affect the heterotrimerization and all of the C-termini in the heterotrimer are exposed on the same side of the ring. Therefore, we can assemble a bacterial P450 and its two electron transfer-related proteins, ferredoxin and ferredoxin reductase, to form a heterotrimeric complex (*P*CNA-*u*tilized *p*rotein complex of P450 and its *e*lectron *t*ransfer-related proteins, PUPPET) through self-assembly of *S. solfataricus* PCNAs by, respectively, fusing the three proteins to the C-termini of the three PCNAs (Fig. 1). Compared to an equimolar mixture of the three proteins, the resulting multienzyme complex showed great enhancement in the P450's activity due to efficient electron transfer in the complex. This chapter describes how to prepare and evaluate a multienzyme complex composed of a bacterial P450 and its redox partner proteins through the self-assembly of *S. solfataricus* PCNAs.

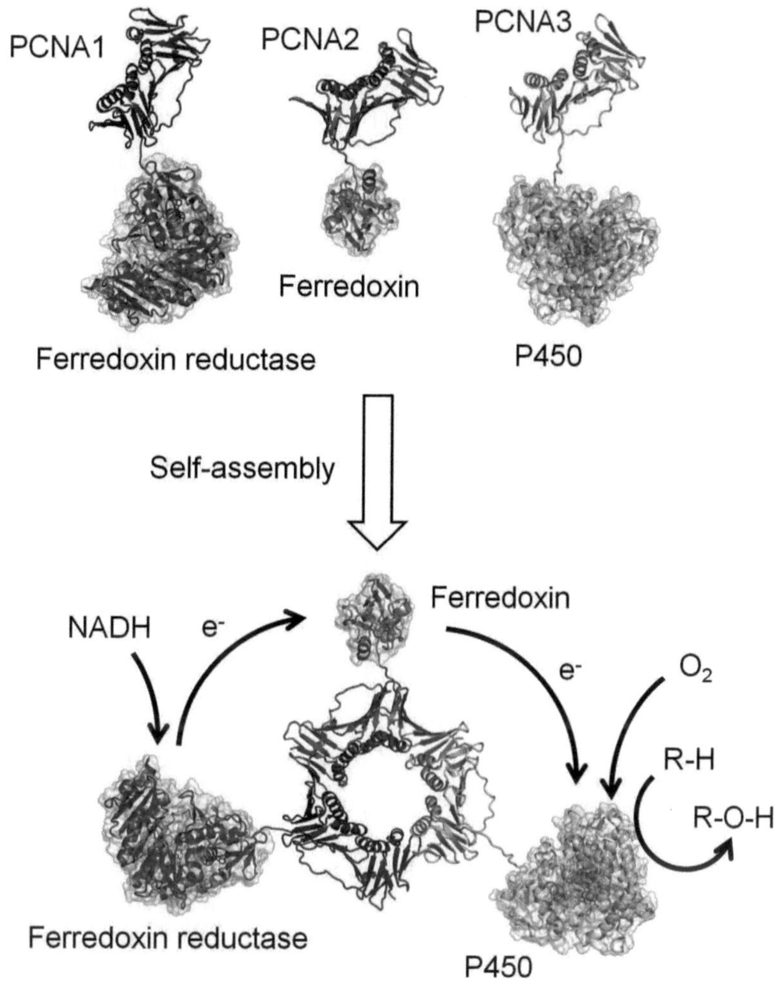

Fig. 1. Multienzyme complex formation of ferredoxin reductase, ferredoxin, and P450 using *S. solfataricus* PCNAs.

2. Materials

2.1. Protein Expression

1. *S. solfataricus* PCNA genes: Synthetic genes that are codon-optimized for expression in *E. coli* (see Note 1).
2. A fusion protein of PCNA1 and PdR (PCNA1-PdR) is constructed by genetically linking the C-terminus of PCNA1 and the N-terminus of PdR with a GGGGSGGGGS linker. A PCNA1-PdR-encoding gene is cloned into pET15b (Merck, Darmstadt, Germany) between *Nde*I and *Bam*HI sites.
3. A fusion protein of PCNA2 and PdX (PCNA2-PdX) is constructed by genetically linking the C-terminus of PCNA2 and

the N-terminus of the C73S/C85S mutant of PdX (see Note 2) with a GGGGSLVPRGSGGGGS linker. A PCNA2-PdX-encoding gene is cloned into pET15b between *Nde*I and *Bam*HI sites.

4. A fusion protein of PCNA3 and P450cam (PCNA3–P450cam) is constructed by genetically linking the C-terminus of PCNA3 and the N-terminus of P450cam with a GGS linker. A PCNA3-P450cam-encoding gene is cloned into pET15b between *Nde*I and *Bam*HI sites.
5. *E. coli* T7 Express I^q (New England Biolabs, Ipswich, MA).
6. 5-Aminolevulinic acid hydrochloride (COSMO BIO, Tokyo, Japan).
7. LB (+ glucose) agar plate: Dissolve 10 g of Bacto Tryptone (Becton, Dickson and Company, Sparks, MD), 5 g of Bacto Yeast Extract (Becton, Dickson and Company), 15 g of agar (Wako Pure Chemical, Osaka, Japan), and 10 g of NaCl (Wako Pure Chemical) in 900 mL of deionized water, and adjust at pH 7.4. Dissolve 10 g of glucose (Wako Pure Chemical) in 100 mL of deionized water. Autoclave, allow liquids to cool for less than 60°C, add 100 mg ampicillin (Wako Pure Chemical), and then mix those. Pour the media into sterile petri dishes and leave the plates to dry and cool overnight.
8. LB (+ glucose) medium: Dissolve 10 g of Bacto Tryptone, 5 g of Bacto Yeast Extract, and 10 g of NaCl in 900 mL of deionized water, and adjust at pH 7.4. Dissolve 10 g of glucose (Wako Pure Chemical) in 100 mL of deionized water. Autoclave, allow liquids to cool for less than 60°C, and then mix those.
9. Terrific Broth medium: Dissolve 12 g of Bacto Tryptone, 24 g of Bacto Yeast Extract, and 4 mL of glycerol in 900 mL of deionized water. Dissolve 12.54 g of K_2HPO_4 and 2.31 g of KH_2PO_4 in 100 mL of deionized water. Autoclave, allow liquids to cool for less than 60°C, and then mix those.

2.2. Protein Purification

1. *d*-Camphor (Wako Pure Chemical) (see Note 3).
2. Buffer A: 20 mM potassium phosphate, pH 7.4, 20 mM imidazole, 150 mM KCl.
3. Buffer B: 20 mM potassium phosphate, pH 7.4, 150 mM KCl.
4. Buffer C: 50 mM potassium phosphate, pH 7.4, 150 mM KCl.
5. Buffer D: 20 mM potassium phosphate, pH 6.0.
6. Buffer E: 50 mM potassium phosphate, pH 7.4, 500 mM KCl.
7. Buffer F: 20 mM potassium phosphate, pH 7.4, 20 mM imidazole, 150 mM KCl, 5 mM *d*-camphor (see Note 2). Add 4 mL of DMSO containing 1.25 M *d*-camphor to vigorously stirred buffer A.

8. Buffer G: 20 mM potassium phosphate, pH 7.4, 150 mM KCl, 5 mM *D*-camphor (see Note 2).
9. Buffer H: 5 mM potassium phosphate, pH 7.4, 5 mM *D*-camphor (see Note 2).
10. Buffer I: 50 mM potassium phosphate, pH 7.4, 150 mM KCl, 5 mM *D*-camphor (see Note 2).
11. HisTrap FF crude column (1.6×2.5 cm, GE Healthcare, Buckinghamshire, UK).
12. HiTrap Q FF column (1.6×2.5 cm, GE Healthcare).
13. HiTrap DEAE FF column (1.6×2.5 cm, GE Healthcare).
14. HiTrap SP FF column (1.6×2.5 cm, GE Healthcare).
15. HiTrap Desalting column (1.6×2.5 cm, GE Healthcare).
16. HiLoad 16/600 Superdex 75 pg column (1.6×60 cm, GE Healthcare).
17. Amicon Ultra-15 Centrifugal Filter Device (30,000 NMWL, Millipore, Billerica, MA).
18. Amicon Ultra-15 Centrifugal Filter Device (50,000 NMWL, Millipore).

2.3. Preparation of Protein Complexes

1. Superdex 200 10/300 GL column (1.0×30 cm, GE Healthcare).

2.4. Determination of FAD-Containing Protein's Concentration

1. 10% SDS solution.
2. Millex-GV syringe-driven filter unit (0.22 μm PVDF membrane filter, Millipore).

2.5. Pyridine Hemochromogen Assay

1. Solution A: Mix 2 mL of pyridine, 1 mL of 1 N NaOH, and 5 mL of deionized water.
2. Saturated sodium dithionate solution in water.

2.6. Enzyme Activity Assay

1. NADH (Sigma-Aldrich, St. Louis, MO): Prepare 10 mM solution in buffer C and store in aliquots at −80°C.
2. Cytochrome c from equine heart (Sigma-Aldrich): Prepare 1 mM solution in buffer C just before use.

3. Methods

A P450 from *Pseudomonas putida*, P450cam, which is a well-studied bacterial P450, catalyzes the hydroxylation of *d*-camphor to 5-exo-hydroxy camphor (12). The catalytic cycle requires electrons from NADH through putidaredoxin reductase (PdR) and putidaredoxin (PdX). Here, we describe the construction of PdR–PdX complex (PCNA1-PdR and PCNA2-PdX heterodimer) and

PdR–PdX–P450cam complex (PUPPET) using self-assembly of *S. solfataricus* PCNAs and the activity assays of those.

There fusion protein, PCNA1-PdR, PCNA2-PdX, and PCNA3, are separately expressed in *E. coli* and purified. PCNA1-PdR and PCNA2-PdX heterodimer is prepared from equimolar mixture of PCAN1-PdR and PCNA2-PdX. The heterodimer works as a CPR-like electron transfer protein from NADH to P450 due to the stable complex formation. The electron transfer activity of the heterodimer is spectroscopically evaluated from cytochrome c reduction, because reduced PdX reduces cytochrome c and, by contrast, PdR cannot reduce it. PUPPET is prepared from a mixture of equimolar concentrations of PCAN1-PdR and PCNA2-PdX and an excess concentration of PCNA3-P450cam. PUPPET and the remaining PCNA3-P450cam are easily separated by size-exclusion chromatography. The *d*-camphor hydroxylation activity of PUPPET is evaluated from the consumption rate of O_2 using liquid-phase Clark-type oxygen electrode, subtracting the value in the absence of *d*-camphor from that in the presence of *d*-camphor.

3.1. Protein Expression

1. *E. coli* T7 Express I^q is transformed with an expression plasmid (see Note 4).
2. Each transformation is spread on LB (+ glucose) agar plate containing 100 μg/mL ampicillin and the plates are incubated at 37°C overnight.
3. A single colony is inoculated in 5 mL of LB (+ glucose) medium containing 100 μg/mL ampicillin and cells are grown at 37°C for 12 h.
4. The grown cells are added to 1L of TB medium containing 100 μg/mL ampicillin and cells are grown at 37°C.
5. When OD at 600 nm reaches a value of 0.6, 1 mmol of IPTG (and 1 mmol of 5-aminolevulinic acid hydrochloride for expression of PCNA3–P450cam) is added and the temperature is lowered at 27°C.
6. After overnight culture, the cells are harvested by centrifugation at 6,000 ×*g* for 20 min.

3.2. Protein Purification

3.2.1. Purification of PCNA1-PdR

1. The harvested cells are resuspended with buffer A and disrupted by 15 repeats of ultrasonication (SONIFIER 250, duty cycle 30 %, output control 8) for 1 min with a 3-min interval, with cooling in an ethanol/ice bath.
2. Cell debris is removed by centrifugation at 22,000 ×*g* for 30 min and the supernatant is loaded on a HisTrap FF crude column pre-equilibrated with buffer A.
3. The column is washed with 50 mL of buffer A.
4. The protein is eluted from the column with a 20–250 mM imidazole gradient. Subsequently, the colored fractions are

loaded on a HiTrap Q FF column pre-equilibrated with buffer B and the column is washed with 50 mL of buffer B.

5. The protein is eluted with a 150–500 mM KCl gradient. The highest purity fractions harboring the lowest Abs_{280}/Abs_{455} ratio are combined and concentrated with an Amicon Ultra-15 Centrifugal Filter Device (50,000 NMWL).
6. The concentrated protein is loaded on a Superdex 75 pg column pre-equilibrated with buffer C. Proteins are eluted with buffer C at 1.6 mL/min.
7. The fractions harboring an Abs_{280}/Abs_{455} ratio of less than 7.4 are combined and concentrated with an Amicon Ultra-15 Centrifugal Filter Device (50,000 NMWL).
8. The concentration of the purified protein is determined on the basis of FAD released from the protein after SDS denaturation (see Subheading 3.5).
9. The purified PCNA1-PdR is stored at −80°C for long-term storage.

3.2.2. Purification of PCNA2-PdX

1. PCNA2-PdX is partially purified using a HisTrap FF crude column and a HiTrap Q FF column as described above.
2. The fractions harboring an Abs_{412}/Abs_{280} ratio of higher than 0.24 are combined and concentrated with an Amicon Ultra-15 Centrifugal Filter Device (30,000 NMWL).
3. The concentrated protein is loaded on a HiTrap Desalting column pre-equilibrated with buffer D and the protein is eluted with buffer D.
4. The eluted protein is loaded on a HiTrap SP FF column pre-equilibrated with buffer D and the column is washed with 50 mL of buffer D.
5. The protein is eluted from the column with a linear change from buffer D to buffer E. The fractions harboring an Abs_{412}/Abs_{280} ratio of higher than 0.25 are combined and concentrated with an Amicon Ultra-15 Centrifugal Filter Device (30,000 NMWL).
6. The concentrated protein is loaded on a HiLoad 16/600 Superdex 75 pg column pre-equilibrated with buffer C. Proteins are eluted with buffer C at 1.6 mL/min.
7. The fractions harboring an Abs_{412}/Abs_{280} ratio of higher than 0.30 are combined and concentrated with an Amicon Ultra-15 Centrifugal Filter Device (30,000 NMWL).
8. The concentration of the purified protein is calculated from $\varepsilon_{412} = 11.0\ mM^{-1}\ cm^{-1}$.
9. The purified PCNA2-PdX is stored at −80°C for long-term storage.

3.2.3. Purification of PCNA3-P450cam

1. PCNA3-P450cam is partially purified using a HisTrap FF crude column and a HiTrap Q FF column as described above except that buffer F and G are used instead of buffer A and B, respectively.
2. The fractions harboring Abs_{392}/Abs_{280} ratio of higher than 1.0 are combined and concentrated with an Amicon Ultra-15 Centrifugal Filter Device (50,000 NMWL).
3. The concentrated protein is loaded on a HiTrap desalting column pre-equilibrated with buffer H and the protein is eluted with buffer H.
4. The eluted protein is loaded on a HiTrap DEAE FF column pre-equilibrated with buffer H and the column is washed with 50 mL of buffer H.
5. The protein is eluted with a 5–500 mL potassium phosphate gradient. The highest purity fractions harboring the highest Abs_{392}/Abs_{280} ratio are combined and concentrated with an Amicon Ultra-15 Centrifugal Filter Device (50,000 NMWL) (see Note 5).
6. The concentrated protein is loaded on a HiLoad 16/600 Superdex 75 pg column pre-equilibrated with buffer I. Proteins are eluted with buffer I at 1.6 mL/min.
7. The fractions harboring an Abs_{392}/Abs_{280} ratio of less than 1.20 are combined and concentrated with an Amicon Ultra-15 Centrifugal Filter Device (50,000 NMWL) (see Note 5).
8. The concentration of the purified protein is determined by pyridine hemochromogen assay (see Subheading 3.5).
9. The purified PCNA3-P450cam is stored at −80°C for long-term storage.

3.3. Preparation of Protein Complexes

3.3.1. PCNA1-PdR:PCNA2-PdX Heterodimer

1. A mixture of 150 μM PCNA1-PdR and 150 μM PCNA2-PdX is incubated in buffer C at 4°C for 1 h.
2. The mixture is loaded on a Superdex 200 10/300 GL column pre-equilibrated with buffer C. Proteins are eluted with buffer C at 0.75 mL/min. The fractions containing all of the two fusion proteins are combined and concentrated with an Amicon Ultra-15 Centrifugal Filter Device (50,000 NMWL).
3. The concentration of the heterodimer is determined on the basis of FAD released from the protein after SDS denaturation (see Subheading 3.4).
4. The purified heterodimer is stored at −80°C for long-term storage.

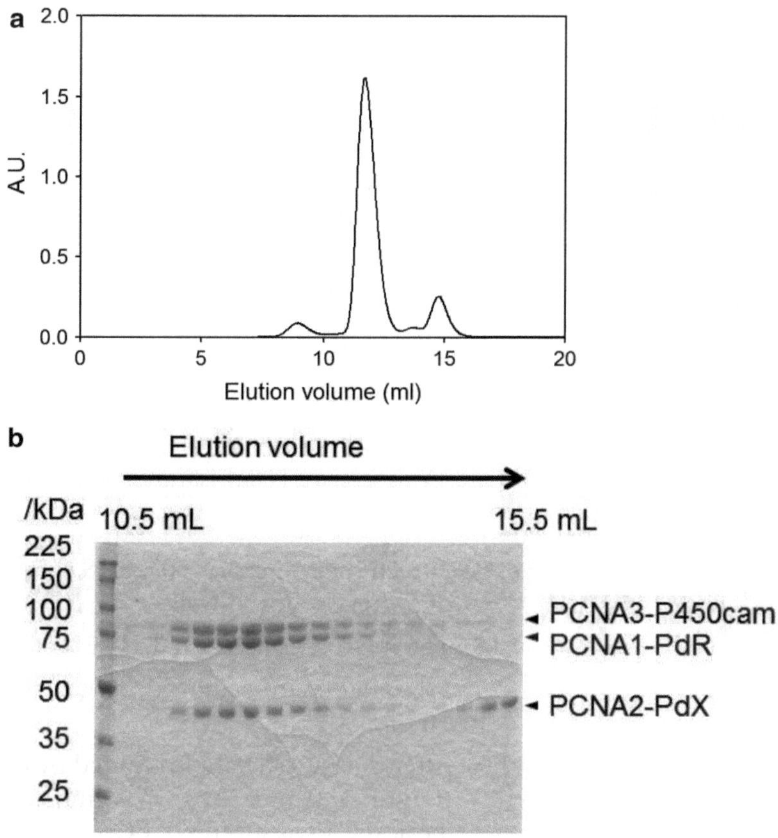

Fig. 2. Purification of PUPPET by size-exclusion chromatography; (**a**) an elution profile from a Superdex 200 10/300 GL column and (**b**) SDS-PAGE analysis of fractions. A main peak contains equimolar amounts of PCNA1-PdR, PCNA2-PdX, and PCNA3-P450cam.

3.3.2. PUPPET

1. A mixture of 150 μM PCNA1-PdR, 150 μM PCNA2-PdX, and higher than 150 μM PCNA3-P450cam (see Note 6) is incubated in buffer I at 4°C for 1 h.
2. The mixture is loaded on a Superdex 200 10/300 GL column pre-equilibrated with buffer I. Proteins are eluted with buffer I at 0.75 mL/min. The fractions containing all of the three fusion proteins (Fig. 2) are combined and concentrated with an Amicon Ultra-15 Centrifugal Filter Device (50,000 NMWL).
3. The concentration of PUPPET is determined by pyridine hemochromogen assay (see Subheading 3.6).
4. Purified PUPPET is stored at −80°C for long-term storage.

3.4. Determination of FAD-Containing Protein's Concentration

1. A dilution series of FAD-containing protein sample is prepared. The absorption at 455 nm for each sample is measured at 25°C.

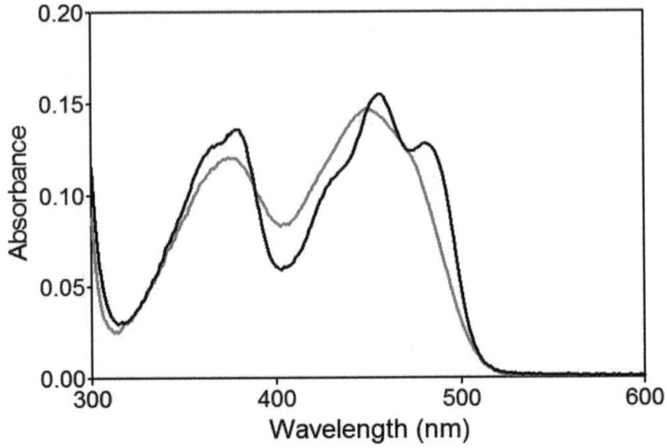

Fig. 3. UV-vis spectra of PCNA1-PdR (*black*) and denatured PCNA1-PdR (*grey*).

2. Fifteen microliters of 10% SDS solution is mixed with 750 μL of protein solution well.
3. The mixture is incubated at 45°C for 1 h.
4. Precipitants are removed by centrifugation at 36,000×*g* for 10 min.
5. After the supernatant is filtrated by 0.22 mm filter, the absorption at 450 nm (Abs_{450}) is immediately measured at 25°C. Typical UV-vis spectra of PCNA1-PdR and denatured PCNA1-PdR are shown in Fig. 3.
6. FAD concentrations are calculated as follows (see Note 7):

$$[\text{FAD}](\mu\text{M}) = \frac{\text{Abs}_{450}}{11.3} \times \frac{100}{97} \times \frac{785}{750}.$$

7. The molecular extinction coefficient is determined by plotting the absorbance at 455 nm for each protein sample versus FAD concentration.

3.5. Pyridine Hemochromogen Assay

1. A dilution series of heme-containing protein sample is prepared. A Soret peak absorbance at 392 nm (Abs_{392}) for each sample is measured at 25°C.
2. Mixture of 0.6 mL of solution A and 0.3 mL of protein solution is put into 10 mm path length quartz cuvette. The absorptions at 541 nm ($Abs_{541}{}^{ox}$) and 557 nm ($Abs_{557}{}^{ox}$) are measured at 25°C.
3. Ten microliters of saturated sodium dithionate solution is mixed with the above mixture well. The absorptions at 541 nm ($Abs_{541}{}^{red}$) and 557 nm ($Abs_{557}{}^{red}$) are immediately measured at 25°C. Typical UV-vis spectra change of PCNA3-P450cam induced by sodium dithionate is shown in Fig. 4.

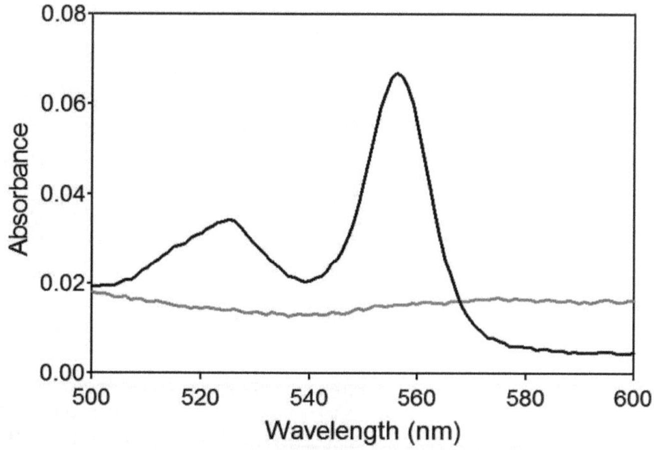

Fig. 4. UV-vis spectra of PCNA3-P450cam before (*grey*) and after (*black*) adding sodium dithionite.

4. Heme concentrations are calculated as follows:

$$[\text{Heme}](\mu\text{M}) = \frac{(\text{Abs}_{557}{}^{\text{red}} - \text{Abs}_{557}{}^{\text{ox}}) - (\text{Abs}_{541}{}^{\text{red}} - \text{Abs}_{541}{}^{\text{ox}})}{0.0207} \times 3.$$

5. The molecular extinction coefficient (ε_{392}) is determined by plotting a Soret peak absorbance for each protein sample versus heme concentration.

3.6. Enzyme Activity Assay

3.6.1. Cytochrome c Reduction by PCNA1-PdR:PCNA2-PdX Heterodimer

1. Put 1.94 mL of buffer C into a standard rectangular quartz cuvette (10 mm path length, 2 mL working volume) equipped with a magnetic stir bar. The reaction mixture is stirred at 400 rpm at 25°C.
2. Add 10 μL of 10 mM NADH solution to the cuvette.
3. Add 40 μL of 1 mM cytochrome *c* solution to the cuvette.
4. Wait until the absorption at 550 nm reaches a plateau, and then add 10 μL of protein solution to the cuvette.
5. Measure the increase rate of the absorption ($\Delta\text{Abs}_{550}/\Delta t$).
6. Reaction rate (V_c) is calculated as follows:

$$V_c(\mu\text{M}/\text{min}) = \frac{\Delta\text{Abs}_{550}/\Delta t(\text{min}^{-1})}{21.1 \times 10^{-3}(\mu\text{M}^{-1}\text{cm}^{-1})}.$$

3.6.2. O_2 Consumption by PUPPET

1. PUPPET is loaded on a HiTrap desalting column pre-equilibrated with buffer C and the protein is eluted with buffer C to remove *d*-camphor (see Note 8).
2. The eluted PUPPET is immediately concentrated with an Amicon Ultra-15 Centrifugal Filter Device (50,000 NMWL).

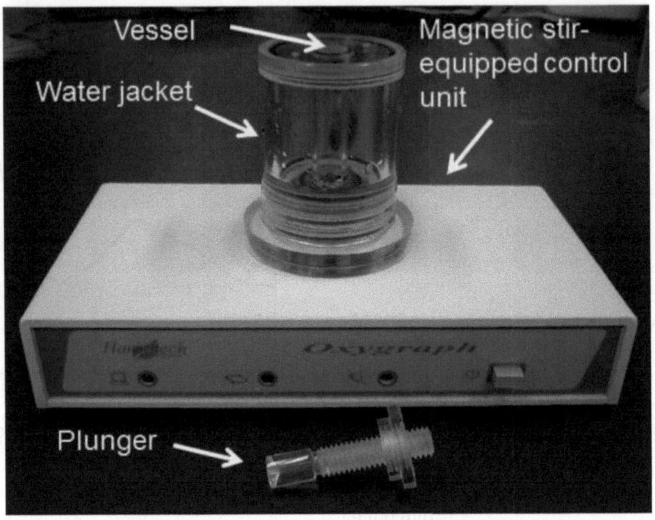

Fig. 5. Clark-type oxygen electrode system composed of a magnetic stir-equipped control unit, a water jacket, a glass reaction vessel, a plunger, and an electrode disc (hidden).

3. Twenty microliters of the concentrated PUPPET is mixed with 780 μL of buffer I. The absorption at 392 nm (Abs_{392}) is immediately measured at 25°C.
4. The concentration of the concentrated PUPPET is calculated as follows:

$$[\text{PUPPET}](\mu\text{M}) = \frac{\text{Abs}_{392}}{\varepsilon_{392}(\mu\text{M}^{-1}\text{cm}^{-1})} \times \frac{800}{20}.$$

5. Put 880 μL of buffer C into a glass reaction vessel of Oxygraph oxygen electrode system (Hansatech Instruments, Norfolk, UK) equipped with a magnetic stir bar (Fig. 5). The reaction mixture is stirred at 50 rpm at 25°C.
6. Add 100 μL of 5 mM *d*-camphor solution (buffer I) to the vessel.
7. Add 10 μL of 10 mM NADH solution to the vessel and insert a plunger into that.
8. Wait until O_2 concentration becomes steady, remove the plunger, add 10 μL of PUPPET solution to the vessel, and then insert a plunger again. An example of real-time monitoring of O_2 consumption is shown in Fig. 6.
9. Measure the decrease rate (V_p) of O_2 concentration.
10. Measure the decrease rate (V_a) of O_2 consumption in the absence of *d*-camphor.
11. Reaction rate (V_o) is calculated as follows:

$$V_o(\mu\text{M}/\text{min}) = V_p(\mu\text{M}/\text{min}) - V_a(\mu\text{M}/\text{min})$$

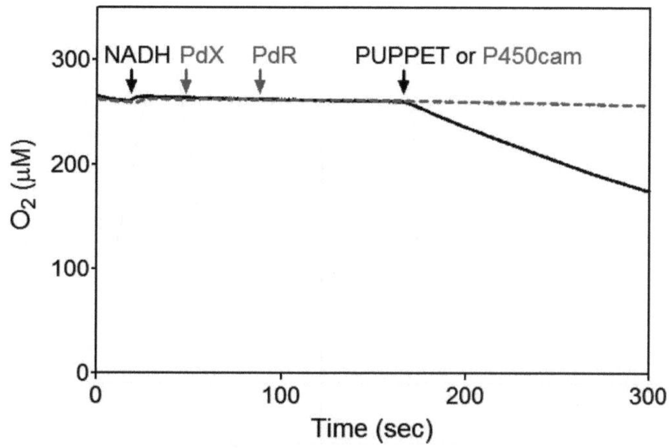

Fig. 6. O_2 consumptions by PUPPET (*solid back*) and an equimolar mixture of PdR, PdX, and P450cam (*broken grey*) in the presence of *d*-camphor. *Black arrows* indicate the time when 10 mL of NADH (10 mM) and 10 mL of PUPPET (9 mM) or P450cam (9 mM) were added. *Gray arrows* indicate the time when PdX (9 mM) and PdR (9 mM) were added for the mixture.

4. Notes

1. PCNA genes cloned directly from *S. solfataricus do not express well in E. coli* because codon usage of *S. solfataricus* is different from that of *E. coli.*
2. The wild-type PdX is highly sensitive to O_2 and easily denatured in aerobic condition. Although the mutation of C73S/C85S in PdX partially decreases the electron transfer activity, it remarkably improves the stability (13). We usually use this mutant as component of P450cam/PdX/PdR system.
3. It takes long time to dissolve *d*-camphor due to its hydrophobicity. *d*-Camphor is once dissolved in minimum volume of DMSO and then added to vigorously stirred buffer.
4. PCNA1-PdR and PCNA2-PdX can be expressed using *E. coli* BL21 (DE3), BL21 (DE3) pLysS, BL21 Star (DE3), and BL21 Star (DE3) pLysS as well as T7 Express I^q. However, expression level of PCNA3-P450cam is not high in the BL21 strains compared to T7 Express I^q.
5. PCNA3-P450cam partially forms homooligomer(s) at the high concentration. Avoid concentrating the protein too much, because the homooligomer(s) does not form a heterotrimer with PCNA1-PdR and PCNA2-PdX.
6. An excess amount of PCNA3-P450cam compared to PCNA1-PdR is necessary to avoid the formation of heterodimer of

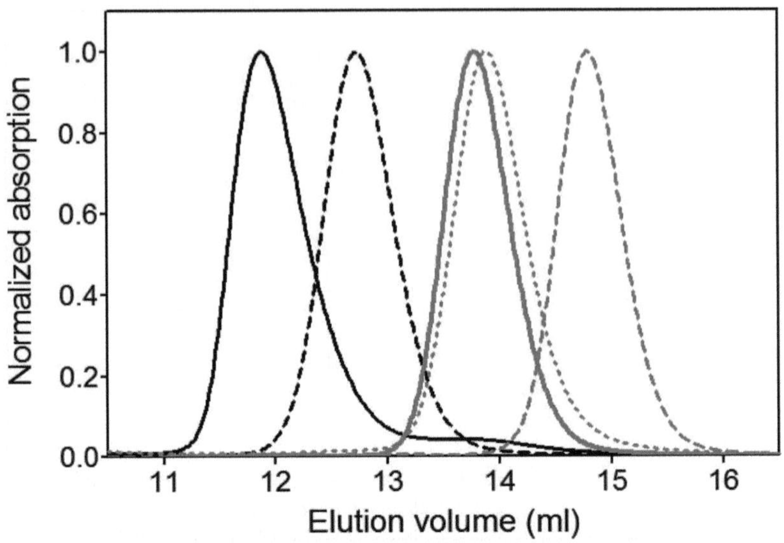

Fig. 7. Size-exclusion chromatographic profiles for purified proteins; PUPPET (*solid black*), heterodimer of PCNA1-PdR and PCNA2-PdX (*broken black*), PCNA1-PdR (*solid gray*), PCNA2-PdX (*broken gray*), and PCNA3-P450cam (*dotted gray*).

PCNA1-PdR and PCNA2-PdX that cannot be completely separated from PUPPET by size-exclusion chromatography (Fig. 7).

7. A correction is made for 3% loss of FAD due to thermal destruction.
8. *d*-Camphor is removed just before activity assay because P450cam easily is inactivated in the absence of *d*-camphor.

Acknowledgements

We gratefully acknowledge the support from a Grant-in-Aid for Young Scientists (B) from the Ministry of Education, Culture, Sports, Science and Technology (MEXT) and the support from the Center for NanoBio Integration (Research and Development in a New Converting Field Based on Nanotechnology and Materials Science, MEXT).

References

1. Jørgensen K, Rasmussen AV, Morant M, Nielsen AH, Bjarnholt N, Zagrobelny M, Bak S, Møller BL (2005) Metabolon formation and metabolic channeling in the biosynthesis of plant natural products. Curr Opin Plant Biol 8:280–291
2. Fontes CMGA, Gilbert HJ (2010) Cellulosomes: highly efficient nanomachines designed to deconstruct plant cell wall complex carbohydrates. Annu Rev Biochem 79: 655–681

3. Schoffelen S, van Hest JCM (2012) Multi-enzyme systems: bringing enzymes together in vitro. Soft Matter 8:1736–1746
4. Ortiz de Montellano PR (ed) (2005) Cytochrome P450: structure, mechanism, and biochemistry, 3rd edn. Plenum Publishers, New York
5. Heyman A, Barak Y, Caspi J, Wilson DB, Altman A, Bayer EA, Shoseyov O (2007) Multiple display of catalytic modules on a protein scaffold: nano-fabrication of enzyme particles. J Biotechnol 131:433–439
6. Mitsuzawa S, Kagawa H, Li Y, Chan SL, Paavola CD, Trent JD (2009) The rosettazyme: a synthetic cellulosome. J Biotechnol 143:139–144
7. Men D, Guo YC, Zhang ZP, Wei HP, Zhou YF, Cui ZQ, Liang XS, Li K, Leng Y, You XY, Zhang XE (2009) Seeding-induced self-assembling protein nanowires dramatically increase the sensitivity of immunoassays. Nano Lett 9:2246–2250
8. Leng Y, Wei H-P, Zhang Z-P, Zhou Y-F, Deng J-Y, Cui Z-Q, Men D, You X-Y, Yu Z-N, Luo M, Zhang X-E (2010) Integration of a fluorescent molecular biosensor into self-assembled protein nanowires: a large sensitivity enhancement. Angew Chem Int Ed 49:7243–7246
9. Moraïs S, Heyman A, Barak Y, Caspi J, Wilson DB, Lamed R, Shoseyov O, Bayer EA (2010) Enhanced cellulose degradation by nano-complexed enzymes: synergism between a scaffold-linked exoglucanase and a free endoglucanase. J Biotechnol 147:205–211
10. Hirakawa H, Nagamune T (2010) Molecular assembly of P450 with ferredoxin and ferredoxin reductase by fusion to PCNA. Chembiochem 11:1517–1520
11. Dionne I, Nookala RK, Jackson SP, Doherty AJ, Bell SD (2003) A heterotrimeric PCNA in the hyperthermophilic archaeon *Sulfolobus solfataricus*. Mol Cell 11:275–282
12. Gunsalus IC, Wagner GC (1978) Bacterial P-450cam methylene monooxygenase components: cytochrome m, putidaredoxin, and putidaredoxin reductase. Methods Enzymol 52:166–188
13. Sevrioukova IF, Garcia C, Li H, Bhaskar B, Poulos TL (2003) Crystal Structure of Putidaredoxin, the [2Fe–2S) Component of the P450cam Monooxygenase System from *Pseudomonas putida*. J Mol Biol 333:377–392

Chapter 12

Gene Synthesis by Assembly of Deoxyuridine-Containing Oligonucleotides

Romualdas Vaisvila and Jurate Bitinaite

Abstract

Gene synthesis is an invaluable technique in synthetic and molecular biology for synthesis of artificial genes, operons, and even genomes. In many cases the traditional methods for obtaining functional DNA sequences through cloning are not applicable due to the novelty of genetic material. Here, we describe the simple and economical DNA synthesis method based on USER™ technology. The method consists of (1) synthesis of building blocks up to 500 bp; (2) assembly of genes up to 3 kb; (3) error correction reassembly; and (4) assembly of operons up to 15 kb if needed.

Key words: Gene synthesis, Polymerase chain reaction, Uracil excision, USER cloning, Synthetic biology

1. Introduction

Gene synthesis is a foundation for enzyme and metabolic engineering as well as synthetic biology. In many cases the traditional methods for obtaining functional DNA sequences are not applicable due to the novelty of genetic material. Numerous protocols have been published in the past 15–20 years, starting from elegant work of Stemmer (1), where an entire plasmid was assembled from an oligonucleotide pool. Another milestone accomplishment utilized programmable DNA microchips for gene synthesis, where a one-step DNA polymerase assembly multiplexing reaction assembled all 21 genes that encode the proteins of the *Escherichia coli* 30S ribosomal subunit (2). In the last 5 years we have experienced an explosion of various gene synthesis methods and improvements because of the significant reduction in the cost of oligonucleotide synthesis (3–5). Many companies offer gene synthesis services, but

James C. Samuelson (ed.), *Enzyme Engineering: Methods and Protocols*, Methods in Molecular Biology, vol. 978,
DOI 10.1007/978-1-62703-293-3_12, © Springer Science+Business Media New York 2013

the price range is still too high for most molecular biology labs. We have recently described an economical long DNA synthesis method based on USER technology (6). The method consists of (1) synthesis of building blocks up to 500 bp; (2) assembly of genes up to 3 kb; (3) error correction reassembly; and (4) assembly of operons up to 15 kb if needed. The advantages of this method include the minimization of restriction endonuclease digestion and ligation steps, ability to produce long and accurate DNA sequences in a very short time (1–2 weeks), and ease of use in any molecular biology laboratory. To date, we have used this method to assemble over 100 kb of synthetic genes ranging from 0.2 to >6 kb in length. Our failure rate in gene synthesis has been negligible and we conclude that the method is suitable for the synthesis and assembly of most genes in any molecular biology lab.

2. Materials

2.1. Codon Optimization and Oligonucleotide Design

1. DNAWorks: http://helixweb.nih.gov/dnaworks/.
2. GeneDesign: http://genedesign.thruhere.net/gd/.

2.2. Reagents

1. Oligonucleotides (any supplier is good, we usually order from Sigma or IDT).
2. Phusion® DNA Polymerase or Q5 High-Fidelity DNA polymerase (NEB), both work well.
3. PfuTurbo Cx DNA polymerase (Agilent Technologies).
4. USER enzyme (NEB).
5. QIAquick Gel Extraction Kit (Qiagen) for purification of the assembly reaction product from multiple nonspecific PCR products (if necessary).
6. Gibson Assembly™ Master Mix (NEB) for improved PCR reactions on high GC templates (if necessary).
7. pNEB206A (NEB).
8. DpnI (NEB).
9. NEB5-alpha competent cells or equivalent.

3. Methods

3.1. Codon Optimization and Oligonucleotide Design

Design overlapping oligonucleotides, generally 55–60 bases in length, encompassing sense and antisense strands of the entire gene of interest (see Note 1). Design the overlap with no less than

15 bases (T_m 62–63°C). Use DNAWorks (7, 8) or Gene Design (9) software: DNAworks: http://helixweb.nih.gov/dnaworks/ or GeneDesign: http://genedesign.thruhere.net/gd/. Order oligonucleotides from a commercial supplier. Standard oligo purification (desalting) gives good results, so there is no need for PAGE purification.

3.2. Design of 5′ and 3′ PCR Amplification Primers to Incorporate Uracil (dU) for Assembling Longer DNAs by the USER Method

The uracil excision-based assembly (USER) method (6) is used to simultaneously stitch together multiple ~500 bp "blocks" synthesized from oligonucleotides. Briefly, choose an 8–12 bp sequence that begins at A and ends at T (5′–3′). This sequence is the overlap between two blocks. Simply, replace the 3′- T in the overlap with U when ordering assembly oligonucleotides. Removing the uracil from the PCR products with the USER enzyme will generate a one-nucleotide gap on each product enabling the short piece of DNA to fall off, leaving compatible, cohesive 8–12 bp ssDNA 3′ overhangs. Ligase is not needed for assembly. Alternatively, the GeneDesign software (http://genedesign.thruhere.net/gd/) may be used to design primers for USER junctions.

3.3. Building Block Amplification

Dilute oligonucleotides to 10 μM in deionized H_2O, mix well, and store at −20°C. Prepare building block amplification mix by combining 10 μl of each of the oligonucleotide (this will be referred to as 1×). Then dilute this mix 1:2 (0.5×) and 1:4 (0.25×) in deionized H_2O. This step is to optimize the building block amplification of the required product, which sometimes varies from gene to gene. Set up two different reactions, one for 0.5× dilution and another for 0.25× dilution of the mix, to determine which gives the best yield and the least background. Use the first 5′ forward oligo and the last 3′ reverse oligo as PCR primers. The amplification primers for PCR reactions must be standard oligonucleotides, without deoxyuridine (see Note 2).

PCR amplification of assembled oligonucleotides with Phusion® DNA polymerase or Q5 High-Fidelity DNA polymerase:

Component	Stock	Amount per reaction
dH_2O		20.1 μl
Reaction buffer	5×	6 μl
dNTPs	10 mM each	0.6 μl
Template	0.5× and 0.25× dilutions	1 μl
5′ forward oligo (standard)	10 μM	1 μl
3′ reverse oligo (standard)	10 μM	1 μl
DNA polymerase	2,000 U/ml	0.3 μl
Total reaction volume		30 μl

Cycling conditions for Phusion or Q5 High-Fidelity DNA polymerase:

No. of cycles	Temperature (°C)	Duration
1	98.0	30 s
30	98.0 63.0 72.0	10 s 10 s 20 s
1	72.0	5 min

Analyze 5 μl of each PCR reaction by agarose gel electrophoresis. If the building block PCR reaction fails, consider using the Gibson Assembly™ Master Mix to improve building block amplification (see Notes 3 and 4). If amplification yielded only one band, or background is not significant, dilute the PCR product 50×, and use 2 μl for amplification with uracil-modified primers and PfuTurbo Cx polymerase (Subheading 3.4). If you observe multiple bands, purify the correct band from an agarose gel with a QIAquick gel purification kit. Then perform PCR with the modified primers and PfuTurbo Cx polymerase.

3.4. Introducing Uracils into Building Blocks for USER Assembly

Perform PCR amplification with the PfuTurbo Cx polymerase as follows:

Component	Stock	Amount per reaction
dH_2O		22.8 μl
PfuTurbo Cx reaction buffer	10×	3 μl
dNTPs	10 mM each	0.6 μl
Template	50× from first PCR	1 μl
5′ dU-modified oligo	10 μM	1 μl
3′ dU-modified oligo	10 μM	1 μl
Pfu Turbo Cx polymerase	2,500 U/ml	0.6 μl
Total reaction volume		30 μl

Cycling conditions for PfuTurbo Cx Polymerase:

No. of cycles	Temperature (°C)	Duration
1	95.0	2 min
30	95.0 55.0 72.0	30 s 30 s 40 s
1	72.0	5 min

Analyze 5 μl of each PCR reaction by agarose gel electrophoresis. PCR product purification is not needed for the full-length gene assembly reaction. Before performing the gene assembly reaction, DpnI digestion should be performed to destroy any plasmid template DNA (see Note 5).

3.5. Full-Length Gene Assembly Reaction

Perform the full-length gene assembly reaction:
Mix the following components on ice or RT:

2–5 μl (20–50 nM) of each building block PCR product amplified with PfuTurbo Cx.

1–2 μl of linear pNEB206A or linear USER vector of choice (see Note 6).

1 μl of USER enzyme (NEB).

0.5× PfuTurbo Cx reaction buffer to 30 μl.

Incubate reaction for 15 min at 37°C, and then let cool down for 15 min to RT. Transform 50 μl NEB5-alpha with 3–5 μl of assembly mixture. Follow the manufacturer's instructions to achieve optimal transformation results. Characterize 3–6 clones by sequence analysis (see Notes 7 and 8).

3.6. Error Correction/ Reassembly

In some cases characterized clones will contain an undesired mutation in one of the building blocks. To correct this error, choose a correct template for the required building block and perform PCR amplification with PfuTurbo Cx polymerase as described in Subheading 3.4. Analyze 5 μl of the PCR reaction by agarose gel electrophoresis. If PCR amplification is successful, mix 3–5 μl of each PCR product with 1× PfuTurbo Cx buffer to a total volume of 30 μl and add 1 μl (20 U) DpnI restriction endonuclease. Incubate for 30 min at 37°C and then heat inactivate DpnI for 20 min at 80°C. Add 1–2 μl of linearized pNEB206A vector and repeat the transformation step. Characterize two clones by sequence analysis.

4. Notes

1. An important parameter for building block assembly is the length of oligonucleotides. We found that the optimal length is 55–60 bp. Ordering oligonucleotides longer than 60 bp increases the probability of errors in the building blocks. If you design overlapping oligonucleotides manually, use the T_m calculator from New England Biolabs (http://www.neb.com/nebecomm/tech_reference/TmCalc/Default.asp). Choose the Phusion or Q5 DNA polymerase option, and then type in the overlapping region. Design the overlap with no less than 15 bp bases (T_m 63–65°C).

2. Taq and Pfu Turbo Cx polymerases can read through uracil (dU). Other archaeal polymerases stall on dU and remain bound to primers for the remainder of the PCR run resulting in no amplification of DNA. PfuTurbo Cx has an error rate comparable to Pfu, so PfuTurbo Cx is the preferred polymerase.
3. If the assembly was not successful, the Gibson Assembly™ Master Mix (NEB) may be used to improve building block amplification. Prepare building block assembly mix by combining 10 μl of each oligonucleotide. Final concentration of each oligonucleotide in the mix will be somewhere between 1 μM (for ten oligonucleotides) and 0.5 μM (for 20 oligonucleotides) depending on the total number of oligonucleotides mixed. Heat the oligonucleotide mix at 95°C for 5 min and then slowly cool down to room temperature. Mix 1 μl oligonucleotide mix with 9 μl deionized H_2O and 10 μl 2× Gibson Master Mix. Incubate the reaction at 60°C for 1 h. Use 1 μl of Gibson reaction product as a template to perform PCR amplification with Phusion® DNA polymerase (NEB) or Q5 High-Fidelity DNA polymerase (NEB). Use the first 5′ forward oligo and the last 3′ reverse oligo as PCR primers.
4. If you use the Gibson Assembly™ Master Mix (NEB) to improve building block amplification, remember that the reaction temperature for oligonucleotide assembly is different from fragment assembly reactions (50°C). For oligonucleotide assembly, always incubate Gibson assembly reactions at 60°C.
5. Remember to treat PCR products with DpnI (NEB) to remove any plasmid template. Otherwise there will be high background of incorrect plasmid. Also remember to purify your plasmid template from a dam + strain (such as NEB5-alpha).
6. pNEB206A comes with premade USER-generated ends. If you wish to clone your synthesized gene into an alternative vector, you must design the appropriate ends, amplify the plasmid backbone, and treat the PCR product with USER enzyme mix. Alternatively, restriction sites may be incorporated into the 5′ and 3′ ends of your gene. Then the gene may be cloned using a traditional digestion and ligation method.
7. There is no need to order extra primers for sequence verification. Simply use the first 5′ oligo and the last 3′ oligo from each building block to sequence USER-assembled clones.
8. If you are constructing a gene or operon greater than 3–4 kb, simply synthesize 3–4 kb blocks and repeat the USER assembly reaction with these blocks.

Acknowledgments

The authors are grateful to Peter Weigele for valuable suggestions. We also thank Laurie Mazzola for DNA sequencing and John Buswell for oligonucleotide synthesis. We are especially thankful for the environment and support provided by Don Comb.

References

1. Stemmer WP, Crameri A, Ha KD, Brennan TM, Heyneker HL (1995) Single-step assembly of a gene and entire plasmid from large numbers of oligodeoxyribonucleotides. Gene 164:49–53
2. Tian J, Gong H, Sheng N, Zhou X, Gulari E, Gao X, Church G (2004) Accurate multiplex gene synthesis from programmable DNA microchips. Nature 432:1050–1054
3. Czar MJ, Anderson JC, Bader JS, Peccoud J (2009) Gene synthesis demystified. Trends Biotechnol 27:63–72
4. Tian J, Ma K, Saaem I (2009) Advancing high-throughput gene synthesis technology. Mol Biosyst 5:714–722
5. Hughes RA, Miklos AE, Ellington AD (2011) Gene synthesis: methods and applications. Methods Enzymol 498:277–309
6. Bitinaite J, Rubino M, Varma KH, Schildkraut I, Vaisvila R, Vaiskunaite R (2007) USER friendly DNA engineering and cloning method by uracil excision. Nucleic Acids Res 35:1992–2002
7. Hoover DM, Lubkowski J (2002) DNAWorks: an automated method for designing oligonucleotides for PCR-based gene synthesis. Nucleic Acids Res 30:e43
8. Hoover D (2012) Using DNAWorks in designing oligonucleotides for PCR-based gene synthesis. Methods Mol Biol 852: 215–223
9. Richardson SM, Nunley PW, Yarrington RM, Boeke JD, Bader JS (2010) GeneDesign 3.0 is an updated synthetic biology toolkit. Nucleic Acids Res 38:2603–2606

Chapter 13

Protein Engineering: Single or Multiple Site-Directed Mutagenesis

Pei-Chung Hsieh and Romualdas Vaisvila

Abstract

Site-directed mutagenesis techniques are invaluable tools in molecular biology to study the structural and functional properties of a protein. To expedite the time required and simplify methods for mutagenesis, we recommend two protocols in this chapter. The first method for single site-directed mutagenesis, which includes point mutations, insertions, or deletions, can be achieved by an inverse PCR strategy with mutagenic primers and the high-fidelity Phusion® DNA Polymerase to introduce a site-directed mutation with exceptional efficiency. The second method is for engineering multiple mutations into a gene of interest. This can be completed in one step by PCR with mutagenic primers and by assembling all mutagenized PCR products using the Gibson Assembly™ Master Mix. This method allows multiple nucleotides to be changed simultaneously, which not only saves time but also reagents compared to traditional methods of mutagenesis.

Key words: Site-directed mutagenesis, Inverse PCR mutagenesis, Gibson assembly, Synthetic biology

1. Introduction

In the past 30 years, site-directed mutagenesis methods have played an important role in understanding the structure and function of proteins and elements of DNA (such as promoter regions). This technique has allowed scientists to change the amino acid of a protein through DNA manipulation and has facilitated the creation of engineered proteins or "biocatalysts" that can perform specific reactions in the field of modern biotechnology. This basic method of gene engineering evolved from the use of phage M13-based vectors where single-stranded templates could be easily generated. Gillam et al. (1) and Zoller et al. (2) demonstrated that chemically synthesized oligonucleotides, carrying a base mismatch, could be

James C. Samuelson (ed.), *Enzyme Engineering: Methods and Protocols*, Methods in Molecular Biology, vol. 978, DOI 10.1007/978-1-62703-293-3_13, © Springer Science+Business Media New York 2013

annealed to single-stranded template DNA and then extended with DNA polymerase to complete the second strand followed by *E. coli* transformation.

Later, several strategies were developed along with the availability of PCR technology (3). Hemsley (4) introduced the concept of inverse PCR, in which both primers (with desired mutations) are designed in a tail-to-tail manner to amplify the whole plasmid DNA containing the gene of interest. Higuchi (5) and Ho (6) also developed a method using four PCR primers to do mutagenesis. Briefly, a designed mutation is planted in two of the four PCR primers, and the sequences of these mutation-encoded primers are complementary to each other. In the first round of PCR, fragments are amplified separately and rejoined again in the second round of PCR due to the presence of two overlap regions. The final products are then purified and ligated to a cloning vector. The second round of PCR can be avoided if a restriction enzyme site can be silently introduced in the primer sequences used to join the two fragments together. If a restriction enzyme site cannot be introduced in the complementary region, one can use a Type IIS restriction enzyme which cuts downstream of the recognition sequence (7). An alternative way to avoid two rounds of PCR is suggested by Landt et al. (8) and Ke et al. (9) who modified the previous method by using three primers in one round of PCR to achieve similar results. This method is known as "megaprimer PCR mutagenesis."

While PCR-based site-directed mutagenesis methods are convenient, exponential amplification of DNA with Taq DNA polymerase may result in undesired nucleotide changes due to the inherent error rate of the polymerase. In addition, subcloning of these mutated fragments usually takes several additional steps with potential technical difficulties. Several companies have developed non-PCR strategies that involve using a high-fidelity T4 DNA polymerase to reduce the risk of introducing spurious mutations and an *E. coli* strain to select the resulting heteroduplex plasmids. For example, Clontech Laboatories, Inc./Takara Bio Inc. developed the Transformer™ Site-Directed Mutagenesis Kit. This technique, requiring two oligonucleotides (one for the purpose of mutagenesis and one for selection), takes advantage of (1) a unique restriction enzyme site in the wild type but not in the mutant and (2) an *E. coli* host with a *mutS* gene deletion, which allows mutants to survive but not wild-type DNA after restriction enzyme selection. Similarly, the GeneEditor™ system from Promega also uses two primers, and the newly synthesized DNA strand will result in new antibiotic resistance. These methods do require more starting DNA template, compared to PCR methods, and rely on T4 DNA polymerase for DNA synthesis.

The emergence of high-fidelity DNA polymerases (such as Phusion® DNA Polymerase) enabled the development of high efficiency PCR mutagenesis methods, which do not require subcloning or ssDNA isolation. For example, the GeneTailor™

Site-directed Mutagenesis System (Life Technologies) claimed greater than 80% mutagenesis efficiency for a range of plasmid sizes and mutation types. In this technique the parental plasmid DNA is methylated in vitro followed by PCR with a mutagenic primer. The resulting products are transformed into wild-type *E. coli* with *Mcr*BC endonuclease restriction activity, which eliminates methylated plasmid DNA and allows *E. coli* with the mutated plasmid DNA to survive. Stratagene/Agilent Technologies developed the ExSite™ PCR-Based Site-Directed Mutagenesis Kit and more recently the QuikChange® Site-Directed Mutagenesis Kit to generate mutations with greater than 85% efficiency in a single reaction. Briefly, instead of modifying the plasmid in vitro in the GeneTailor™ protocol, the Stratagene methods cleverly use a restriction enzyme, *Dpn*I, to remove the adenine-methylated parental plasmids. These plasmids are sensitive to *Dpn*I treatment if they are isolated from standard *E. coli* strains with a Dam modification system. While QuikChange® Site-Directed Mutagenesis is a popular method for mutagenesis, the problem of primer dimerization may still occur even when primers were properly designed (10). Zheng et al. (11) developed an improved method of primer design not only solving the inconsistency of the primer dimerization problem but also improving the PCR amplification efficiency.

In this chapter, we summarize two methods of site-directed mutagenesis. For single site-directed mutagenesis, including point mutation, insertion, deletion, and multiple nearby substitutions (accomplished by one megaprimer), we outline Hemsley's (4) method with some modifications, which will benefit researchers in terms of cost savings and time savings. For mutagenesis of multiple sites (hundreds of base pairs apart), we have combined Higuchi's (5), Jones' (12), and Gibson's (13, 14) methods. This multiple site method offers a simplified design but also circumvents the need for restriction enzymes during subcloning. In addition, all desired mutations can be completed together with great efficiency using the Gibson Assembly™ Master Mix, which directs the assembly of multiple mutagenic PCR fragments together in one isothermal step.

2. Materials

2.1. Single Site-Directed Mutagenesis

1. A plasmid DNA with gene of interest.
2. Plasmid DNA Miniprep Kit (Qiagen kit or equivalent).
3. Oligonucleotides (forward and reverse primers) with and without mutagenic sequences (see Note 1).
4. A thermocycler machine.

5. Phusion® Hot Start Flex DNA Polymerase (or other high-fidelity DNA polymerase) and dNTP nucleotide mix. Alternatively, Phusion® Hot Start Flex 2× Master Mix.
6. Agarose.
7. 1× TBE buffer (89 mM Tris–HCl, 89 mM boric acid, 2 mM EDTA, pH 8.3).
8. DNA fragment gel extraction kit.
9. PCR fragment purification kit.
10. 37°C water bath.
11. T4 polynucleotide kinase.
12. T4 DNA ligase supplied with 10× T4 DNA ligase buffer.
13. *Dpn*I.
14. NEB5-alpha chemical competent *E. coli* cells or equivalent (high efficiency >10^9 cfu/microgram DNA). Alternatively, NEB10-beta chemical competent *E. coli* cells or equivalent. Alternatively, NEB10-beta electrocompetent *E. coli* cells or equivalent.
15. Appropriate DNA restriction enzymes and buffers.

2.2. Multiple Site-Directed Mutagenesis

1. A plasmid DNA with gene of interest.
2. Plasmid DNA Miniprep Kit (Qiagen kit or equivalent).
3. Oligonucleotides (forward and reverse primers) with and without mutagenic sequences (see Note 1).
4. A thermocycler machine.
5. Phusion® Hot Start Flex DNA Polymerase (or other high-fidelity DNA polymerase) and dNTP nucleotide mix. Alternatively, Phusion® Hot Start Flex 2× Master Mix.
6. Agarose.
7. 1× TBE buffer (89 mM Tris–HCl, 89 mM boric acid, 2 mM EDTA, pH 8.3).
8. DNA fragment gel extraction kit.
9. PCR fragment purification kit.
10. 37°C water bath.
11. 2× Gibson Assembly™ Master Mix.
12. *Dpn*I.
13. NEB5-alpha chemical competent *E. coli* cells or equivalent (high efficiency >10^9 cfu/microgram DNA). Alternatively, NEB10-beta chemical competent *E. coli* cells or equivalent. Alternatively, NEB10-beta electrocompetent *E. coli* cells or equivalent.
14. Appropriate DNA restriction enzymes and buffers.

2.3. *E. coli* Transformation

1. SOC outgrowth medium: 2% (w/v) vegetable peptone, 0.5% (w/v) yeast extract, 10 mM NaCl, 2.5 mM KCl, 10 mM $MgCl_2$, 10 mM $MgSO_4$, 20 mM glucose, pH 7.0.
2. Bucket of ice.
3. Water bath set at 42°C.
4. Shaking 37°C incubator.
5. Sterile cell spreader.
6. LB agar plates with appropriate antibiotic.
7. LB medium: 1% (w/v) Bacto Tryptone, 0.5% (w/v) yeast extract, 1% (w/v) NaCl, pH 7.0.

2.4. Colony PCR for Clone Screening

1. Bacterial colonies on a transformation plate.
2. OneTaq® Quick-Load® 2× Master Mix with standard Buffer or equivalent.
3. Primers for gene-specific PCR.
4. Sterile pipette tips to pick colonies.
5. A thermocycler machine.
6. PCR tubes.

3. Methods

3.1. Design of Single Site-Directed Mutagenesis

3.1.1. Strategy for Point Mutagenesis

1. A general scheme of creating a point mutation is shown on Fig. 1. Primer A (with a designed mutation marked "x") and primer B are used to produce linear PCR products with the desired mutation, which must be phosphorylated and ligated prior to transformation into *E. coli*. *Dpn*I digestion eliminates the original plasmid template.
2. Order both oligonucleotides with the designed sequences (primer A and B). Primer A and primer B are designed tail-to-tail (5' ends) in order to direct the polymerization of DNA resulting in an inverse PCR product. Place the mutation 9 nts away from the 5' end and 18–27 nts from 3′ end. The 3' exact match region is used for determining the T_m for PCR (Fig. 2). The T_m for both gene-specific primers can be calculated from the NEB website (see Note 3).
3. Go to Subheading 3.1.4 for PCR setup.

3.1.2. Strategy for Insertion Mutagenesis

1. A general scheme of creating insertion mutations is shown in Fig. 3. Primer A (designed according to the length of the insertion) and primer B are used to produce linear PCR products with the desired mutation. The linear PCR product must be phosphorylated and ligated prior to transformation into *E. coli*. *Dpn*I digestion eliminates the original plasmid template.

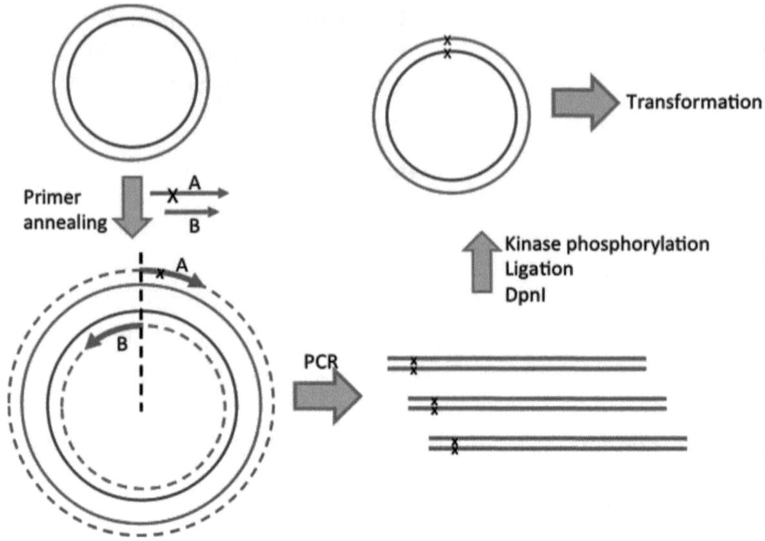

Fig. 1. Schematic representation of the point mutation method.

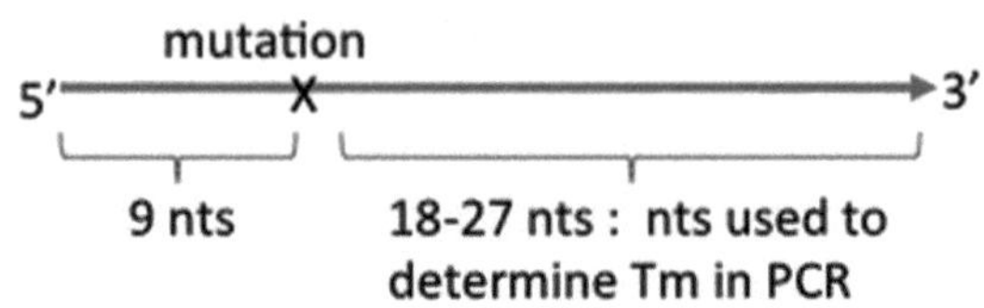

Fig. 2. Design of primer a for creating a point mutation.

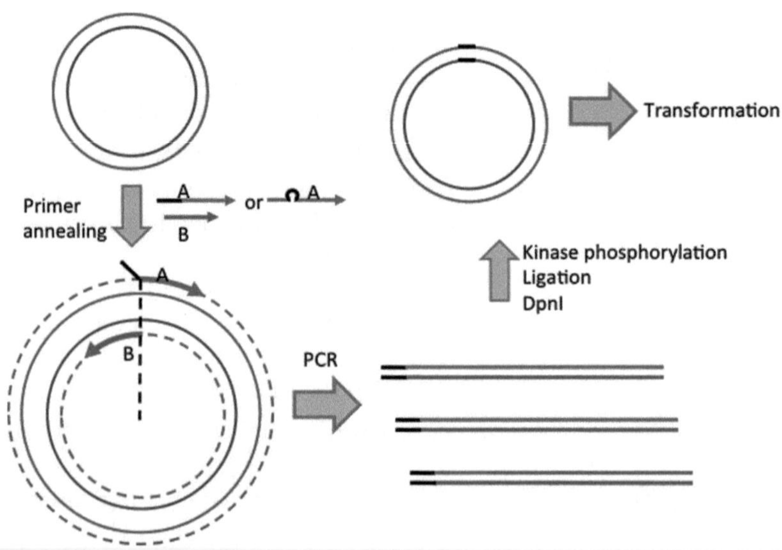

Fig. 3. Schematic representation of the method for creating insertion mutations.

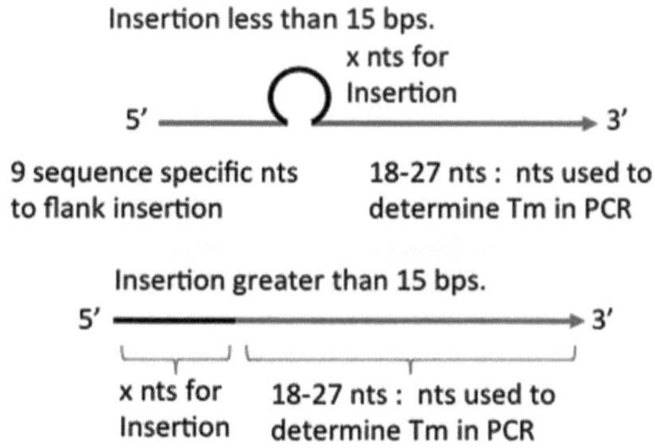

Fig. 4. Two options for designing primer A for insertion mutations.

2. Order both oligonucleotides with the designed sequences (primer A and B, Fig. 4). Primer A and primer B are designed tail-to-tail (5′ ends) in order to direct the polymerization of DNA resulting in an inverse PCR product. If the insertion is greater than 15 bp, place the "insertion region" at the 5′ end of the primer A with 18–27 matching nt at 3′ end. If the insertion is less than 15 bp, place the insertion 9 nt away from the 5′ end and 18–27 nt from 3′ end. The 3′ exact match region is used for determining the T_m for PCR. T_m for both gene-specific primers can be calculated from NEB website (see Note 3).
3. Go to Subheading 3.1.4 for PCR setup.

3.1.3. Strategy for Deletion Mutagenesis

1. A general scheme of creating deletion mutations is shown in Fig. 5. Primer A and primer B are designed so that ultimate ligation of the 5′ ends of the inverse PCR product results in a deletion of the gene of interest. The linear PCR product must be phosphorylated and ligated prior to transformation into *E. coli*. The T_m of primer A and B can be calculated from the NEB website (see Note 3).

3.1.4. PCR Setup

1. Add the following components to a 0.2 ml PCR reaction tube:

PCR component	50 ml PCR reaction	Final concentration
Phusion® Hot Start Flex 2× Master Mix	25 μl	1×
10 μM forward primer	2.5 μl	0.5 μM
10 μM reverse primer	2.5 μl	0.5 μM
Template DNA	Variable	50 pg (plasmid) to 400 ng (genomic DNA)
Nuclease-free water	Up to 50 μl	

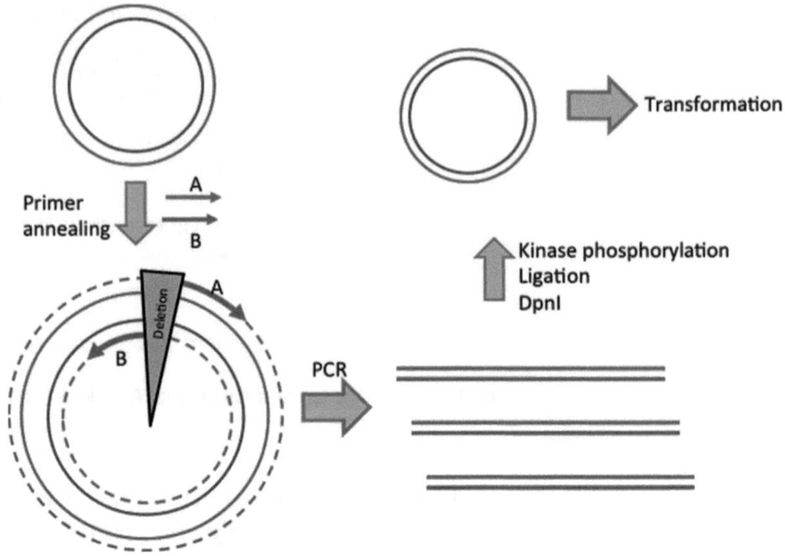

Fig. 5. Schematic representation of the method for creating insertion mutations.

2. Gently mix the reaction. If necessary, collect all liquid to the bottom of the tube by a quick spin. Overlay the sample with mineral oil if using a PCR machine without a heated lid.
3. Transfer PCR tubes to a PCR machine with the block preheated to 98°C and start the cycling program.

   Cycling protocol:

Cycling step	Temperature	Time
1 cycle of		
Initial denaturation	98°C	30 s
30 cycles of		
Denaturation	98°C	10 s
Annealing	60–65°C	20 s
Extension	72°C	30 s/kb
1 cycle of		
Final extension	72°C	5 min

3.1.5. Isolation of Designed Single-Site Mutagenesis Clones

1. Run 5–10 μl of PCR products on an agarose gel. No PCR product purification is necessary if the desired PCR products are greater 80% pure (see Note 3).
2. Digest PCR product with *Dpn*I to degrade circular plasmid template. Add 1 μl of *Dpn*I to the PCR product and incubate at 37°C for 30 min.

3. Mix the following components in a 1.5 ml reaction tube:

Reaction component	Volume (μl)
Nuclease-free water	40.5
10× T4 DNA ligase buffer	5
PCR product (>5 ng/μl)	3
T4 polynucleotide kinase	1
T4 DNA ligase (see Note 4)	0.5
Total	50

4. Mix well by pipetting gently up and down, and incubate at room temperature for 30 min.

3.1.6. Transformation

1. Thaw a tube of NEB5-α Competent *E. coli* cells on ice for 10 min. Add 3 μl of the phosphorylation/ligation mixture directly from step in Subheading 3.1.5 to the competent cells. Carefully flick the tube few times to mix cells and DNA. Do not vortex.
2. Place the mixture on ice for 30 min.
3. Heat shock at 42°C for 30 s.
4. Place on ice for 2 min.
5. Pipette 950 μl of room temperature SOC medium into the mixture.
6. Incubate in a 37°C shaker for 60 min.
7. Warm selection plates to 37°C.
8. Spread 100 μl onto a selection plate and incubate overnight at 37°C.

3.1.7. Isolate Plasmid DNA

1. Inoculate single colonies into individual culture tubes containing 5 ml LB. If the inverse PCR product was pure, then four colonies should be analyzed. If there were additional PCR products and the fragment of interest was not gel purified, then screen several colonies.
2. Incubate LB cultures with shaking (220 rpm) at 37°C.
3. Purify plasmid DNA according to kit manufacturer's protocol.
4. Sequence the plasmid insert to confirm engineered mutations.

3.2. Design of Multiple Site-Directed Mutagenesis

In this section, we outline an example to create three point mutations at the same time in the *lacZ* gene. We utilize the Gibson Assembly™ Master Mix, which contains enzymes capable of mediating the assembling several DNA fragments. The end result is that multiple gene alterations can be created simultaneously. Briefly, this approach includes four primary steps:

Table 1
Primers used in multiple site-directed mutagenesis

Primers	Primer sequence
LacZ-F1 (see Note 6)	TTTAAGAAGGAGATATACATATGACCATGATTACGGATTC (see Note 7)
LacZ-R1 (see Note 6)	CACATCTG**GAA**TTCAGCCTCCAGTACAGC (see Notes 8 and 9)
LacZ-F2	AGGCTGAA**TTC**CAGATGTGCGGCGAGTT
LacZ-R2	GGCCTGAT**GAA**TTCCCCAGCGACCAGAT
LacZ-F3	CTGGGGAA**TTC**ATCAGGCCACGGCGC
LacZ-R3	ACACTGAG**GAA**TTCCGCCAGACGCCA
LacZ-F4	TGGCGGAA**TTC**CTCAGTGTGACGCTCCC
LacZ-R4	TTTGTTAGCAGCCGGATCTCATTTTTGACACCAGACCAACT
T7-For	TGAGATCCGGCTGCTAACAAAG
T7-Rev	ATGTATATCTCCTTCTTAAAGTTAAACAAAAT

3.2.1. Primer Design

See Subheadings 3.1.1–3.1.3 for guidance on designing primers for point mutations, insertions or deletions, respectively.

1. Primers used in this multiple site mutagenesis example are described in Table 1. The corresponding regions of each primer in the pET21a–lacZ plasmid are shown in Fig. 6, where the gray bar is *lacZ* gene and the backbone is pET21a.

3.2.2. Fragment Generation

1. Prepare plasmid DNA with *lacZ* gene in pET21a vector.
2. 5 PCR reactions (50 μl each) were set up separately to create five fragments with the designed mutations using primer pairs of lacZ-F1/R1, lacZ-F2/R2, lacZ-F3/R3, lacZ-F4/R4, and T7-For/T7-Rev as depicted in Fig. 6.
3. Prepare five 0.2 ml PCR reaction tubes to amplify each fragment separately. Add the following components sequentially to the tube:

PCR component	50 μl PCR reaction	Final concentration
Phusion® Hot Start Flex 2× Master Mix	25 μl	1×
10 μM forward primer	2.5 μl	0.5 μM
10 μM reverse primer	2.5 μl	0.5 μM
Template DNA	1 μl	50 pg
Nuclease-free water	19 μl	

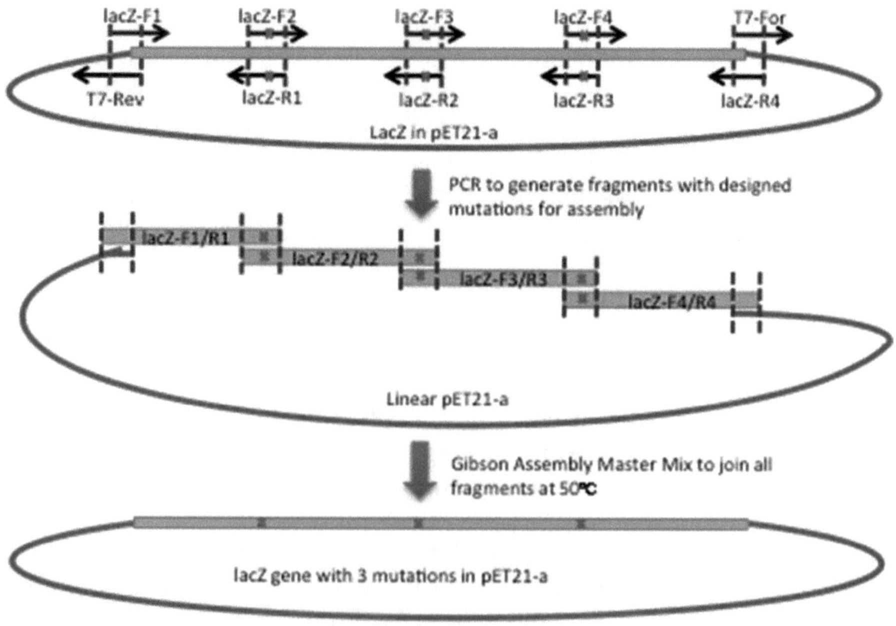

Fig. 6. The general scheme of multiple site-directed mutagenesis.

4. Gently mix the reaction. If necessary, collect all liquid to the bottom of the tube by a quick spin. Overlay the sample with mineral oil if using a PCR machine without a heated lid.
5. Transfer PCR tubes to a PCR machine and start the cycling program.

Cycling program:

Cycling step	Temperature	Time
1 cycle of		
Initial denaturation	98°C	1 min
30 cycles of		
Denaturation	98°C	10 s
Annealing	55°C	15 s
Extension	72°C	40 s[a]
1 cycle of		
Final extension	72°C	5 min

[a]Linear pET21a vector was PCR amplified using T7-For and T7-Rev as primers and with a modified program by changing the extension time at 72°C for 3 min instead of 40 s.

6. Add 1 μl of *Dpn*I to the PCR product and incubate at 37°C for 30 min. DNA templates from PCR fragments were purified using a QIAquick® PCR purification kit and eluted in 25 μl TE buffer.
7. Run 5 μl of PCR product on an agarose gel to confirm adequate DNA yield.

3.2.3. Fragment Assembly

1. Determine the concentration of each fragment using a Nanodrop spectrophotometer or estimate the amount from an agarose gel relative to the DNA ladder. With a 5-fragment assembly, we use 0.05 pmol of each fragment and adjust the total DNA volume to 10 μl (see Note 9).
2. Keep the dsDNA solution on ice. Thaw the 2× Gibson Assembly™ Master Mix at room temperature and vortex it thoroughly. Remove 10 μl of 2× Gibson Assembly Master Mix and mix it with the dsDNA fragments from step 1.
3. Transfer the PCR tube to a thermocycler and incubate at 50°C and for 1 h.
4. Gibson assembly samples can be stored at −20°C until the transformation step. Alternatively, the assembled product can be use as a template for PCR or rolling circle amplification.

3.2.4. Transformation

1. Thaw a tube of NEB5-alpha competent *E. coli* cells on ice for 10 min. Add 2 μl of assembly mix from step in Subheading 3.2.3 to the competent cells. Carefully flick the tube a few times to mix cells and DNA. Do not vortex.
2. Place the mixture on ice for 30 min.
3. Heat shock at 42°C for 30 s.
4. Place on ice for 2 min.
5. Pipette 950 μl of room temperature SOC medium into the mixture.
6. Leave the cells for recovery in the 37°C shakers for 60 min.
7. Warm selection plates to 37°C.
8. Spread 40–100 μl onto a selection plate and incubate overnight at 37°C.

3.2.5. Colony PCR (Optional)

1. Prepare 0.2 ml PCR tubes to analyze colonies for the presence of the desired DNA clone. Add the following components sequentially to the tube:

PCR component	50 μl PCR reaction	Final concentration
OneTaq Quick-Load 2× Master Mix	25 μl	1×
10 μM T7 sequencing primer	1 μl	0.2 μM
10 μM T7 terminator primer	1 μl	0.2 μM
Nuclease-free water	23 μl	

2. Set up ten individual PCR reactions. Pick single colonies (~1 mm diameter) with a pipette tip or toothpick and add each colony into a different PCR tube.

3. Colony PCR cycling protocol:

Cycling step	Temperature	Time
1 cycle of		
Initial denaturation	94°C	4 min
32 cycles of		
Denaturation	94°C	30 s
Annealing	55°C	30 s
Extension	68°C	3.5 min
1 cycle of		
Final extension	72°C	5 min

3.2.6. Isolate Plasmid DNA

1. Inoculate the 5 ml LB medium with a single *E. coli* colony using an inoculation loop.
2. Incubate with shaking (220 rpm) at 37°C.
3. Purify plasmid DNA according to manufacturer's protocol.
4. Sequence the plasmid DNA to confirm engineered mutations.

4. Notes

1. Standard desalting of the oligonucleotides is sufficient for the purpose of the outlined mutagenesis methods. Reverse-phase high-performance liquid chromatography (RP-HPLC) or polyacrylamide gel electrophoresis (PAGE) purification is not necessary.
2. 5′-Phosphorylation is not necessary if the T4 polynucleotide kinase reaction is carried out prior to or at the same time as the ligation step.
3. http://www.neb.com/nebecomm/tech_reference/TmCalc/Default.asp.
4. PCR products vary in purity, length, GC content, and secondary structure. Before adding T4 DNA ligase, gel analysis should be conducted on each mutagenesis reaction in order to determine the yield and specificity of the desired PCR product. A single mutagenesis reaction should then be selected and subsequently ligated and used for transforming an appropriate host. If necessary, one can purify the desired PCR product from an agarose gel after electrophoresis.
5. We use high-concentration T4 DNA ligase (2,000,000 units/ml).
6. LacZ-Fx: forward primer x; LacZ-Rx: reverse primer x.

7. Underlined nucleotides indicate the overlapping region where two fragments can join together in the presence of Gibson Assembly Master Mix.
8. Nucleotides in bold indicate the designed mutations, which result in an *Eco*RI site for restriction enzyme analysis.
9. If the mutation is located in the overlapping region, both primers should have the same mutations. Mutations can be located in the nonoverlapping region as long as there are enough matching nucleotides at the 3′ end to anneal to the template during PCR.
10. This website (http://www.endmemo.com/bio/dnacopynum.php) offers a tool to convert unit between "weight" and "mole."

Acknowledgments

The authors would like to thank Dr. Daniel Gibson for his advice on Gibson Assembly applications.

References

1. Gillam S, Astell CR, Smith M (1980) Site-specific mutagenesis using oligodeoxyribonucleotides: isolation of a phenotypically silent phi X174 mutant, with a specific nucleotide deletion, at very high efficiency. Gene 12:129–137
2. Zoller MJ, Smith M (1983) Oligonucleotide-directed mutagenesis of DNA fragments cloned into M13 vectors. Methods Enzymol 100:468–500
3. Scharf SJ, Horn GT, Erlich HA (1986) Direct cloning and sequence analysis of enzymatically amplified genomic sequences. Science 233:1076–1078
4. Hemsley A, Arnheim N, Toney MD, Cortopassi G, Galas DJ (1989) A simple method for site-directed mutagenesis using the polymerase chain reaction. Nucleic Acids Res 17:6545–6551
5. Higuchi R, Krummel B, Saiki RK (1988) A general method of in vitro preparation and specific mutagenesis of DNA fragments: study of protein and DNA interactions. Nucleic Acids Res 16:7351–7367
6. Ho SN, Hunt HD, Horton RM, Pullen JK, Pease LR (1989) Site-directed mutagenesis by overlap extension using the polymerase chain reaction. Gene 77:51–59
7. Ko JK, Ma J (2005) A rapid and efficient PCR-based mutagenesis method applicable to cell physiology study. Am J Physiol Cell Physiol 288:C1273–C1278
8. Landt O, Grunert HP, Hahn U (1990) A general method for rapid site-directed mutagenesis using the polymerase chain reaction. Gene 96:125–128
9. Ke SH, Madison EL (1997) Rapid and efficient site-directed mutagenesis by single-tube 'megaprimer' PCR method. Nucleic Acids Res 25:3371–3372
10. Williams M, Louw AI, Birkholtz LM (2007) Deletion mutagenesis of large areas in Plasmodium falciparum genes: a comparative study. Malar J 6:64
11. Zheng L, Goddard JP, Baumann U, Reymond JL (2004) Expression improvement and mechanistic study of the retro-Diels-Alderase catalytic antibody 10F11 by site-directed mutagenesis. J Mol Biol 341:807–814
12. Jones DH (1994) PCR mutagenesis and recombination in vivo. PCR Methods Appl 3:S141–S148
13. Gibson DG (2009) Synthesis of DNA fragments in yeast by one-step assembly of overlapping oligonucleotides. Nucleic Acids Res 37:6984–6990
14. Gibson DG, Young L, Chuang RY, Venter JC, Hutchison CA 3rd, Smith HO (2009) Enzymatic assembly of DNA molecules up to several hundred kilobases. Nat Methods 6:343–345

Chapter 14

Gene Assembly and Combinatorial Libraries in *S. cerevisiae* via Reiterative Recombination

Nili Ostrov, Laura M. Wingler, and Virginia W. Cornish

Abstract

While mutagenesis of single genes is now common practice in molecular biology, engineering multiple target genes still requires complex cloning techniques and thus is limited to expert laboratories. Here, we describe "Reiterative Recombination," a user-friendly DNA assembly technique in *Saccharomyces cerevisiae* for the integration of an indefinite number of DNA fragments sequentially into the yeast genome. The high efficiency of chromosomal integration can further be utilized for the assembly of large combinatorial libraries for metabolic engineering.

Key words: DNA assembly, Homologous recombination, Combinatorial libraries, Biosynthetic pathways, Homing endonuclease, *Saccharomyces cerevisiae*, Cell engineering

1. Introduction

Here we describe "Reiterative Recombination," a technically simple, robust method for in vivo DNA assembly on the chromosome of *Saccharomyces cerevisiae* (1). Harnessing the endogenous homologous recombination process in yeast, Reiterative Recombination offers high-efficiency DNA integration, making it especially attractive for nonexpert investigators in the field of yeast genetics. The protocol described below enables (1) cloning of multiple DNA fragments in yeast, with indefinite potential for elongation of the same construct (10s–100s of kbs), (2) construction of stable strains by integration of all DNA constructs (e.g., entire multicomponent pathways) into the yeast chromosome at a predefined location, and (3) cloning of large combinatorial DNA libraries.

James C. Samuelson (ed.), *Enzyme Engineering: Methods and Protocols*, Methods in Molecular Biology, vol. 978, DOI 10.1007/978-1-62703-293-3_14, © Springer Science+Business Media New York 2013

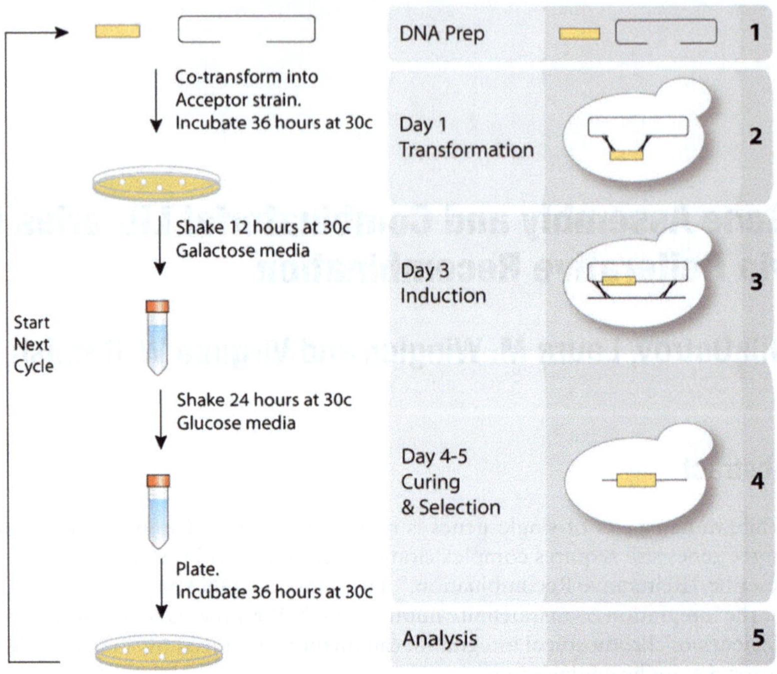

Fig. 1. Reiterative Recombination protocol overview. The DNA of interest (*yellow*) is incorporated into the chromosome using a donor plasmid in a 5-day cycle. Homologous recombination between DNA fragment, donor plasmid, and genomic DNA occurs entirely in vivo, requiring only the exchange of selective media.

Reiterative Recombination utilizes pairs of alternating, orthogonal endonucleases and selectable markers as well as the efficient process of in vivo endonuclease-stimulated homologous recombination to integrate and elongate a construct of interest. The system consists of two modules: a "donor" plasmid carrying the DNA to be assembled and an "acceptor" strain carrying a predefined target locus for assembly. As shown in Fig. 1, each cycle consists of four steps: (1) *DNA preparation*—the DNA fragment to be assembled is PCR amplified to add homology regions on both ends. (2) *Transformation*—co-transformation of the PCR product and a linearized donor plasmid into an acceptor strain. (3) *Induction*—galactose-induced expression of an endonuclease triggers homologous recombination, resulting in integration of the PCR product and an auxotrophic marker into the acceptor strain chromosome. (4) *Curing and selection*—cells in which successful integration events occurred are isolated by selecting for the new auxotrophic marker. The acceptor strain is cured from excess donor plasmid by counter selection, and recombinants containing the correct construct are used as acceptor strain for the next cycle of elongation. A single cycle takes 5 days to complete. Importantly, the use of only two recyclable auxotrophic selective markers (histidine and leucine metabolic genes, Fig. 2) provides a generic selection independent of the nature of DNA that is being assembled. At each cycle of assembly, one or

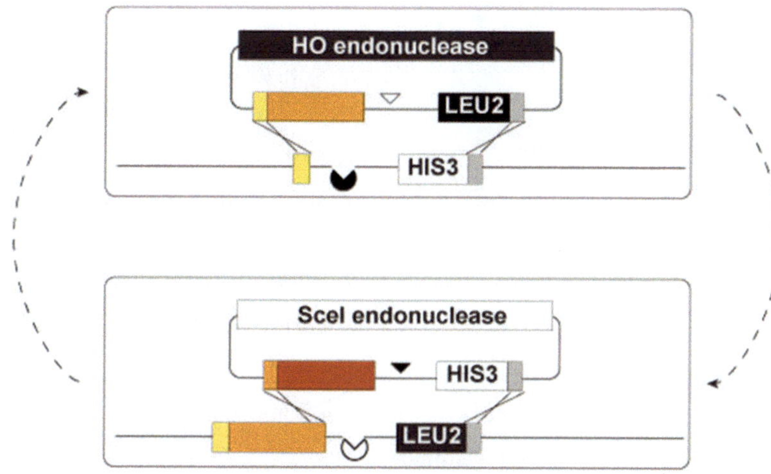

Fig. 2. General scheme of Reiterative Recombination, showing endonucleases and selective markers for each cycle of assembly. Assembled DNA fragments are in (*orange*) and (*brown*). *Triangles* indicate endonuclease cleavage site. Generic homology sequences are indicated in gray.

more DNA fragments can be integrated, and the order of fragment assembly is determined by the design of appropriate homology on both ends of the fragments that is incorporated (see Fig. 3).

Each cycle of Reiterative Recombination assembly yields ~10^5/mL recombinant colonies. Considering yeast transformation (Fig. 1, step 2 and methods section 3.2 below) should yield ~10^6–10^8 cells, the efficiency of DNA integration with Reiterative Recombination is estimated to be 1–10% of all transformed, induced cells. It is important to note, however, that the number of recombinant colonies drops significantly if one attempts to co-transform multiple DNA fragments in a single cycle.

Significantly, all that is required to adapt Reiterative Recombination as a tool for library mutagenesis is to use a pool of mutagenized DNA fragments as opposed to a single DNA sequence during the transformation step (figure 1, step 2). At the end of the cycle each of the ~10^5/mL cells will contain a different DNA mutant integrated at the acceptor site in the chromosome. The mutant displaying the desired phenotype can be isolated using a suitable screen or selection.

While generic Reiterative Recombination acceptor strains are readily available (1), any strain can be converted to be an acceptor for DNA assembly. Two cloning steps are necessary, both performed using classical integration vectors (see Subheading 3 below): (1) knock out of the endogenous HO endonuclease cleavage site on the acceptor strain, to avoid unwanted double-strand breaks during assembly and (2) integration of the "acceptor module" at the HO endonuclease gene locus of the strain. It is important to note that the acceptor strain must have *his3*, *trp1*, *ura3*, and *leu2* auxotrophies (Table 1).

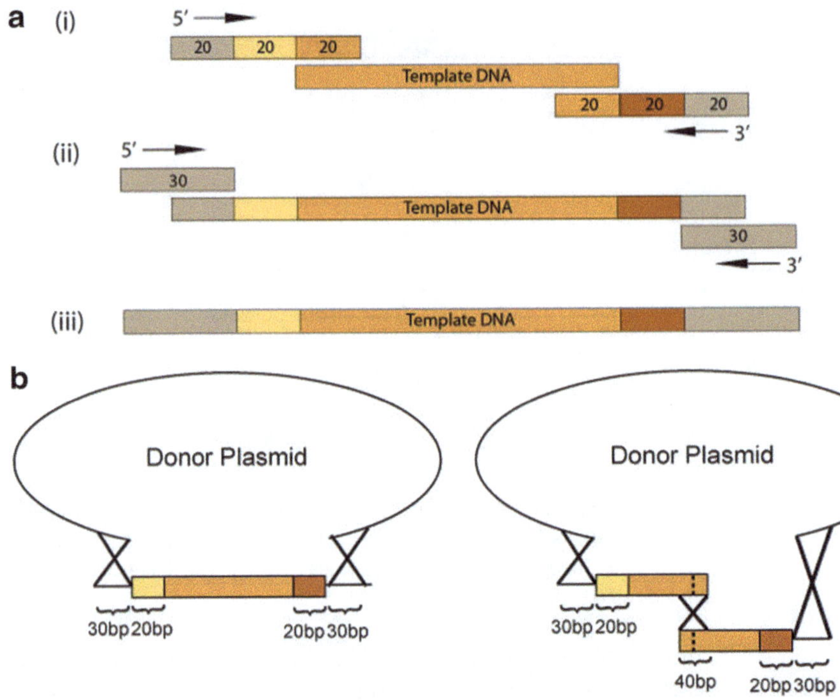

Fig. 3. Primer design. (**a**) PCR overview. Each DNA fragment is added with homology to allow its recombination with the donor plasmid and other fragments previously assembled on the chromosome. (i) Primers are designed to add homology to preceding assembly fragments (*yellow*), next assembly fragment (*brown*), and donor plasmid (*gray*) see 'PCR1' in Table 4 and Note 1–2. (ii) Generic primers extend the homology to donor plasmid and make the final fragment (iii) see 'PCR2' in Table 4, Note 1–3. (**b**) Assembly of DNA fragments with a donor plasmid. Single (*left*) or multiple (*right*) DNA fragments can be used at each Reiterative Recombination cycle, provided each is added with 30–40 bp of overlapping sequence.

Table 1
Plasmid list

Plasmid	Description
pLMW2594	Donor plasmid—cycle 1 (integrating selective marker: leucine)
pLMW2593	Donor plasmid—even cycles (integrating selective marker: histidine)
pLMW2592	Donor plasmid—odd cycles (integrating selective marker: leucine)
pLMW2588	Integration plasmid containing non-cleavable allele "*MATa-inc*" to be used in 'a' acceptor strain
pLMW2586	Integration plasmid containing non-cleavable allele "*MATα-inc*" to be used in 'alpha' acceptor strain
pLMW2590	Integration plasmid for acceptor module

2. Materials

All solutions should be prepared using deionized water and sterile techniques. All reagents should be prepared and stored at room temperature (unless indicated otherwise).

2.1. Yeast Media Components

1. Amino acid mix (500 mL, not including histidine, tryptophan, leucine, and uracil): Weigh and combine the amino acids listed in Table 2 (except for aspartic acid and threonine) with 500 mL water in a 1 L flask. Autoclave 15 min. When cooled to about 50°C, add aspartic acid and threonine. Using sterile technique, filter sterilize into autoclaved bottles.
2. 40% Glucose (500 mL): Autoclave 300 mL water in a 1 L flask. When cooled to about 50°C, add 200 g glucose and shake or stir until fully dissolved (keep sterile). Add autoclaved water to reach 500 mL total volume. Filter sterilize into autoclaved bottles.
3. 30% Galactose (500 mL): Same protocol as glucose, using 150 g galactose.
4. 20% Raffinose (500 mL): Same protocol as glucose, using 100 g raffinose.

Table 2
Amino acid mix (no histidine, tryptophan, uracil, leucine)

Compound name	Amount for 500 mL (g)
Adenine sulfate	0.2
Arginine HCl	0.2
*Aspartic acid	1.0
Glutamic acid	1.0
Isoleucine	0.3
Lysine HCl	0.3
Methionine	0.2
Phenylalanine	0.5
Serine	4.0
*Threonine	2.0
Tyrosine	0.3
Valine	1.5

*Amino Acid added after autoclave. See section 2.1

Table 3
SC (synthetic complement) media[b]

Component	For 250 mL media (mL)
SD (synthetic-defined) media	125
40% glucose[a]	12.5
Amino acid mix (HUTL-)	12.5
1% Histidine	0.5
1% Tryptophan	0.5
0.2% Uracil	2.5
1% Leucine	2.25
Sterile H_2O (for liquid) or 4% agar (for plates)	250

[a]Glucose is replaced with galactose and raffinose for SC–galactose media (see 2.1.9 below)
[b]For counter selection (SC-FOA) plates see section 2.18

5. Amino acids: For the following amino acid solutions, add the indicated amino acid into 200 mL water in 250 mL bottle and autoclave. 1% tryptophan, 2 g tryptophan; 0.2% uracil, 0.4 g uracil; 1% histidine, 2 g histidine; 1% leucine, 2 g leucine.
6. 4% Agarose: Mix 8 g Bacto-agar with 200 mL water. Autoclave. Microwave before use.
7. SD (synthetic-defined) media (YNB w/ammonium sulfate, 500 mL): Dissolve 6.7 g in 500 mL water. Autoclave.
8. Yeast SC-glucose (synthetic complete) dropout liquid media: Mix components as described in Table 3 using histidine/tryptophan/leucine/uracil, as required. Use autoclaved water in the last step. For 0.1% FOA plates, add 0.1% g/volume 5-FOA (5-fluoroorotic acid) to media components; shake at 37°C until fully dissolved. Filter sterilize before use.
9. Yeast SC–galactose dropout liquid media: Mix components as described in Table 3 replacing the 40% glucose solution with 16.6 mL 30% galactose and 25 mL 20% raffinose. Add sterile water to 250 mL.
10. Yeast SC-glucose (synthetic complete) dropout plates: Same as dropout liquid media above, using 4% agar in the last step. (Dissolve in microwave if solid).
11. 15 mL culture tubes.
12. Sterile inoculation loops.
13. QIAquick Gel Extraction Kit.
14. ZymoPrep Genomic DNA Extraction Kit.

2.2. Yeast Transformation Components

1. E-buffer (250 mL, keep at 4°C): Weigh 46.1 g sucrose and 0.78 g Tris-HCl. Add to 500 mL sterile water and stir to dissolve. Adjust pH to 7.5. Add $MgCl_2$ to final concentration 1 mM and filter sterilize before use. (Final concentration: 10 mM Tris-HCl pH 7.5, 270 mM sucrose, 1 mM $MgCl_2$).
2. DTT (1,4-dithiothreitol) solution: Prepare 1 M Tris-HCl, pH 8. Immediately before use, dissolve DTT to a final concentration of 2.5 M, vortex, and filter sterilize.
3. Yeast complete media (YPD): Weigh 10 g Bacto-Yeast extract and 20 g Bacto-peptone. Add 500 mL water in 1 L bottle. Autoclave. When cooled, add 25 mL of 40% glucose solution (*see* Subheading 2.1 above).
4. Pellet Paint co-precipitant—for concentrating DNA samples (Novagen).
5. Autoclaved 250 mL flask.
6. Electroporation cuvettes.

3. Methods

Carry out all procedure using sterile technique (2, 3). Fig. 1 depicts each step.

3.1. Preparation of DNA for Assembly

It is critical to follow the primer design instructions for PCR1 (see Note 1 and Fig. 3a).

3.1.1. PCR of DNA Fragments for Assembly

1. Prepare PCR1 mix (see Note 1 and Fig. 3): Add 0.5–1 μL template DNA (containing your fragment of interest), 0.2 μL 5′-primer (100 μM stock), 0.2 μL 3′-primer (100 μM stock), 0.5 μL VENT polymerase, 0.5 μL dNTP mix, and 5 μL ThermoPol buffer, and add sterile water to final volume of 50 μL. PCR amplify (see Note 2). Run the PCR product on agarose gel; verify the expected fragment size, and purify the fragment from gel.
2. Prepare PCR2 mix (see Fig. 3a and Note 3): 1 μL of purified PCR1 template, 0.2 μL of 5′-PCR2 primer (100 μM stock), 0.2 μL of 3′-PCR2 primer (100 μM stock), 0.5 μL VENT polymerase (NEB), 0.5 μL dNTP mix (NEB), and 5 μL ThermoPol buffer (NEB) and add sterile water to final volume of 50 μL. PCR amplify and purify the resulting fragment from gel.
3. Measure the DNA concentration of all purified PCR products.

3.1.2. Linearization of Donor Plasmid

The donor plasmid for the first cycle, pLW2594, is used below. For each assembly cycle, choose the suitable donor plasmid and its respective restriction enzyme using Table 4. Note that odd and even donor plasmids are alternated for all cycles except for the first cycle.

Table 4
Cycle-specific DNA preparation

Cycle	PCR1: Add sequence to 5′ end of 5′-primer	PCR1: Add sequence to 5′ end of 3′-primer	PCR2: 5′-primer	PCR2: 3′-primer	Donor plasmid	Choose restriction enzyme
Cycle 1	5′-AAAATTGTGCCTTTGG ACTTAAAATGGCGT	5′-CTTAGGGATAA CAGGGTAAT	No PCR2 primers required. (Use PCR1 primer)	LW367	pLW2594	*Sma*I, *Xma*I, *Tsp*MI, *Bso*BI, *Ava*I. (1,500 bp KanMX gene will be lost upon digestion)
Even cycle	5′-GGACGCTCGAAGGCTTT	5′-CTGTTGCGGAA AGCTGAAA	LW374	LW375	pLW2593	*Sph*I, *Sal*I, *Tsp*MI, *Xma*I, *Sma*I, *Eco*53kI, *Sac*I, *Eco*RI
Odd cycle	5′- GGACGCTCGAAGGCTTT	5′-CTTAGGGATAA CAGGGTAAT	LW374	LW367	pLW2592	*Hind*III, *Bsa*BI, *Not*I, *Eag*I, *Ale*I, *Eco*53kI, *Sac*I

1. Mix 1–2 μg donor plasmid pLW2594, 2 μL of restriction enzyme SmaI, 5 μL of NEB buffer 4, and sterile water to final volume of 50 μL in a microcentrifuge tube. Incubate at 25°C for 6 h.
2. Run the products on agarose gel (compare to a non-digested plasmid sample control). Purify the linearized plasmid DNA from gel.
3. Measure the DNA concentration of the purified linear plasmid.

3.2. Yeast Transformation

We use a high-efficiency yeast electroporation protocol with slight modifications (4). The media described below is specific for the first cycle of DNA assembly in the parent acceptor strain (1). Ensure that the appropriate media is used for the particular cycle of assembly (see Table 5). The resulting colonies after this step will be referred to as "transformants."

1. Day before transformation: Use a patch of the acceptor strain on SC-glucose(H-) to inoculate a 5 mL SC-glucose(H-) liquid culture in 15 mL culture tube. Shake at 30°C overnight.
2. Day of transformation: Measure OD_{600} of the acceptor strain culture.
3. Inoculate 100 mL YPD media using the overnight culture to reach final $OD_{600}=0.1$. Grow cells in 30°C shaker until $OD_{600}\approx 0.8$ (approx. ~7 h). While waiting, prepare DNA mix to be transformed (see Note 4) and concentrate DNA mix to 4 μL using Pellet Paint co-precipitant (see Subheading 2).
4. When culture reaches $OD_{600}\approx 0.8$, add 1 mL filter-sterilized DTT solution (see Subheading 2) and incubate for another 20 min in 30°C shaker.
5. *From now on, try to keep the cells on ice and work quickly to maintain the cell competence.* Harvest the cells (5 min, 2,000×*g*, 4°C, in two 50 mL Falcon tube swinging bucket centrifuge). Discard the supernatant and resuspend cells using 25 mL cold E-buffer (see Subheading 2) by pipetting. Harvest the cells again.
6. Discard the supernatant and resuspend cells in 1 mL of cold E-buffer and transfer them to a sterile 1.5 mL microcentrifuge tube. Harvest the cells (1 min, 20,000×*g*).
7. Discard the supernatant and resuspend the cells using 40 μL cold E-buffer (see Note 5). Aliquot 60 μL of cells into sterile 1.5 mL microcentrifuge tubes containing the DNA and donor plasmid for assembly. Mix by gently pipetting, and incubate for 10 min on ice.

Table 5
Cycle-specific media

	Transformation (overnight culture)	Transformation media	After electroporation	Induction media	Curing		Inoculate next cycle
	(liquid)	(liquid)	(plates)	(liquid)[a]	(liquid)	(plates)	(liquid)
Cycle 1	[b]SC-glucose/H-	SC-glucose/H-	SC-glucose/HU-	SC–galactose/U-	SC-glucose/L-	SC-glucose/L- 0.1% FOA	SC-glucose/L-
Even cycle	SC-glucose/L-	SC-glucose/L-	SC-glucose/LU-	SC–galactose/U-	SC-glucose/H-	SC-glucose/H-0.1% FOA	SC-glucose/H-
Odd cycle	SC-glucose/H-	SC-glucose/H-	SC-glucose/HU-	SC–galactose/U-	SC-glucose/L-	SC-glucose/L-0.1% FOA	SC-glucose/L-

[a]It is highly recommended to perform a negative control by culturing in SC–glucose (noninductive), to determine the induction efficiency with galactose.
[b]"SC–glucose" media refers to glucose media, unless indicated otherwise

8. Set the gene pulser to 25 μF, 540 V, 0 Ω. Transfer the cells to a 0.2 cm electroporation cuvette and pulse (see Note 6). Immediately after electroporation, add 1 mL warm YPD and let shake at 30°C for 1 h.
9. Plate 1:10 and $1{:}10^3$ dilutions (see Note 7) on selective plates SC-glucose(HU-). Incubate plates at 30°C for 2 days or until colonies are visible (see Note 8).

3.3. Induction of DNA Integration into the Chromosome

The resulting colonies after this step are referred to as "induced recombinants."

1. Using a sterile inoculation loop, pick 5–10 transformant colonies from SC-glucose(HU-) plate and resuspend them in 1 mL sterile water in 1.5 mL microcentrifuge tube (see Note 9). Spin down in tabletop centrifuge (5 min at 20,000×*g*) and carefully discard the supernatant (a cell pellet should be clearly visible, otherwise repeat this step by picking more colonies).
2. Resuspend cell pellet in 1 mL sterile water; aliquot 500 μL of cell suspension into two sterile microcentrifuge tubes. One tube is used as a negative control and labeled "Glucose (control)" and the other for induction, labeled "Galactose." Spin down both samples (5 min at 20,000×*g*) and carefully discard the supernatant. Cell pellet should be visible.
3. Resuspend the "Glucose" control sample with 1 mL of SC–glucose(U-) and the "Galactose" sample with SC–Galactose(U-). Transfer each 1 mL sample into a 15 mL culture tubes and incubate in 30°C shaker for 12 h (see Note 10).
4. Optional: To evaluate the efficiency of the induction process, plate 100 μL of each cell culture immediately after 12 h induction on SC-glucose(L-) plates, incubate at 30°C, and compare the number of colonies (see Note 11). Otherwise, skip straight to curing step.

3.4. Curing and Selection for Recombinants

The resulting colonies after this step are referred to as "cured recombinants."

1. To cure cell from donor plasmid, take 100 μL of induced cell culture from SC–galactose(U-) and add 900 μL SC-glucose(L-) media in a 15 mL culture tube. Incubate the samples for 24 h in 30°C shaker (see Note 12).
2. After 24 h, take 100 μL of the cured sample into 900 μL sterile water, mix, and plate 100 μL on SC-glucose(L-/0.1% FOA) selective plates. It is recommended to plate several dilutions, as necessary (see Notes 13 and 8). Incubate at 30°C for 2–3 days or until colonies are visible.

Table 6
Primer sequences

Primer name	Priming location	Sequence
VC1052	Reverse primer, CYC terminator	GGGACCTAGACTTCAGGTTG
VM172	Reverse primer, end of HO gene	TTAGCAGATGCGCGCAC
LMW274	Reverse Primer, 76 bp after end of MAT locus	CATTTGTCATCCGTCCCGTATA
LMW308	Downstream of assembly construct	CAGCCGAACGACCGAGCGCAGCG AGTCAGTGATCTAGAATGTC TAAAGGTGAAGAATTAT
LMW309	Downstream of assembly construct	GACAACACCAGTGAATAATTCTTCAC CTTTAGACATTGTGATGATGTTTTA
LMW317	Forward primer, upstream of HO gene (1820bp prior to start codon) HO gene	CTTTGGACTTAAAATGGCGT
LMW318	Reverse primer, 1200bp into HO gene ORF HO gene	GTGAAGTTGTTCCCCCAG
LMW319	Internal MATa primer	TTAGAAGAAAGCAAAGCCTTA
LMW320	Internal MATα primer	CCTGTTCCTTCCTCTCGA
LMW367	Reverse primer, Downstream of integrated fragment for odd cycles	TCAGTACAATCTTAGGGATAACA GGGTAAT
LMW374	Forward Primer, Upstream of integrated fragments for even and odd cycles	TGAGAAGGTTTTGGGACG CTCGAAGGCTTT
LMW375	Reverse Primer, Downstream of integrated fragment for even cycles	GCACAGTTATACTGTTGCGGAAAG CTGAAA

3.5. Analysis of Assembly Products

1. Patch a few cured recombinant colonies on SC-glucose(L-/0.1% FOA) plate and incubate plates 2–3 days in 30°C incubator (to produce more cells for analysis).
2. Perform colony PCR (see step 5 of Subheading 3.7.1 for protocol and Table 6 for primers. see Note 14)
3. Analyze the resulting DNA fragment on an agarose gel to verify the correct assembly of DNA.

3.6. Starting the Next Cycle of Assembly

1. Pick a single colony from the previous cycle (confirm by PCR or sequencing) and inoculate 5 mL SC-glucose(L-) media in a 15 mL culture tube. Incubate in 30°C shaker overnight, and use as acceptor strain for the next cycle of assembly (starting Subheading 3.1 above). (For libraries assembly, see Note 15).

3.7. Construction of Acceptor Strain (This Step Is Optional, See Note 16)

3.7.1. For MATa Strains: Replace MATa Locus with Non-cleavable Allele "MATa-inc"

1. Mix 1–2 μg plasmid pLW2588, 2 μL NEB Buffer 3 and 2 μL restriction enzyme BglII (total volume of 20 μL). Incubate the reaction at 37°C for 3 h.
2. Purify the resulting linear plasmid from agarose gel (see Subheading 2) and determine DNA concentration.
3. Plasmid "pop-in"—transform >500 ng of purified linear plasmid (see Subheading 3.2 above). Plate the transformants on SC-glucose(U-) plates and incubate at 30°C for 2 days.
4. Plasmid "pop-out"—use one or more "pop-in" colonies colony to inoculate 5 mL YPD media, shake at 30°C for 1 day. Dilute the culture (1:100–1:1,000) and plate on SC-glucose(0.1% FOA) plates. Incubate at 30°C for 2 days (see Note 17).
5. Plasmid "pop-out" analysis by *colony PCR*—patch a few of the colonies and perform colony PCR to confirm correct DNA integration: Using an inoculation loop or sterile tip, pick some of the patch cells and suspend them in 30 μL 0.2% SDS solution in a PCR tube. Boil the sample for 5 min at 95°C in a PCR block. Spin down the sample 10 min at 2,000×*g* (swinging bucket centrifuge) and incubate 30 min at RT.
6. Perform PCR: Mix 1–2 μL colony/SDS supernatant, 0.2 μL primer LMW274 (100 μM), 0.2 μL primer LMW319 (100 μM), 0.2 μL VENT polymerase, 0.2 μL dNTPs, 4 μL polymerase buffer, and 13.4 μL sterile water. Expected band size is 650 bp (see Table 6 for primer sequences).
7. Digest PCR product: Mix 10 μL PCR product with 1.5 μL Buffer 3, 1 μL restriction enzyme AciI, and 2.5 μL sterile water. Incubate at 37°C for 3 h. Run on agarose gel. Correct integration of *MATa-inc* allele results in digestion product 357 and 293 bp. Wild-type *MATa* results in digestion products 293, 225, and 132 bp.

3.7.2. For MATα Strains: Replace MATα Locus with Non-cleavable Allele "MATα-inc"

1. Mix 1–2 μg plasmid pLW2586, 2 μL NEB buffer 2, 2 μL BSAx 10, and 2 μL restriction enzyme NheI (total volume of 20 μL). Incubate the reaction at 37°C for 3 h.
2. Purify the resulting linear plasmid from agarose gel by kit (see Subheading 2) and determine DNA concentration.
3. Plasmid "pop-in"—transform >500 ng of purified linear plasmid into the strain (see Subheading 3.2 above). Plate the transformants on SC-glucose(U-) plates and incubate at 30°C for 2 days.
4. Plasmid "pop-out"—use one colony to inoculate 5 mL YPD media, shake at 30°C for 1 day. Dilute the culture (1:100–1:1,000) and plate on SC-glucose(0.1% FOA) plates. Incubate at 30°C for 2 days (see Note 17).
5. Plasmid "pop-out" analysis—patch a few of the colonies and perform *colony PCR* primers LMW274 and LMW320 to

confirm correct DNA integration (see step 5 of Subheading 3.7.1 for colony PCR protocol). Expected band size is 675 bp (see Table 6 for primer sequences).

6. Digest PCR product: Mix 10 μL PCR product with 1.5 μL buffer 4, 1.5 μL BSA buffer 10 (NEB), 1 μL restriction enzyme HhaI, and 1 μL sterile water (total volume 15 μL). Incubate at 37°C for 3 h. Run on agarose gel. Correct integration of *MATα-inc* allele results in 675 bp (PCR is not digested). Wild-type *MATα* results in digestion products 423 and 252 bp.

3.7.3. Integration of Acceptor Module at HO Gene Locus

1. Mix 1–2 μg plasmid pLW2590, 3 μL NEB buffer 4, and 2 μL restriction enzyme SpeI (total volume 20 μL). Incubate the reaction at 37°C for 3 h.
2. Purify the resulting linear fragment from agarose gel and determine DNA concentration.
3. Transform 2 μg of purified linear plasmid into the acceptor strain (see Subheading 3.2). Plate the transformants on SC-glucose(H-) plates, and incubate at 30°C for 2 days.
4. Plasmid integration analysis: patch a few of the colonies and perform colony PCR (see step 5 of protocol section 3.7.1 above) using primers LMW309 and LMW318. Expected band for successful integration is 913 bp. No amplification if wild type (i.e., if integration unsuccessful). For further verification, purify the genomic DNA (using kit, see Subheading 2) and PCR amplify using the following primers: (1) primers LMW309 and LMW319, expected band for successful integration is 913 bp. No amplification if unsuccessful. (2) Primers LMW309 and VM172, expected band for successful integration is 1,458 bp. No amplification for unsuccessful. (3) Primers LMW317 and LMW308, expected band for successful integration is 1,710 bp. No amplification if unsuccessful.
5. After acceptor module integration was confirmed by restriction analysis (preferably by sequencing as well), inoculate the confirmed strain for use in the first cycle of DNA assembly.

4. Notes

1. Primer design is critical for successful DNA assembly, as the integration of new fragments is guided in vivo solely by homologous sequences. Each DNA fragment is amplified using primers that add short regions of homology (30–40 bp) on both ends (a) to the preceding and next piece of the growing assembly and (b) to the donor plasmid.

To design primers for PCR1, start by determining the sequences for PCR amplification of your template of interest (~20 bp). Then, as shown in Fig. 3a(i), add the 5′ end of your 5′-primer with 20 bp of overlap with the preceding DNA assembly fragment (i.e., last 20 bp of the last fragment assembled in the previous cycle of assembly). Similarly, add the 5′ end of your 3′-primer with 20 bp of overlap with the next DNA assembly fragment (i.e., the first 20 bp of the next fragment to be inserted in the next cycle of assembly). Lastly, to add donor plasmid homology ("gray" sequences in Fig. 3a), add the sequences specified in Table 4 at the respective ends. Perform PCR1. PCR1 products are then purified and used as template for PCR2, to add further homology to the donor plasmid. The primers for PCR2 are generic and do not require any adjustments, and they are specified in Table 4. *We want to emphasize that each cycle requires specific pairs of PCR1 primers unique to the specific pathway you desire to integrate into the yeast chromosome and that PCR1 primers change at every cycle for each new DNA fragment.*

2. Both annealing temperature and elongation time should be adjusted according to the Tm of each primer pair and the length of the template DNA, respectively. (Suggested PCR cycle—using the following cycle: 95°C for 5 min, 95°C for 30 s, 50°C for 30 s,72°C for 2 min, repeat steps 2–4 29 times, 72°C for 10 min). In case multiple fragments are assembled in a single cycle, PCR1 should be performed to add homology to each fragment.
3. PCR2 primers add homology to the donor plasmid. Therefore, in case multiple DNA fragments are assembled in a single cycle (see Fig. 3b), PCR2 should only be performed on the outermost 5′ and 3′ fragments of PCR1.
4. We recommend using a molar ratio of 1:100 plasmid: PCR product for yeast transformation. In case the volume of total DNA to be transformed (donor plasmid + PCR fragments) is higher than 10 μL—pellet the DNA using Pellet Paint co-precipitant and resuspend pellet in 4 μL (see Subheading 2).
5. This last resuspension step aims to form a paste of cells, rather than a liquid cell culture, for electroporation. Thus it might be necessary to use less E-buffer at this stage to achieve the required consistency, depending on the amount of cells.
6. Electroporation time constant should be ~18 ms or lower.
7. Dilution 1:10 means plating 100 μL of 1 mL YPD culture. $1:10^2$ means plating 100 μL of a 1:10 diluted 1 mL culture, etc.
8. Cultures can be kept at 4°C for up to 2 days and replated, in case a different dilution is required.

9. The number of colonies used for induction depends on the size of the colonies; therefore in the case of very small colonies, it is recommended to use more colonies or wait 1–2 more days of plate incubation. Cell pellet should be visible after spinning down.
10. In our hands, Induction works better if it is initialized at exponential phase. Therefore, if the induction culture seems to have a high OD (visibly white and cloudy)—it is recommended to dilute the sample 1:10 (100 μL sample + 900 μL appropriate media to final OD_{600} = ~0.1–0.2) to improve recombination efficiency.
11. For control purposes, we recommend plating a sample of the induced culture immediately after 12 h of induction on selective plates with the opposite marker (histidine/leucine) and with *no FOA*. This allows better observation of the difference between induced (galactose) and uninduced (glucose) samples than if plated after the curing step.
12. In our hands, a 1-day period was found to be sufficient for curing a significant number of cells of their donor plasmid. However, the curing period can be extended if necessary.
13. In order to get single colonies of cured recombinants, it might be necessary to plate higher dilutions. We recommend plating 1:10, 1:10^3, 1:10^5 dilutions the first time and adjust as necessary.
14. In case colony PCR does not work, especially for long DNA fragments (above 2 kb), purify genomic DNA (see Subheading 2) and use it as template for PCR analysis.
15. *Assembly of DNA Libraries*: In the case of DNA libraries, a pool of excess mutant DNA is co-transformed with the donor plasmid, and each recombinant cell encodes a unique DNA mutant at the end of the assembly cycle:
    - Mutagenesis technique—any technique can be used to construct the library, as long as all DNA variants carry homology on both sides to enable assembly by Reiterative Recombination, as described for standard non-library assembly. Common methods for DNA mutagenesis include error-prone PCR and DNA shuffling.
    - Tips for improved transformation—since library experiments require a large number of cells (i.e., a large number of transformant colonies) to cover the total number of variants, it is critical to carefully execute the yeast transformation step in order to maintain high number of transformants: Use fresh patch to inoculate the cultures, prewarm YPD media to 30°C prior to sample inoculation,

and perform the protocol quickly, keeping the cells on ice as much as possible.

- Recover more recombinants—in order to carry all library variants to the next cycle of assembly, use a large (150 mm) plate in the last curing step to recover more recombinant colonies. Then, scrape the entire plate by spreading ~10 mL sterile water and gently lifting all the colonies into a microcentrifuge tube. Use the entire pool of colonies as an acceptor strain for the next cycle of assembly.

16. The integration of acceptor module is optional, since a genetic acceptor strain is available (1). Integration of acceptor module should be performed in cases where a specific host strain is required. The generic acceptor strain carries the acceptor module at the HO gene locus, knocking out the function of the endogenous HO endonuclease gene. The acceptor strain also carries non-cleavable allele at the MATa locus.
17. It is highly recommended to perform the pop-out step for multiple colonies of the previous step, to increase the chances of isolating a successful integration event.

Acknowledgments

This research was partially supported by National Institutes of Health (NIH) grant R01 GM062867 and GM96064, NSF grant CHE-09-5756.

References

1. Wingler LM, Cornish VW (2011) Reiterative recombination for the in vivo assembly of libraries of multigene pathways. Proc Natl Acad Sci USA 108:15135–15140
2. Adams A, Gottschling D, Kaiser C et al (eds) (1998) Methods in yeast genetics. Cold Spring Harbor Laboratory Press, Plainview, NY
3. Ausubel F, Brent R, Kingston R et al (1995) Current protocols in molecular biology. Wiley, New York
4. Wittrup KD (2007) http://openwetware.org/wiki/Wittrup:_Yeast_Transformation Accessed 1 Feb 2012

Chapter 15

Promiscuity-Based Enzyme Selection for Rational Directed Evolution Experiments

Sandeep Chakraborty, Renu Minda, Lipika Salaye, Abhaya M. Dandekar, Swapan K. Bhattacharjee, and Basuthkar J. Rao

Abstract

Error-prone PCR, DNA shuffling, and saturation mutagenesis are techniques used by protein engineers to mimic the natural "evolutionary walk" that conjures new enzymes. Rational design is often critical in efforts to accelerate this "random walk" into a "resolute sprint." Previous work by our group established a computational method for detecting active sites (CLASP) based on spatial and electrostatic properties of catalytic residues, and a method to quantify promiscuous activities in a wide range of proteins (PROMISE). Here, we describe a rational design flow (DECAAF) based on the PROMISE methodology to choose a protein which, when subjected to minimal mutations, is most likely to mirror the scaffold of a desired enzymatic function. Modeling the diversity in catalytic sites and providing precise user control to guide the search is a key goal of our implementation. The flow details have been worked out in a real-life example to select a plant protein to substitute for human neutrophil elastase in a chimeric antimicrobial enzyme designed to bolster the innate immune defense system in plants.

Key words: Directed evolution, Computational biology, Promiscuous active sites, Active site prediction, Finite difference Poisson Boltzmann, Proteases, Pathogenesis-related protein, Human neutrophil elastase

1. Introduction

The remarkable ability of enzymes to selectively catalyze chemical reactions between compounds from the cellular soup is essential for the proper functioning of most pathways in biological systems (1, 2). Promiscuity, the catalysis of reactions distinct from the one the protein has evolved to perform using the same domain, is an apparent contradiction to this specificity (3–5). Jensen boldly hypothesized that in pristine life, enzymes were few and promiscuous (6) and that subsequent gene duplication and specialization

James C. Samuelson (ed.), *Enzyme Engineering: Methods and Protocols*, Methods in Molecular Biology, vol. 978,
DOI 10.1007/978-1-62703-293-3_15,

formed the basis of the evolution of complex organisms (7, 8). This hypothesis provides a rationalization for the specificity/promiscuity dichotomy.

Directed evolution is a generic term for methods that mimic and accelerate this natural process by random mutations (9), in vitro recombination (10–13), continuous evolution (14, 15), and high-throughput screening (16). The fact that mutations near the active site are likely to yield faster results (17) has led to more directed approaches (18–20). Rational design takes the logical next step of identifying preexisting scaffolds in proteins (21–25), which are often secondary activities under neutral drift (26–28). Other de novo approaches have also succeeded in endowing function in proteins from scratch (29–33). These developments have been reviewed recently at length (34–37).

We previously demonstrated conservation of electrostatic potential differences in cognate pairs of residues in a wide range of related proteins (38). We supplemented this electrostatic correlation with spatial information to establish a computational method (CLASP) to detect active sites (38) (Fig. 1) and quantified promiscuous activities in a wide range of proteins (PROMISE) (39). We showed that active site residues in a heme cytochrome C peroxidase are a good spatial and electrostatic match with residues in the active site of a Zn^{2+} carboxypeptidase A (Table 1). This existing scaffold can provide a starting point for engineering heme-binding

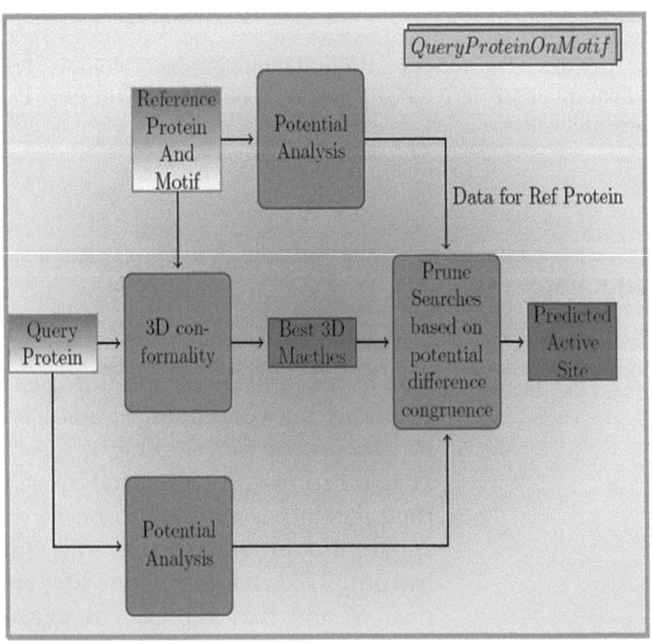

Fig. 1. CLASP Flow: A motif from a protein with known function and 3D structure is used to detect a similar scaffold in an unknown protein based on spatial and electrostatic properties. (Figure reproduced from ref. 38).

Table 1
Predicted residues, pairwise distances, and potential differences in carboxypeptidase A and cytochrome c peroxidases using the motif (Ala48, His52, Trp191) from a cytochrome C peroxidase (PDBid: 1DJ1). Table reproduced from ref. 39

	Predicted residues			Pairwise distance			Pairwise potential difference		
	a	b	c	ab	ac	bc	ab	ac	bc
1DJ1	Ala48	His52	Trp191	6.3	11.4	13.8	329.1	317.2	11.9
5CPA	Ala143	His69	Trp63	7.5	11.5	13.3	321.1	255.3	65.7

sites in this protein, since the carboxypeptidase gains pyruvate oxidase activity on replacement of Zn^{2+} with Cu^{2+} (40). We have presented a rational flow to help enzyme engineers select a protein on which to endow an enzymatic activity, and predict subsequent mutations based on the superimposition of the matching scaffolds (DECAAF) (41). Here, we detail the steps in DECAAF intended to choose a target protein based on the presence of a significant subset of the desired catalytic scaffold as a promiscuous domain in its structure.

Similar rational approaches based on identifying enzymes with a partial catalytic structure have led to successful efforts at inducing or strengthening the desired enzymatic function (42–44). The advantage of the CLASP algorithm lies in its use of "exact" electrostatic properties early in the search, combined with structural properties. Other methods have examined binding energy and energy minimization at a later stage of the search (21, 22, 24). Electrostatic congruence of a few active site residues implies a favorable milieu for the desired catalytic activity and thus encodes residues in the close vicinity as well. This allows CLASP to filter out unfeasible configurations at a much lower computational cost.

Another aspect of catalysis important to computational modeling is the flexibility and diversity observed in the active site scaffold of related enzymes. CLASP and PROMISE are designed with this heterogeneity in mind. While stereochemical equivalence can be hardwired for amino acids with similar properties, there are instances where residues with different properties occupy the same sequence and spatial location and perform the same functionality. A well-known example is the equivalence of Ser130 and Tyr150 in Class A and C β-lactamases, respectively (45). CLASP allows the user to explicitly specify the group of residues that can match a particular residue from the input motif. A major source of overfitting in motif searches (leading to pessimistic results) is that all pairs are not equally significant in a given set of atoms. The weight given to such pairwise deviations, equal by default, can be varied in CLASP to handle this redundancy. Finally, while choosing a site in a protein as a possible scaffold, it is sometimes useful to relax the constraint that the site must be close to the active site. Such relaxations can allow identification of moonlighting domains as possible targets for mutagenesis (46). Modulating the radius defining the "active site vicinity" has helped us characterize properties of residues that determine promiscuity (39).

We describe detailed steps in choosing a plant protein to substitute for human neutrophil elastase (47) in a chimeric antimicrobial enzyme designed to bolster the innate immune defense system in plants (48). In this example, P14A is identified as a very significant match (49). P14A is a member of the PR-1 group of pathogenesis-related proteins (50). While P14A is not known to have elastase activity, it is structurally homologous to a snake venom protein that was previously identified as an elastase (51). Furthermore, the residues in the predicted catalytic triad in P14A (Ser49, His48, and Tyr36) are conserved and hypothesized to be critical for catalysis: "His48, Ser49, and His93 are in close proximity, so that the conservation of this group of residues could be taken as an indication of it being important as an active site in P14a" (49).

The complete scaffold of an elastase does not exist in the P14A protein. While Ser195, His57, and Gly193 from the input motif have a highly matching scaffold in P14A, the spatial position of the elastase Asp102 is close to Asn35 and Ser39 in P14A when the proteins are superimposed based on the matching scaffolds. We intend to express this protein and test for elastase activity. If we find no detectable activity, we believe that an Asn35Asp or Ser39Asp P14A mutant will gain elastase function and thus validate the rational design flow DECAAF.

2. Materials

1. The CLASP and PROMISE packages can be downloaded from http://www.sanchak.com/clasp/.
2. Adaptive Poisson-Boltzmann Solver (APBS) and PDB2PQR packages were used to calculate the potential difference between the reactive atoms of the corresponding proteins and need to be installed (52, 53).
3. We have extensively integrated and used the freely available BioPerl (54) modules and Emboss (55) tools.
4. System requirements: CLASP and PROMISE packages are written in Perl for Linux.
5. Hardware requirements are modest—all results here are from a simple workstation (8 GB RAM).
6. Protein structures were rendered by PyMol (http://www.pymol.org/).

3. Methods

1. Choose structure and active sites: We first chose the PDB structure (PDB id: 1B0F) and active site residues Ser195, His57, Asp102, Ser214, and Gly193 of the known enzymatic function elastase.
2. Choose target set of proteins: We chose the set of proteins from which a target protein(s) is to be selected using a keyword search for "plants" in http://www.pdb.org/, which was then pruned for redundancy based on a 40% sequence similarity. This yielded 288 proteins (http://www.sanchak.com/elastase/targetset.html).
3. Do electrostatic analysis: We did electrostatic analysis for the elastase and all 288 proteins using APBS (52) (see Note 1). Each atom of each protein now has a potential associated with it (see Note 2).
4. Create partial motif library: We created subset motifs from active site residues; by default, each position in the match can be occupied by the same amino acid. An example motif file can be found at http://www.sanchak.com/elastase/M1.txt and in Fig. 2. A utility script generated all possible motifs of size n (three in this example) given a set of N (five in this case) active site residues. Each residue was represented by its reactive atom (see Note 3).
   (a) M1 = (Ser195, His57, Ser214)
   (b) M2 = (Ser195, His57, Gly193)
   (c) ...
   (d) M10 = (Ser195, His57, Asp102)
5. Run each motif on the target list: We applied each motif (M1, M2, ... M10) on the target list and obtained best matches. Each run of a single motif on the list of 288 proteins takes less than 30 min on a simple workstation (8 GB RAM). The best matches with M1 and M2 were determined (Fig. 2).
6. Choose best consensus match: We found that the P14A protein (PDB id: 1CFE) had a good match with most of these motifs.
7. Allow stereochemical equivalence to increase congruence: Subsequently, we analyzed the mismatches, and figured whether expanding the stereochemical equivalence group increased the congruence. In M3 = (Ser195, His57, Ser214), the position in Ser214 could be matched by either a Ser or a Tyr (see Note 4). P14A protein, where a Tyr36 matches the Ser214, now scored better (Fig. 2). Tyr36 had a much better spatial orientation than Ser120, although the potential differences were almost similar (Table 2).

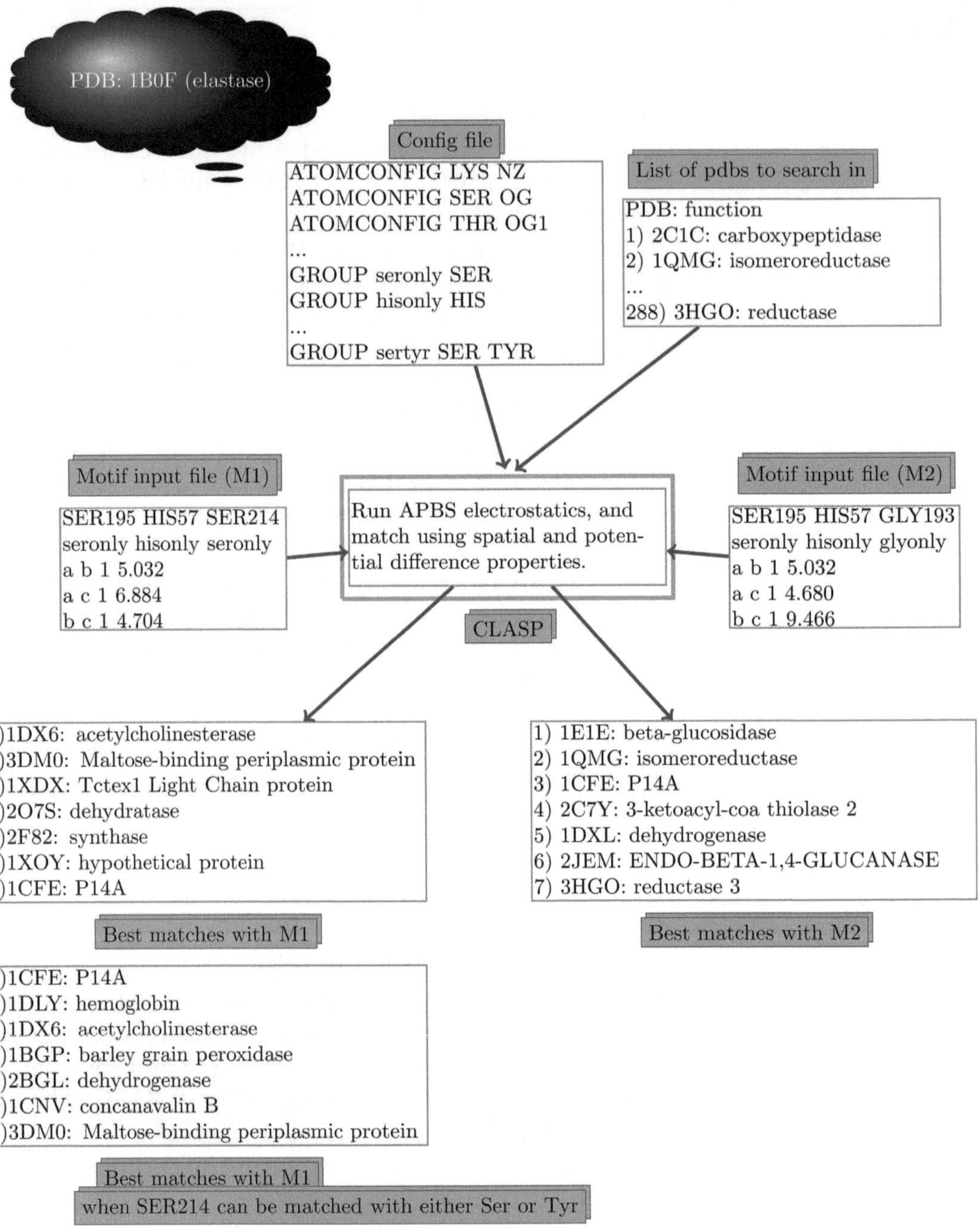

Fig. 2. The flow adopted to select a plant protein with a significantly elastase-like scaffold. The input configuration files, list of proteins on which the search was performed. The input motifs (M1, M2 …) can be viewed at http://www.sanchak.com/elastase/. A pathogenesis-related protein (PDB id: 1CFE) was selected as the best possible candidate. The ability to specify a set of residues to match a particular position allows this protein (PDB id: 1CFE) to score much better when a Tyr (Tyr36) is matched instead of a Ser (Ser120). Figure reproduced from ref. 41.

Table 2
Expanding the set of residues that can match a single position in the input motif. When the position Ser214 in the input motif (Ser195, His57, Ser214) can be matched by a Ser or a Tyr, Tyr36 has a much better spatial orientation with respect to Ser49 and His48 than Ser120 in the P14A protein (PDBid: 1CFE). Table reproduced from ref. 41

	Predicted residues			Pairwise distances in Å			Pairwise potential difference			
PDB	a	b	c	ab	ac	bc	ab	ac	bc	Score
1B0F	Ser195	His57	Ser214	5	6.8	4.7	53.7	105.7	51.9	0
1CFE (Ser214 can be Ser only)	Ser49	His48	Ser120	0.1	−0.2	1.9	15.3	83.9	68.6	0.111
1CFE (Ser214 can be Ser or Tyr)	Ser49	His48	Tyr36	0.1	0.3	−0.3	15.3	88.6	73.3	0.023

8. Increase the motif size: We relaxed the constraint that the motif size should be three, and based on matches in M1, M2 … M10, we chose a larger set of four residues in the elastase that had a very significant match in the P14A protein (Table 3). This table shows that the Gly193 in the elastase matches with the stereochemically equivalent Ala51 in P14A, increasing the odds of obtaining elastase activity in the P14A protein, either in the wild type or by directed evolution.
9. Identify possible outliers: The difficulty comes with Asp102. The P14A protein is not a good match when the input scaffold includes Asp102, leaving open the possibility that it does not have elastase activity. If so, directed evolution techniques would be necessary to impart elastase function to the protein. However, many unconventional serine proteases with variations in the catalytic triad have been observed (56). It is possible that P14A is one of them.
10. Figure out the possible mutations that need to be applied to take care of the outliers: Finally, we superimposed the elastase and P14A based on the matching scaffold to determine possible mutations to apply. We applied linear and rotational transformations on both proteins using the first three atoms of the match. After the transformations, (Ser195/OG, His57/ND1, Ser214/OG) and (Ser49/OG, His48/ND1, Tyr36/OH) lie on the same plane ($z=0$ for all atoms), Ser195/OG and Ser49/OG are at the center of the coordinate axis, and His57/ND1 and His48/ND1 lie on the x–y-axis ($y=0$). The initial and

Table 3
Extending the partial match: Spatial and electrostatic potential difference (PD) congruence in cognate pairs in the human neutrophil elastase (PDB id: 1B0F) and the pathogenesis-related P14A protein (PDB id: 1CFE) for an extended set of four residues in the active site. Table reproduced from ref. 41

	a	b	c	d
1B0F motif	SER195	HIS57	SER214	GLY193
1CFE motif	SER49	HIS48	TYR36	ALA51

	ab	ac	ad	bc	bd	cd
PDB	Pairwise distances					
1B0F	5	6.8	4.6	4.7	9.4	11
1CFE	4.9	6.8	4.5	5	9.1	10
	Pairwise potential difference					
1B0F	53.7	105.7	-111.1	51.9	-164.9	-216.8
1CFE	15.3	88.6	-58.1	73.3	-73.4	-146.8

Table 4
Superimposing proteins based on partial matches: Applying linear and rotational transformations on both proteins such that all three atoms lie on the same plane ($z=0$), the first atom lies on the center of the coordinate system ($x=0$), and the first two atoms lie on the *x–y*-axis. Table reproduced from ref. 41

	Atom	Before			After		
	Coordinates	*X*	*Y*	*Z*	*X*	*Y*	*Z*
PDB: 1B0F (template)	Ser195/OG	64.4	57	53.8	0	0	0
	His57/ND1	63.3	54.8	58.2	5	0	0
	Ser214/OG	63.6	50.6	56.1	5	4.7	0
PDB: 1CFE (target)	Ser49/OG	8.8	-6.3	-4.8	0	0	0
	His48/ND1	9.2	-2.4	-1.8	4.9	0	0
	Tyr36/OH	13.7	-1.6	-3.7	4.7	5	0

transformed coordinates for both the proteins are listed (Table 4) and the superimposed motifs can be drawn (Fig. 3). The spatial position of Asp102 in P14A is occupied by Asn35 or Ser39 (Fig. 3), so an Asn35Asp or a Ser39Asp mutant might gain elastase function.

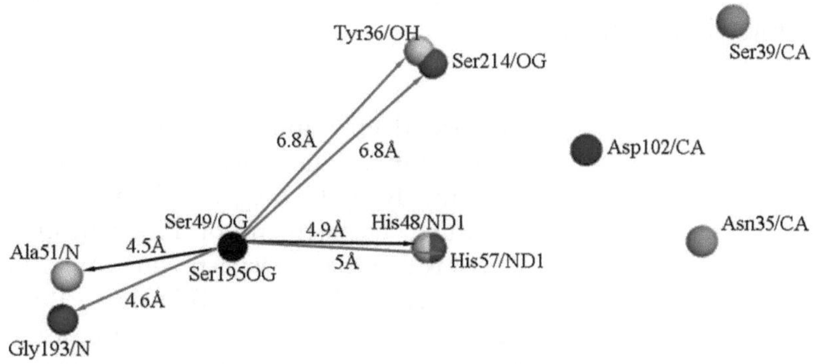

Fig. 3. Superimposing proteins based on partial matches (Ser195/OG, His57/ND1, Ser214/OG from PDBid: 1B0F) and (Ser49/OG, His48/ND1, Tyr36/OH from PDBid: 1CFE:). Applying linear and rotational transformations on both proteins such that all three atoms lie on the same plane ($z=0$), the first pair lies on the center of the coordinate system ($x=0$, $y=0$, $z=0$), and the second pair lie on the x–y-axis ($y=0$). Figure reproduced from ref. 41.

4. Notes

1. The APBS parameters are set as follows: Solute dielectric, 2; solvent dielectric, 78; solvent probe radius, 1.4 Å; temperature, 298 K; and ionic strength, 0.
2. APBS writes out the electrostatic potential in dimensionless units of kT/e where k is Boltzmann's constant, T is the temperature in K, and e is the charge of an electron.
3. Each amino acid is represented by one atom. Usually this is the reactive atom, but the user can specify it to be any atom, as described in the configuration file (Fig. 2 and http://www.sanchak.com/elastase/config.txt).
4. While matching, the set of amino acids that are equivalent for each position in the motif can be specified by the user. Note that this flexibility comes at a computational cost, quickly reaching prohibitive proportions for more than three or four amino acids for each position. These groups are specified in the configuration file and must be modified in the motif-matching file (Fig. 2, http://www.sanchak.com/elastase/config.txt and http://www.sanchak.com/elastase/M1.txt).

Acknowledgements

We are deeply indebted to J. M. Frere (Centre for Protein Engineering, Universite de Liege, Institut de Chimie B6, Sart Tilman, B-4000 Liege, Belgium), Bjarni Asgeirsson (Science Institute, Department of Biochemistry, University of Iceland), Masataka Oda (Department of Microbiology, Faculty of Pharmaceutical Science, Tokushima Bunri University, Japan), and Felix M. Goni (Unidad de Biofisica (CSIC-UPV/EHU) and Departamento de Bioquimica, Universidad del Pais Vasco, Bilbao, Spain), for technical discussions, suggestions, and support. AMD acknowledges support received from the California Department of Food and Agriculture's Pierce's Disease Board to conduct this collaborative research. BJR acknowledges a J.C. Bose award fellowship grant.

References

1. Lehninger A, Nelson DL, Cox MM (2008) Lehninger principles of biochemistry, 5th edn. W. H. Freeman, New York
2. Koshland DE (1958) Application of a theory of enzyme specificity to protein synthesis. Proc Natl Acad Sci U S A 44:98–104
3. Brien PJ, Herschlag D (1999) Catalytic promiscuity and the evolution of new enzymatic activities. Chem Biol 6:R91–R105
4. Hult K, Berglund P (2007) Enzyme promiscuity: mechanism and applications. Trends Biotechnol 25:231–238
5. Khersonsky O, Tawfik DS (2010) Enzyme promiscuity: a mechanistic and evolutionary perspective. Annu Rev Biochem 79:471–505
6. Jensen RA (1976) Enzyme recruitment in evolution of new function. Annu Rev Microbiol 30:409–425
7. Lewis EB (1951) Pseudoallelism and gene evolution. Cold Spring Harb Symp Quant Biol 16:159–174
8. Tawfik DS (2010) Messy biology and the origins of evolutionary innovations. Nat Chem Biol 6:692–696
9. Cirino PC, Mayer KM, Umeno D (2003) Generating mutant libraries using error-prone PCR. Methods Mol Biol 231:3–9
10. Hall BG, Zuzel T (1980) Evolution of a new enzymatic function by recombination within a gene. Proc Natl Acad Sci U S A 77:3529–3533
11. Stemmer WP (1994) Rapid evolution of a protein in vitro by DNA shuffling. Nature 370:389–391
12. Zhao H, Giver L, Shao Z et al (1998) Molecular evolution by staggered extension process (StEP) in vitro recombination. Nat Biotechnol 16:258–261
13. Kolkman JA, Stemmer WP (2001) Directed evolution of proteins by exon shuffling. Nat Biotechnol 19:423–428
14. Esvelt KM, Carlson JC, Liu DR (2011) A system for the continuous directed evolution of biomolecules. Nature 472:499–503
15. Johns GC, Joyce GF (2005) The promise and peril of continuous in vitro evolution. J Mol Evol 61:253–263
16. Goddard JP, Reymond JL (2004) Enzyme assays for high-throughput screening. Curr Opin Biotechnol 15:314–322
17. Morley KL, Kazlauskas RJ (2005) Improving enzyme properties: when are closer mutations better? Trends Biotechnol 23:231–237
18. Reetz MT, Carballeira JD (2007) Iterative saturation mutagenesis (ISM) for rapid directed evolution of functional enzymes. Nat Protoc 2:891–903
19. Climie S, Ruiz-Perez L, Gonzalez-Pacanowska D et al (1990) Saturation site-directed mutagenesis of thymidylate synthase. J Biol Chem 265:18776–18779
20. Reetz MT, Carballeira JD, Peyralans J et al (2006) Expanding the substrate scope of enzymes: combining mutations obtained by CASTing. Chemistry 12:6031–6038
21. Zanghellini A, Jiang L, Wollacott AM et al (2006) New algorithms and an in silico

benchmark for computational enzyme design. Protein Sci 15:2785–2794

22. Dahiyat BI, Mayo SL (1997) De novo protein design: fully automated sequence selection. Science 278:82–87
23. Malisi C, Kohlbacher O, Hocker B (2009) Automated scaffold selection for enzyme design. Proteins 77:74–83
24. Georgiev I, Lilien RH, Donald BR (2008) The minimized dead-end elimination criterion and its application to protein redesign in a hybrid scoring and search algorithm for computing partition functions over molecular ensembles. J Comput Chem 29:1527–1542
25. Lovell SC, Word JM, Richardson JS et al (2000) The penultimate rotamer library. Proteins 40:389–408
26. Kimura M (1986) DNA and the neutral theory. Philos Trans R Soc Lond B Biol Sci 312: 343–354
27. Amitai G, Gupta RD, Tawfik DS (2007) Latent evolutionary potentials under the neutral mutational drift of an enzyme. HFSP J 1:67–78
28. Wroe R, Chan HS, Bornberg-Bauer E (2007) A structural model of latent evolutionary potentials underlying neutral networks in proteins. HFSP J 1:79–87
29. Bolon DN, Mayo SL (2001) Enzyme-like proteins by computational design. Proc Natl Acad Sci U S A 98:14274–14279
30. Jiang L, Althoff EA, Clemente FR et al (2008) De novo computational design of retro-aldol enzymes. Science 319:1387–1391
31. Faiella M, Andreozzi C, de Rosales RT et al (2009) An artificial di-iron oxo-protein with phenol oxidase activity. Nat Chem Biol 5:882–884
32. Siegel JB, Zanghellini A, Lovick HM et al (2010) Computational design of an enzyme catalyst for a stereoselective bimolecular Diels-Alder reaction. Science 329:309–313
33. Rothlisberger D, Khersonsky O, Wollacott AM et al (2008) Kemp elimination catalysts by computational enzyme design. Nature 453:190–195
34. Nannemann DP, Birmingham WR, Scism RA et al (2011) Assessing directed evolution methods for the generation of biosynthetic enzymes with potential in drug biosynthesis. Future Med Chem 3:809–819
35. Dalby PA (2011) Strategy and success for the directed evolution of enzymes. Curr Opin Struct Biol 21:473–480
36. Lutz S (2010) Beyond directed evolution-semi-rational protein engineering and design. Curr Opin Biotechnol 21:734–743
37. Antikainen NM, Martin SF (2005) Altering protein specificity: techniques and applications. Bioorg Med Chem 13:2701–2716
38. Chakraborty S, Minda R, Salaye L et al (2011) Active site detection by spatial conformity and electrostatic analysis—unravelling a proteolytic function in shrimp alkaline phosphatase. PLoS One 6:e28470
39. Chakraborty S, Rao BJ (2012) A measure of the promiscuity of proteins and characteristics of residues in the vicinity of the catalytic site that regulate promiscuity. PLoS One 7:e32011
40. Yamamura K, Kaiser ET (1976) Studies on the oxidase activity of copper(ii) carboxypeptidase A. J Chem Soc, Chem Commun: 830–831
41. Chakraborty S (2012) An automated flow for directed evolution based on detection of promiscuous scaffolds using spatial and electrostatic properties of catalytic residues. PLoS One PLoS One. 2012;7(7):e40408. doi: 10.1371/journal.pone.0040408.
42. Savile CK, Janey JM, Mundorff EC et al (2010) Biocatalytic asymmetric synthesis of chiral amines from ketones applied to sitagliptin manufacture. Science 329:305–309
43. Chen CY, Georgiev I, Anderson AC et al (2009) Computational structure-based redesign of enzyme activity. Proc Natl Acad Sci U S A 106:3764–3769
44. Sandstrom AG, Wikmark Y, Engstrom K et al (2012) Combinatorial reshaping of the Candida antarctica lipase A substrate pocket for enantioselectivity using an extremely condensed library. Proc Natl Acad Sci U S A 109:78–83
45. Lobkovsky E, Moews PC, Liu H et al (1993) Evolution of an enzyme activity: crystallographic structure at 2 Å resolution of cephalosporinase from the ampC gene of Enterobacter cloacae P99 and comparison with a class A penicillinase. Proc Natl Acad Sci U S A 90:11257–11261
46. Jeffery CJ (2009) Moonlighting proteins—an update. Mol Biosyst 5:345–350
47. Macdonald SJ, Dowle MD, Harrison LA et al (2002) Discovery of further pyrrolidine trans-lactams as inhibitors of human neutrophil elastase (HNE) with potential as development candidates and the crystal structure of HNE complexed with an inhibitor (GW475151). J Med Chem 45:3878–3890
48. Dandekar AM, Gouran H, Ibanez AM et al (2012) An engineered innate immune defense protects grapevines from Pierce disease. Proc Natl Acad Sci U S A 109:3721–3725
49. Fernandez C, Szyperski T, Bruyere T et al (1997) NMR solution structure of the pathogenesis-related protein P14a. J Mol Biol 266:576–593

50. Stintzi A, Heitz T, Prasad V et al (1993) Plant 'pathogenesis-related' proteins and their role in defense against pathogens. Biochimie 75: 687–706
51. Bernick JJ, Simpson W (1976) Distribution of elastase-like enzyme activity among snake venoms. Comp Biochem Physiol B 54:51–54
52. Baker NA, Sept D, Joseph S et al (2001) Electrostatics of nanosystems: application to microtubules and the ribosome. Proc Natl Acad Sci U S A 98:10037–10041
53. Dolinsky TJ, Nielsen JE, McCammon JA et al (2004) PDB2PQR: an automated pipeline for the setup of Poisson-Boltzmann electrostatics calculations. Nucleic Acids Res 32: W665–W667
54. Stajich JE, Block D, Boulez K et al (2002) The bioperl toolkit: Perl modules for the life sciences. Genome Res 12:1611–1618
55. Rice P, Longden I, Bleasby A (2000) EMBOSS: The European Molecular Biology Open Software Suite. Trends Genet 16:276–277
56. Ekici OD, Paetzel M, Dalbey RE (2008) Unconventional serine proteases: variations on the catalytic Ser/His/Asp triad configuration. Protein Sci 17:2023–2037

Chapter 16

Rational Protein Sequence Diversification by Multi-Codon Scanning Mutagenesis

Jia Liu and T. Ashton Cropp

Abstract

A new method for protein sequence diversification is based on generating random codon mutations to an encoding DNA. This allows for the scanning of user-defined amino acid changes to any protein of interest, and is an alternative to traditional directed evolution strategies. This chapter describes the procedures required to apply this technology to any protein of interest. The resulting libraries can then be screened for new or improved protein function.

Key words: Directed evolution, Protein mutagenesis, Green fluorescent protein, Library

1. Introduction

Nearly all random protein diversification approaches rely on generating single-nucleotide changes in an encoding DNA sequence. While this is quite easy to do in the lab using methods such as error-prone PCR (1) and/or DNA shuffling (2), such changes are limited by the redundancy of the genetic code. Indeed many nucleotide changes do not result in protein changes and simply dilute out functional diversity. Likewise, there are many amino acid substitutions that require three consecutive nucleotide changes and are thus much less likely to occur during random mutation. We recently reported (3) an alternative approach that randomly changes DNA sequence "codon at-a-time" rather than "nucleotide at-a-time," which ensures a high percentage of functional diversity. Importantly the codon can be chosen as a single amino acid or group to offer precise diversity. Moreover the method is easily adaptable to multiple codons, for example "two codons at-a-time" libraries. The methodology is complementary to other methods and, depending

James C. Samuelson (ed.), *Enzyme Engineering: Methods and Protocols*, Methods in Molecular Biology, vol. 978,
DOI 10.1007/978-1-62703-293-3_16, © Springer Science+Business Media New York 2013

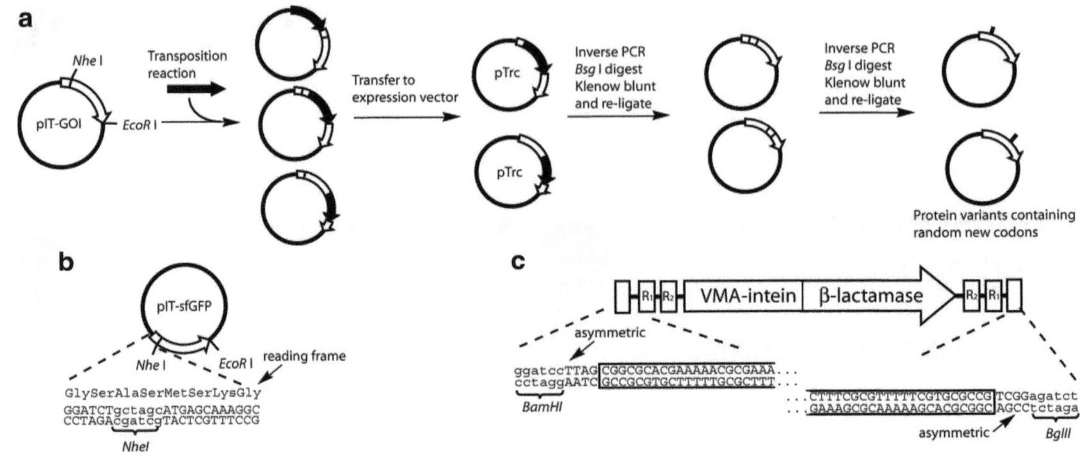

Fig. 1. (**a**) Depiction of the overall procedure described in this protocol. (**b**) pIT-sfGFP (available upon request) is used to generate pIT-GOI by cloning gene of interest into the NheI and EcoRI sites. The reading frame of the target protein must be maintained as shown. (**c**) The transposon (available upon request) is made ready for the reaction by BglII, BamHI digestion.

on the experiment, might even be combined with, for example, DNA shuffling to further expand protein sequence space.

Multi-codon scanning mutagenesis is based on the development of an asymmetric mini-Mu transposon. Transposon mutagenesis is used to essentially deliver randomly placed primer-binding sites to facilitate polymerase chain reactions. The overall method, which is outlined in Fig. 1, consists of cloning a gene of interest into a special targeting plasmid, pIT-GOI. This template is then subjected to a round of transposon mutagenesis to create a library of new plasmids. These then serve as a template for two inverse-PCR reactions which add, and then remove, DNA sequence scars. The final result is a library of new expression vectors, each containing a new gene with one, two, or three new in-frame codon mutations. Depending on the experimental design, generating such libraries can be accomplished in approximately 3 weeks. Below outlines the experimental procedures required using the model protein superfolder green fluorescent protein (sfGFP) as an example.

2. Materials

1. Intein targeting plasmid (pIT) (3) (see Notes 1 and 2).
2. pUC18-transposon (3) (see Note 3).
3. Gene of interest (see Notes 4 and 5).
4. Engineered pTrcHisA vector (Invitrogen Corporation, Carlsbad, CA) (3) or other appropriate expression vector (see Note 6).

5. HyperMu MuA transposase and 10× transposition reaction buffer (Epicentre Biotechnologies, Madison, WI). Store at −20°C.
6. 10 mM Tris–HCl buffer, pH 8.5 at 25°C.
7. TE buffer: 10 mM Tris–HCl, 0.1 mM EDTA, pH 8.5 at 25°C.
8. 500 mM ethylenediaminetetraacetic acid (EDTA), pH 8.0 at 25°C.
9. 50× TAE buffer: 242 g Tris base, 57.1 mL glacial acetic acid, and 100 mL 500 mM EDTA (pH 8.0 at 25°C) and distilled water up to 1 L.
10. Primers:
    (a) Forward and reverse primers for cloning the gene of interest into pIT vector (see Note 7).
    (b) Forward and reverse primers for cloning the gene of interest with random transposon insertions into a desired expression vector.
    (c) Primers for EI-PCR reactions (see Table 1 for sequence).

    Dissolve primer stocks in TE buffer at a final concentration of 100 μM. Store the stock primers at −20°C and dilute to 10 μM prior to use.
11. dNTPs stock solution, 10 mM each of dATP, dTTP, dGTP, and dCTP. Store as 20 μL aliquots at −20°C.
12. Phusion DNA polymerase and 5× HF Phusion buffer (New England Biolabs, Ipswich, MA). Store at −20°C.
13. FastDigest *Bpm* I (Fermentas, Glen Burnie, MD), *Bsg* I (NEB), *Nhe* I, *EcoR* I, and other desired restriction enzymes. Store at −20°C.
14. Large (Klenow) fragment of DNA polymerase I (NEB). Store at −20°C.
15. T4 DNA ligase (Fermentas). Store at −20°C.
16. 1% (w/v) agarose gels: Add 1 g agarose to 100 mL 1× TAE. Microwave to thoroughly melt prior to use (see Note 8).
17. QIAquick PCR purification kit (Qiagen, Valencia, CA).
18. QIAEX II gel extraction kit (Qiagen, Valencia, CA).
19. Chemically or electro-competent *Escherichia coli* DH10B cells. Store at −80°C.
20. 1,000× antibiotic stock solutions: Ampicillin (100 mg/mL) and kanamycin (50 mg/mL). Store as 1 mL aliquots at −20°C.
21. 1,000× inducing reagent for protein expression, such as 1 M isopropyl β-d-1-thiogalactopyranoside (IPTG). Store as 1 mL aliquots at −20°C.
22. Luria Bertani (LB) medium (sterilized): 10 g tryptone, 5 g yeast extract, 10 g sodium chloride, and distilled water up to 1 L.

Table 1
List of primers for EI-PCR reactions

Name	Sequence (*Bsg* I and *Bpm* I sites underlined)	Use
FF1	TACTTTTTCTGGAGCCGTCGGA	First PCR, Fwd, remove one codon
FF2	TTTTTCGTGTGCAGTCGGA	First PCR, Fwd, remove two codons
FR1[a]	MNNAATCAACGACTTTGCGCCGCTAAG	First PCR, Rev, introduce one codon
FR2[a]	MNNMNNAATCAACGACTTTGCGCCGCTAAG	First PCR, Rev, introduce two codons
FR3[a]	MNNMNNMNNAATCAACGACTTTGCGCCGCTAAG	First PCR, Rev, introduce three codons
SF0	TAACGTGCAGTTACAAGTCGTTGATT	Second PCR, Fwd, cleave to new codons
SF1	CATCGTGCAGATGCGCCGCTAAG	Second PCR, Fwd, remove one codon
SR	CTTCGTGCAGTAAATGCGCCGCTAAG	Second PCR, Rev, cleave to target gene

[a]These reverse primers will deliver one to three NNK (N: A, T, G, C; K: G, T) degenerate codons into the gene of interest. MNN (M: A, C) codons, the reverse complementary sequence of NNK, are designed in these primers because the new codons will be delivered in a reverse complementary manner

23. SOC medium (sterilized): 20 g tryptone, 5 g yeast extract, 10 mM sodium chloride, 25 mM potassium chloride, and distilled water up to 1 L. Autoclave and then add sterile magnesium chloride to 10 mM and glucose to 20 mM.
24. 2×YT medium (sterilized): 16 g tryptone, 10 g yeast extract, 5 g sodium chloride, and distilled water up to 1 L, pH 7.2.
25. LB agar plates with appropriate antibiotics (Table 1).

3. Methods

3.1. Generation of pIT-GOI Plasmid Containing Gene of Interest

1. Digest 2 μg of the pIT-sfGFP (3) vector using *Nhe* I/*EcoR* I. Gel-purify the 3.2 Kbp fragment corresponding to pIT backbone using QIAEX II gel extraction kit (see Note 9).
2. Amplify the gene of interest using PCR to introduce *Nhe* I and *EcoR* I sites at the 5′ and 3′ ends, respectively. Ensure that the gene of interest is in-frame with the targeting sequence of pIT-vector backbone, as indicated in Fig. 1.

3. Purify the PCR product using with the QIAquick PCR purification kit.
4. Digest the PCR product using *Nhe* I/*EcoR* I and gel-purify the digestion product using QIAEX II gel extraction kit.
5. Ligate the purified digestion products of pIT vector and gene of interest using T4 DNA ligase (see Note 10). Name the recombinant plasmid as pIT-GOI.

3.2. Transposition Reaction

1. Prepare asymmetric transposon DNA by digesting pUC18-transposon plasmid with *Bgl* II/*Bam*H I as follows: 2 μg DNA, 10 U each of *Bgl* II and *Bam*H I (Fermentas), and 2× buffer Tango (Fermentas) in a 50 μL solution (see Note 11). Incubate the reaction at 37°C for 5 h.
2. Purify the 1.3 Kbp fragment using QIAEX II gel extraction kit (Qiagen) and determine its concentration by comparison with a DNA standard. Store the purified transposon DNA at –20°C prior to use.
3. Perform a 20 μL transposition reaction as follows: 450 ng of pIT-GOI plasmid DNA, 1.3 molar excess of asymmetric transposon DNA (see Note 12), 1 unit of HyperMu MuA transposase (Epicentre), 50 mM Tris–acetate, pH 7.5, 150 mM potassium acetate, 10 mM magnesium acetate, and 4 mM spermidine. Incubate the reaction at 30°C for 4 h. Stop the reaction by addition of 2 μL 0.1% SDS, followed by heat-inactivation at 70°C for 10 min. Cool the reaction on ice.
4. Transform each 1 μL of transposition product into 50 μL of electro-competent *E. coli* DH10B cells (see Notes 13 and 14). Recover the cells in 500 μL SOC at 37°C for 1 h. Plate transformed cells on LB agar supplemented with 50 μg/mL kanamycin and 40 μg/mL ampicillin (see Note 15). Grow the plates at room temperature for 48 h (see Notes 16 and 17).
5. For a gene of *L*bps in length, collect at least 3×(*L*+1,500) colonies from the transposition reaction (see Note 18). Resuspend the collected cells in LB medium supplemented with 50 μg/mL kanamycin and 40 μg/mL ampicillin. Save at least five tubes of 1 mL stock cells with an OD_{600} of ~1.0 in 10% glycerol at –80°C (see Note 19). Extract the plasmid DNA from the remaining cells to obtain the pIT-GOI-transposon library.

3.3. Transfer of Gene of Interest with Random Transposon Insertions into pTrcHisA Vector

1. Digest 2 μg of the transposon library DNA using *Nhe* I/*EcoR* I.
2. Use QIAEX II gel extraction kit to gel-purify the DNA fragment corresponding to the gene of interest with transposon insertions (see Note 20).
3. Ligate the purified DNA fragment into pTrcHisA precut with *Nhe* I/*EcoR* I using T4 DNA ligase.

4. Transform the ligation products into electro-competent *E. coli* DH10B cells. Recover the cells in 500 μL SOC at 37°C for 1 h and then plate the cells on LB agar supplemented with 100 μg/mL ampicillin.
5. Collect 3 × *L* colonies (see Note 21). Resuspend the collected cells in LB medium supplemented with 100 μg/mL ampicillin. Save at least five tubes of 1 mL stock cells with an OD_{600} of ~1.0 in 10% glycerol at −80°C. Extract the plasmid DNA from the remaining cells to obtain the pTrcHisA-GOI-transposon library.

3.4. Introduction of New Codons and "Anchor" Sequence

One, two, and three codon mutations can be generated using the same pTrcHisA-GOI-transposon library DNA as the template but different primer pairs in the PCR amplification. The optimum template concentrations and annealing temperatures have been individually determined for each PCR reaction (3).

1. Set up several 100 μL PCR reactions containing **X** pg/μL pTrcHisA-GOI-transposon library DNA, 0.5 μM each of forward and reverse primers (see below), 0.2 mM dNTPs, 0.005 U/μL of Phusion DNA polymerase (NEB), 1× Phusion HF buffer (NEB, providing 1.5 mM $MgCl_2$), and additional 2 mM $MgSO_4$. **X** indicates the optimum template concentrations and are 100, 125, and 200 pg/μL for one, two, and three codon mutations, respectively (see Note 22). The primer pairs are FF1/FR1, FF2/FR2, and FF2/FR3 for one, two, and three codon mutations, respectively (see Table 1 for sequence).
2. Cycle the PCR reactions using the following conditions: Initial denaturation at 98°C for 2 min, 18 cycles of 98°C for 10 s, **Y**°C for 30 s and 72°C for 2 min (see Note 23), and a final extension at 72°C for 10 min. **Y** represents the optimum annealing temperatures and are 57°C, 59°C, and 65°C for one, two, and three codon mutations, respectively (see Note 22).
3. Purify the PCR product using QIAquick PCR purification kit (Qiagen).
4. Digest the purified PCR product of one codon library with *Bpm* I in a 50 μL reaction containing 500 ng DNA, 1 U FastDigest *Bpm* I (Fermentas), and 1 × FastDigest buffer (Fermentas) (see Note 24). Incubate the reaction at 30°C for 4.5 h. Heat-inactivate the reaction at 65°C for 20 min. Digest the purified PCR product of two and three codon libraries with *Bsg* I in a 50 μL reaction containing 500 ng DNA, 6 U *Bsg* I (NEB), 80 μM *S*-adenosylmethionine (SAM), and 1× buffer 4 (NEB) (see Note 25). Incubate the reaction at 37°C for 4.5 h. Heat-inactivate the reaction at 65°C for 20 min.
5. For each 50 μL of the above heat-inactivated digestion product, add 1.7 μL of 1 mM dNTPs and 0.5 U of Klenow fragment (NEB). Incubate the end-blunting reaction at 25°C for

15 min (see Note 26). Stop the reaction (50 μL) by adding 1.1 μL of 0.5 M EDTA, pH 8.0, followed by heat-inactivation at 75°C for 20 min (see Note 27).

6. Gel-purify the Klenow-treated digestion product using QIAEX II gel extraction kit (Qiagen).
7. Re-circularize the purified DNA into plasmid in a 10 μL ligation reaction containing 2.5 ng/μL DNA, 0.5 Weiss unit/μL T4 DNA ligase (Fermentas), 40 mM Tris–HCl, pH 7.8 at 25°C, 10 mM $MgCl_2$, 10 mM DTT, and 0.5 mM ATP (see Note 28). Incubate the reaction at 16°C overnight. Heat-inactivate the reaction at 70°C for 7 min (see Note 29). Cool the reaction on ice.
8. Transform 1 μL of the ligation product (see Note 30) into 50 μL electro-competent *E. coli* DH10B cells. Plate the transformed cells on LB agar supplemented with 100 μg/mL ampicillin.
9. Collect desired number of colonies to maintain library diversity (see Note 31). Save cell stocks as above and extract the plasmid DNA from the remaining cells to obtain the pTrcHisA-GOI-anchor libraries for each of one, two, and three codon scanning libraries.

3.5. Removal of Primer Sequence and Generation of Seamless, In-frame Codon Mutations

The second EI-PCR/digestion process is conceptually the same as the first one as described above. The goal of this process is to remove the reverse primer sequence ("anchor" sequence) left in the gene of interest from the first PCR, generating seamless codon mutations.

1. Set up several 100 μL PCR reactions containing X2 pg/μL pTrcHisA-GOI-anchor library DNA, 0.5 μM each of forward and reverse primers (see below), 0.2 mM dNTPs, 0.005 U/μL of Phusion DNA polymerase (NEB), 1 × Phusion HF buffer (NEB, providing 1.5 mM $MgCl_2$), and additional 2 mM $MgSO_4$. X2 represents the optimum template concentrations and are 200, 125, and 200 pg/μL for one, two, and three codon scanning libraries, respectively. The primer pairs for one, two, and three codon scanning libraries are SF0/SR, SF0/SR, and SF1/SR, respectively (Table 1).
2. Cycle the PCR reactions using the following conditions: Initial denaturation at 98°C for 2 min, 18 cycles of 98°C for 10 s, Y2 °C for 30 s and 72°C for 2 min (see Note 22), and a final extension at 72°C for 10 min. Y2 represents the optimum annealing temperatures and are 67°C, 63°C, and 66°C for one, two, and three codon scanning libraries, respectively.
3. Purify the PCR product using QIAquick PCR purification kit (Qiagen).

4. Digest the purified PCR product of all the three libraries with *Bsg* I as described above.
5. Treat the digestion product with Klenow fragment as described above.
6. Gel-purify the Klenow-treated digestion product using QIAEX II gel extraction kit (Qiagen).
7. Ligate and transform the gel-purified product as described above.
8. Collect sufficient colonies to build the final library. Save cell stocks as above and extract library DNA for future experiments (Fig. 1).

4. Notes

1. pIT-sfGFP (3) plasmid and other necessary DNA stocks can be sent upon request.
2. pIT vector contains a DNA sequence upstream fused to the gene of interest. This DNA sequence encodes the "head" reporter tatSS signal peptide and the N-terminal domain of *Saccharomyces cerevisiae* vacuolar membrane ATPase (VMA) intein. The tatSS signal peptide will direct a fused peptide or protein into the periplasm of Gram-negative bacteria, which is necessary for the function of many proteins such as β-lactamase. The VMA intein N-terminal domain (VMAI-N) is part of a split intein splicing system, the function of which requires the insertions of asymmetric transposon carrying the VMA intein C-terminal domain (VMAI-C). Importantly, the leading peptide (tatSS + VMAI-N) usually disrupts the native function of a fused protein. Hence, activity assays cannot be directly performed in the context of the pIT vector.
3. Prior to the transposition reaction, the asymmetric transposon is released from the pUC18-transposon vector by *Bgl* II/*Bam*H I digestion. This creates a 5′-GATC overhang at each end of the transposon that can increase the efficiency of transposition (4).
4. *Bpm* I, *Bsg* I, *Nhe* I, and *EcoR* I restriction sites should be removed from the gene of interest to accommodate subsequent processes.
5. It is advisable to optimize the codons in the gene of interest to accommodate protein expression in host cells.
6. This methodology has been successfully applied using the pTrcHisA vector (3). If another expression vector is used, some subsequent experimental conditions, such as PCR reactions, may

need to be re-optimized. *Bsg* I and *Bpm* I sites should be removed from this vector to accommodate subsequent processes.

7. The forward primer should contain an *Nhe* I site for the cloning. The nucleotides in the primers should be adjusted such that the gene of interest will be in-frame fused with the leading peptide. The reverse primer should contain an *EcoR* I site after stop codon.
8. Cautions should be taken if agarose DNA gels are to be stained by ethidium bromide.
9. When gel slices are incubated in solubilization buffer (Qiagen), temperatures above 50°C are not recommended. Higher temperatures denature double-stranded DNA and reduce the yield of gel-purification.
10. It is important to ensure that the gene of interest is placed in frame with the N-terminal fusion peptide carried in the pIT vector.
11. It is not recommended to use excess DNA (e.g., >2 μg) in *Bgl* II/*Bam*H I digestion. In our experience, excess DNA may result in incorrect or incomplete digestion that will reduce the efficiency of transposon integration.
12. Transposon DNA should be maintained at a molar excess (versus pIT-GOI plasmid). Higher transposon concentrations may lead to the formation of unstable pIT-GOI-transposon complexes containing multiple transposon insertions. These unstable complexes will compromise transposition reactions.
13. The time constant for each electroporation experiment should be monitored. A time constant above 4.0 ms is preferred for efficient transformation. Low time constant (below 3.0 ms) is indicative of too little resistance in medium and is frequently caused by too much residual salt. If this is the case the ligation product can be purified by spin column or ethanol precipitation. Alternatively, less ligation product can be used (for example, adding 0.5 μL reaction to per 50 μL competent cells).
14. When using competent cells with a competency of 5×10^7 CFU/μg DNA (colony-forming units, CFU), 20 μL transposition product typically results in >10,000 colonies. If desired, multiple 20 μL transposition reactions can be performed at once, pooled, ethanol precipitated, and then transformed.
15. The asymmetric transposon carries a reading-frame selection marker that consists of VMAI-C and TEM-1 β-lactamase (BLA). This fusion peptide has a stop codon after the BLA coding sequence but does not bear a start codon, such that its expression is strictly dependent on the insertion of transposon into an open reading frame. When the asymmetric transposon inserts in

frame with the gene of interest (as well as with the leading peptide), VMAI-N from the pIT vector will be in the same reading frame with VMAI-C from the asymmetric transposon. In this case, the VMA intein will self-splice and assemble the tatSS signal peptide and BLA, conferring ampicillin resistance.

16. Incubation at a reduced temperature is critical for a non-biased intein-mediated splicing (5). High temperatures may disrupt the folding of VMA intein when a destabilizing fusion peptide is introduced between VMAI-N and VMAI-C by transposon insertion. Compromised intein stability at high temperatures could be the main cause of a biased pattern of mutations.
17. The randomness of transposon insertions can be verified by restriction digestion or sequencing. The asymmetric transposon used in this study contains two closely placed *Mly* I sites at the 3′ end (~50 bp apart). Individual colonies from the transposon library can be digested using *Mly* I together with another unique restriction site on the pIT-GOI plasmid. Depending on the position of transposon insertion, two DNA fragments with variable sizes should be expected (3).
18. To calculate the required colonies for a full coverage library, the number of allowed transposon insertion sites and the site preference of the transposase (6) must be considered. The pIT vector backbone roughly contains 1,500 possible sites for transposon insertions (2,900 of total length minus 600 bps of replication origin and 800 bps of kanamycin resistance gene). There are thus $(L+1,500)$ possible insertion sites in pIT-GOI plasmid. Threefold redundant colonies, $3\times(L+1,500)$, are required to cover 95% of all the possible insertion events, assuming a non-biased insertion pattern (7). To compensate for the site preference of insertions, we assume that additional threefold redundancy is sufficient (8). Therefore, $3\times3\times(L+1,500)=9\times(L+1,500)$ colonies are sufficient to give a 95% coverage of all the transposon insertions. Among these, only one-third is expected to survive ampicillin selection due to the reading-frame selection system. Therefore, $3\times(L+1,500)$ colonies should be collected at this point.
19. For future experiments, the whole tube (1 mL) of stock cells should be used for inoculation to maintain the library diversity.
20. The transposon can insert both inside and outside of the gene of interest. In the former case, *Nhe* I/*EcoR* I digestion will result in two DNA fragments corresponding to pIT vector backbone and gene + transposon. If transposon inserts outside of the gene of interest, pIT + transposon and gene of interest bands are expected.
21. Because transposon insertions outside of the gene of interest have been purged from the library, only $3\times L$ colonies are required at this point to maintain a full coverage library.

22. It is recommended to monitor the quality of the PCR product. Due to the random transposon insertions, the "head" of a PCR product may overlap with the "tail" of another product, resulting in homologous recombination (9). This side product, shown as a smear on agarose DNA gels, should be avoided. In our experience, the side product becomes predominant once the concentration of the desired PCR product exceeds 2 ng/μL.
23. Phusion DNA polymerase requires 15–30 s for the extension of 1 Kbp DNA. The extension time needs to be adjusted according to the length of the gene of interest.
24. Although Fermentas provides standard versions of *Bpm* I (*Gsu* I), in our hands the FastDigest enzyme gives better cleavage efficiency.
25. Fresh SAM is critical for complete *Bsg* I digestion. Liquid SAM stocks can be stored at −80°C for no longer than 3 months.
26. Longer incubation time or higher temperatures are not desired, as overdigestion (such as removal of extra nucleotides) may occur in these cases.
27. It is important to quench the reaction first with EDTA prior to heat-inactivating the Klenow enzyme. Heat-inactivation without quenching frequently results in removal of extra nucleotides.
28. It is recommended to use 0.5 mM ATP in blunt-end ligations for higher end-joining efficiency.
29. Heat-inactivation of T4 DNA ligase has been reported to increase the efficiency of following transformation (10). This is particularly important if electro-transformation is to be used (11). Alternatively, the ligation product can be purified using commercial spin columns.
30. Multiple 10 μL ligation reactions can be performed, pooled, ethanol-precipitated, and then transformed into competent cells.
31. At this stage, the introduced codons and protein size should be taken into account for calculating the library diversity. For example, if NNK codon is being scanned through a protein consisting of 200 amino acids, 3 (redundancy factor) × 32 (NNK codon combination) × 200 = 19,200 colonies should be collected.

Acknowledgments

The authors thank the National Institutes of Health (GM084396) for financial support.

References

1. Cirino PC et al (2003) Generating mutant libraries using error-prone PCR. Methods Mol Biol 231:3–9
2. Stemmer WP (1994) DNA shuffling by random fragmentation and reassembly: in vitro recombination for molecular evolution. Proc Natl Acad Sci U S A 91:10747–10751
3. Liu J, Cropp TA (2012) A method for multi-codon scanning mutagenesis of proteins based on asymmetric transposons. Protein Eng Des Sel 25:67–72
4. Haapa S et al (1999) An efficient and accurate integration of mini-Mu transposons *in vitro*: a general methodology for functional genetic analysis and molecular biology applications. Nucleic Acids Res 27:2777–2784
5. Gerth ML et al (2004) A second-generation system for unbiased reading frame selection. Protein Eng Des Sel 17:595–602
6. Kim YC, Morrison SL (2009) N-terminal domain-deleted Mu transposase exhibits increased transposition activity with low target site preference in modified buffers. J Mol Microbiol Biotechnol 17:30–40
7. Patrick WM et al (2003) User-friendly algorithms for estimating completeness and diversity in randomized protein-encoding libraries. Protein Eng 16:451–457
8. Poussu E et al (2004) Probing the alpha-complementing domain of *E. coli* beta-galactosidase with use of an insertional pentapeptide mutagenesis strategy based on Mu *in vitro* DNA transposition. Proteins 54:681–692
9. Williams R et al (2006) Amplification of complex gene libraries by emulsion PCR. Nat Methods 3:545–550
10. Michelsen BK (1995) Transformation of *Escherichia coli* increases 260-fold upon inactivation of T4 DNA ligase. Anal Biochem 225:172–174
11. Ymer S (1991) Heat inactivation of DNA ligase prior to electroporation increases transformation efficiency. Nucleic Acids Res 19:6960

Chapter 17

Screening Libraries for Improved Solubility: Using *E. coli* Dihydrofolate Reductase as a Reporter

Jian-Wei Liu and David L. Ollis

Abstract

Low protein solubility is a problem in many areas of protein science. Although chemical methods have been developed to solubilize proteins these are not always effective and add to the cost of producing the protein. One way of overcoming these difficulties is to evolve the protein to be more soluble. A major hurdle in this process is the ability to select mutant proteins with enhanced solubility from a large library of randomly mutated proteins. In this article, we describe such a method. The method relies on the fact that increasing the expression of dihydrofolate reductase (DHFR) makes *Escherichia coli* resistant to Trimethoprim (TMP). Proteins fused to DHFR will produce chimeras with altered levels of resistance to TMP. This variation in TMP resistance can be used to identify mutant proteins with enhanced solubility.

Key words: Protein solubility, Directed evolution, Selection, Dihydrofolate reductase, Trimethoprim

1. Introduction

The cheapest and easiest way to produce large quantities of recombinant protein is with *Escherichia coli* (1). However, frequently proteins fail to express in a soluble form and may be incorporated into inclusion bodies (2). There are various approaches to solving this problem (3–5). Here we describe an evolutionary approach that involves screening large mutant libraries for variants with enhanced solubility. The method involves fusing the gene of the target protein to the *folA* gene that encodes for dihydrofolate reductase (DHFR)—a protein that is essential for the survival for *E. coli*. Now, DHFR can be inhibited by Trimethoprim (TMP) so that most strains of *E. coli* will not grow on agar plates containing more than 2 μg/mL of the antibiotic (6, 7). However, expressing DHFR at high levels in *E. coli* confers resistance to TMP—the concentration of DHFR exceeds that

James C. Samuelson (ed.), *Enzyme Engineering: Methods and Protocols*, Methods in Molecular Biology, vol. 978, DOI 10.1007/978-1-62703-293-3_17, © Springer Science+Business Media New York 2013

of TMP so that there is enough active DHFR for the organism to survive. If we fuse a soluble protein to DHFR and express the fusion protein at high levels then the *E. coli* will survive. Conversely, if the target protein is insoluble then the resulting fusing protein will be insoluble and the organism will not survive. Mutant forms of an insoluble target protein will confer resistance to TMP if the mutations improve the solubility. The level of resistance will depend upon the solubility of the mutant protein and can be used in directed evolution experiments to select proteins with enhanced solubility.

In practice a target protein may be partially soluble and may give rise to a TMP resistance level higher than that of wild-type *E. coli*. Trials are used to establish the lowest level at which wild-type colonies are inhibited by TMP. Once this basal level is established then large libraries can be screened at higher levels of TMP. Typically four rounds of evolution are used to produce soluble protein as shown in published studies (8, 9). The size of the libraries to be scanned will depend upon the mutation rate. As with any directed evolution experiment, the mutation rate should be kept low enough so that there is a high probability that a large fraction of the possible library can be screened.

The method has worked well with a small protein and a small domain of a larger protein. Problems were encountered with a larger oligomeric protein—possible due to the DHFR interfering with the oligomeric structure of the target protein. It should be noted that mutants that confer increased solubility might do so at the expense of activity. Mutants that survive the TMP screen should be subjected to a secondary screen for activity as part of each round of evolution. It should also be noted that this screen might select for mutant proteins that display increased expression.

There are other reporter proteins that have been used to screen for improved solubility (10–13). The system described here is relatively insensitive—small changes in solubility will result in small changes in TMP resistance. Significant increases in solubility can be acquired over several rounds of evolution with the mutants produced in each round showing only modest improvements over the previous round. The technique could be automated to screen for very large libraries.

The plasmid pJWL1030folA that is used to form the fusion protein is available from the authors upon request. It gives high level constitutive expression of the target protein attached to the N-terminus of DHFR by a five-residue linker as shown in Fig. 1. This mode of attachment means that mutations that result in a stop codon will not give rise to a viable protein, as would sometimes be the case if the target protein were fused to the C-terminus of DHFR. Furthermore, attachment to the C-terminus of DHFR may facilitate protein folding and enhance the solubility of the fusion protein, but not necessarily the target protein.

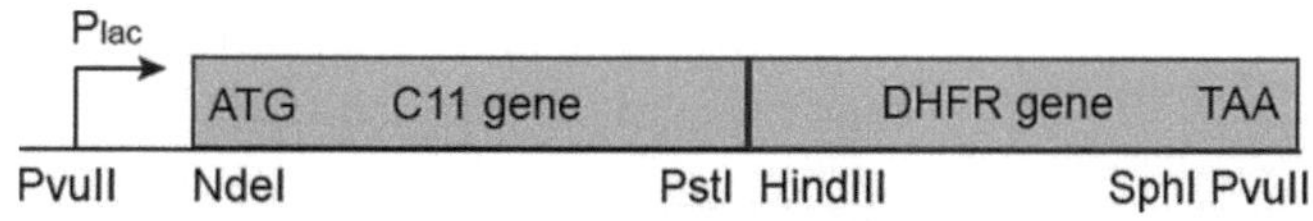

Fig. 1. The section of the pJWL1030folA plasmid with the gene for C11 attached to the *folA* gene. The ATG start codon is half of the site recognized by NdeI. The C11 gene codes for an insoluble protein and can be excised with NdeI and PstI (or HindIII) and replaced with the gene encoding the target protein.

2. Materials

1. *E. coli* strain DH5α (*sup*E44 Δ*lac*U169 ø80 *lac*Ω M15 *hs*dR17 *rec*A1 *end*A1 *gyr*A96 *thi*-1 *rel*A1) (see Note 1).
2. *E. coli* strain BL21(DE3) (F– *dcm ompT hsdS*(rB—mB–) *gal* λ (DE3)).
3. *E. coli* DHFR fusion vector: pJWL1030folA.
4. *E. coli* expression vector: pET vector.
5. Mueller Hinton Agar (MHA) (BD): Add 38 g in 1 L dH_2O, do not adjust pH, autoclave at 121°C for 20 min (see Note 2).
6. Luria–Bertani medium (LB): 10 g tryptone, 5 g yeast extract, 10 g NaCl, adjust pH to 7.5 with NaOH, add 15 g agar for plate, add dH_2O to 1 L, autoclave at 121°C for 20 min.
7. Minimal A medium (MMA): 10.5 g K_2HPO_4, 4.5 g KH_2PO_4, 1 g $(NH)_2SO_4$, 0.5 g sodium citrate·$2H_2O$, add dH_2O to 1 L, autoclave at 121°C 20 min. Then add 1 mL of 20% $MgSO_4$, 0.5 mL of 1% B1 (thiamine hydrochloride), 10 mL of 20% glucose, 20 mL of BME amino acid 50× stock solution (Sigma).
8. 10 mg/mL TMP stock: Dissolve 0.1 g TMP in 10 mL of propylene glycol (Fluka). Store at 4°C (see Note 3).
9. 50 mg/mL Kanamycin stock: Dissolve 0.5 g kanamycin in 10 mL dH_2O. Store at –20°C.
10. High-fidelity DNA polymerase: Phusion DNA polymerase.
11. Non-proofreading DNA polymerase: Taq DNA polymerase.
12. Restriction enzymes NdeI, PstI, HindIII.
13. T4 DNA ligase.
14. Alkaline phosphatase, calf intestinal (CIP).
15. Bugbuster Protein Extraction Reagent.
16. Plasmid mini prep kit, PCR purification kit, and gel purification kit.
17. BioRad Gene Pulser.

18. 96-well deep well growth blocks.
19. 96-well clear plates.
20. 96-well plate reader.
21. French Press Cell, sonicator, or homogenizer.

3. Methods

3.1. Test the Trimethoprim Resistance Level of the DFHR Fusion with the Wild-Type Protein of Interest

1. Amplify the coding region of the protein of interest by PCR with a high-fidelity DNA polymerase. Primers should be used that will result in the addition of NdeI and PstI at the 5′ and 3′ ends of the gene. The PCR product is isolated and digested with restriction enzymes so that it can be ligated into the DHFR fusion vector between NdeI and PstI sites (see Fig. 1, Note 4). Transform the plasmid containing the fusion proteins into *E. coli* DH5α. Plate out transformants onto LB agar plate containing 50 μg/mL of kanamycin and incubate at 37°C overnight. Isolate plasmid DNA and confirm its identity by digestion with NdeI and PstI or DNA sequencing.
2. Patch or streak the *E. coli* DH5α containing pJWL1030folA with the gene for the target protein onto MHA selection plates that contain 50 μg/mL of kanamycin plus 2, 5, 10, or 20 μg/mL of TMP. Incubate the plates at 37°C for 1–2 days. The minimal concentration of TMP that inhibits growth is the basal level and is used to select soluble proteins from the first-generation mutant library (see Note 5).

3.2. Select Soluble Mutants by E. coli DHFR Fusion Reporter

1. Create a mutant library of the target protein by introducing random mutations using a non-proofreading DNA polymerase (see Note 6). As described in the previous section, ligate the mutant library into the DHFR fusion vector between NdeI and PstI sites (see Note 7).
2. Transform the mutant library into the DH5α cells that have been made competent for electroporation. Plate out 50 μL of transformed cells onto LB agar plate containing 50 μg/mL of kanamycin. Determine the library size from the number of kanamycin-resistant colonies (see Note 8).
3. Wash the transformants with dH_2O to remove traces of medium that might contain thymidine. Plate out a volume of transformed cells that is equivalent to 1,000–2,000 kanamycin-resistant colonies onto each MHA agar selection plate. The MHA agar selection plates contain 50 μg/mL of kanamycin and the basal level of TMP in the first round. Subsequent rounds will be screened at higher levels of TMP.

4. The entire library may require 80–100 MHA selection plates which should be left at 37°C for 2–5 days (see Note 9).
5. Set up a 96-well growth block by adding 0.5 mL of MMA containing TMP (see Note 10) and 50 μg/mL of kanamycin to each well. Pick up 10–20 larger sized colonies from each MHA agar plate and inoculate the colonies in the MMA selection medium. Always include DH5α cells expressing fusion proteins of the DHFR and wild-type protein in each one of 96 wells as a control.
6. It may require 10–20 growth blocks to grow all the mutants from all the plates (see Note 11). Incubate the blocks at 37°C in a shaker overnight.
7. Transfer 250 μL of overnight cell culture of each clone from the 96-well growth block to a 96-well clear plate. Measure cell density with a plate reader at OD600.
8. Pick up one to two clones that give the highest cell growth from each growth block (see Note 12). Streak them onto LB agar plate that contains 50 μg/mL of kanamycin. Incubate at 37°C overnight.
9. Inoculate each clone in 2–10 mL LB containing 50 μg/mL of kanamycin and incubate at 37°C overnight. Isolate plasmid DNA and test that it has the correct insert by restriction digestion with NdeI and PstI. The gene should also be sequenced.
10. Combine equal quantities of the DNA of each mutant plasmid DNA and use them as template for next round of selection (see Note 13).
11. The process can be repeated until the desired solubility has been reached or until the evolution has converged.

3.3. Compare the Solubility of Wild-Type and Mutant Proteins

1. False positives may arise (see Note 14) and it may be wise to run an SDS-PAGE gel to inspect the soluble and insoluble fractions of the fusion proteins.
2. Amplify coding region of wild-type or mutant proteins with a high-fidelity DNA polymerase.
3. Clone the coding region of the wild-type or mutant proteins into an expression vector and transform into an *E. coli* expression host (see Note 15).
4. Express wild-type or mutant proteins in 5–100 mL LB containing appropriate antibiotic at 37°C.
5. The cells are disrupted by a French Press Cell, sonication, homogenizer, or with Bugbuster or some other detergent (see Note 16). Separate soluble and insoluble proteins by centrifugation at 30,000 × *g* for 30 min. Analyze soluble and insoluble protein by SDS-PAGE (see Note 17). An example of this type of analysis is given in Fig. 2.

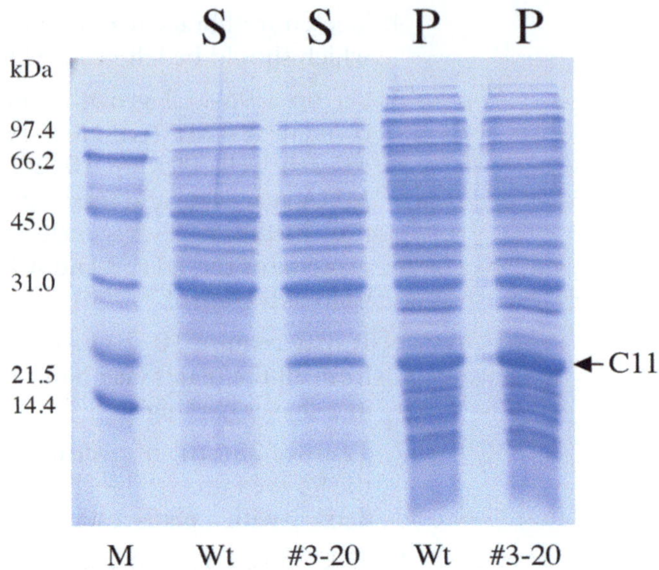

Fig. 2. Shows an SDS-PAGE gel with molecular weight markers in the first lane (M). The next two lanes contain the soluble fractions (S) of cells expressing the wild-type C11 protein (Wt) and the protein after three rounds of evolution (#3–20). There is very little, if any, soluble wild-type protein. Lanes four and five show the insoluble fraction (P) of cells expressing wild type (Wt) and mutant (#3–20). Directed evolution has resulted in an increase in the amount of expression and solubility of the C11 protein (#3–20).

4. Notes

1. *E. coli* DH5α is inhibited by less than 2 μg/mL of TMP. Other strains can be used after the minimal inhibition concentration of TMP has been determined.
2. Inhibition of thymidine production is a major aspect of the activity of TMP. The specialist media MHA is used for selecting colonies resistance to TMP, as Luria–Bertani medium contains thymidine (7).
3. Vortex will help to dissolve TMP in propylene glycol. Alternatively TMP can be dissolved in DMSO.
4. A coding region of a protein can be cloned into the *E. coli* DHFR fusion vector between NdeI and PstI or between NdeI and HindIII. The protein of interest should be in frame with the DHFR. There should be no stop codon between the testing protein and the *E. coli* DHFR.
5. The minimal inhibition concentration of the DHFR fusion with an insoluble protein is typically about 2 μg/mL and it is used to select soluble proteins from the first generation of mutant library. The TMP concentration is increased to 5 μg/mL, 10 μg/mL, 20 μg/mL, 50 μg/mL, and 100 μg/mL in the following generations.

6. Mutant libraries can be created by a number of methods including error-prone PCR, DNA shuffling, and StEP PCR (14).
7. *E. coli* DH5α containing the DHFR fusion vector pJWL-1030folA is resistant to more than 200 μg/mL of TMP. When preparing the DHFR fusion vector for mutant library construction, any incompletely digested vectors will produce false positives. To avoid this, a known insoluble protein, a putative phosphotriesterase C11 (15), is placed into the DHFR fusion vector. DH5α containing this vector is resistant to less than the 2 μg/mL of TMP (8). The insert is removed by digestion with NdeI and PstI and the vectors are dephosphorylated with alkaline phosphatase.
8. High-efficiency competent cells are required. For each generation, a mutant library of at least 100,000–200,000 clones should be selected on MHA selection plates.
9. At low levels of TMP colonies appear within 2 days. Up to 5 days are needed with high concentrations of TMP.
10. TMP selection pressure is increased by a factor of two over the TMP concentration that is used in the MHA agar plate selection.
11. About 1,000–2,000 of larger colonies that are equivalent to 1% of the whole library are picked from the MHA agar selection plates. They are further selected in secondary selection with MMA selection medium in 96-well plates.
12. About 10–40 clones that give the highest cell growth are picked from the 96-well plate MMA selections.
13. Like other protein evolution experiments, a number of mutant libraries are required to be selected sequentially on MHA selection plates with increasing concentration of TMP. TMP resistance of 20–100 μg/mL is a good indication that soluble fusion proteins are present in *E. coli* DH5α cells.
14. False positives can arise. They can occur if a proteolytic cleavage site is produced in the mutant library. Alternatively, if a ribosome-binding site evolves just upstream from a methionine codon then a short peptide fused to DHFR may be produced.
15. A pET/T7 expression vector is recommended to over-express wild-type and mutant proteins in *E. coli* BL21 (DE3) cells (16).
16. Detergent such as BugBuster sometimes causes less stable protein to precipitate. More than one method can be used to release proteins from the *E. coli* cells and some consideration should be given to this aspect of the process.
17. If the soluble protein is too little to be detected with Coomassie blue stain, western blotting should be used to compare the solubility of the wild-type and mutant proteins.

References

1. Graslund S, Nordlund P, Weigelt J et al (2008) Protein production and purification. Nat Methods 5:135–146
2. Peleg Y, Unger T (2012) Resolving bottlenecks for recombinant protein expression in E. coli. Methods Mol Biol 800:173–186
3. Dale GE, Broger C, Langen H et al (1994) Improving protein solubility through rationally designed amino acid replacements: solubilization of the trimethoprim-resistant type S1 dihydrofolate reductase. Protein Eng 7:933–939
4. Sun L, Petrounia IP, Yagasaki M et al (2001) Expression and stabilization of galactose oxidase in Escherichia coli by directed evolution. Protein Eng 14:699–704
5. Leong SR, Chang JC, Ong R et al (2003) Optimized expression and specific activity of IL-12 by directed molecular evolution. Proc Natl Acad Sci U S A 100:1163–1168
6. White PA, McIver CJ, Deng Y et al (2000) Characterisation of two new gene cassettes, aadA5 and dfrA17. FEMS Microbiol Lett 182:265–269
7. Watson M, Liu JW, Ollis D (2007) Directed evolution of trimethoprim resistance in Escherichia coli. FEBS J 274:2661–2671
8. Liu JW, Boucher Y, Stokes HW et al (2006) Improving protein solubility: the use of the Escherichia coli dihydrofolate reductase gene as a fusion reporter. Protein Expr Purif 47:258–263
9. Liu JW, Hadler KS, Schenk G et al (2007) Using directed evolution to improve the solubility of the C-terminal domain of Escherichia coli aminopeptidase P. Implications for metal binding and protein stability. FEBS J 274:4742–4751
10. Maxwell KL, Mittermaier AK, Forman-Kay JD et al (1999) A simple in vivo assay for increased protein solubility. Protein Sci 8:1908–1911
11. Waldo GS, Standish BM, Berendzen J et al (1999) Rapid protein-folding assay using green fluorescent protein. Nat Biotechnol 17:691–695
12. Wigley WC, Stidham RD, Smith NM et al (2001) Protein solubility and folding monitored in vivo by structural complementation of a genetic marker protein. Nat Biotechnol 19:131–136
13. Lesley SA, Graziano J, Cho CY et al (2002) Gene expression response to misfolded protein as a screen for soluble recombinant protein. Protein Eng 15:153–160
14. Neylon C (2004) Chemical and biochemical strategies for the randomization of protein encoding DNA sequences: library construction methods for directed evolution. Nucleic Acids Res 32:1448–1459
15. Horne I, Sutherland TD, Oakeshott JG et al (2002) Cloning and expression of the phosphotriesterase gene hocA from Pseudomonas monteilii C11. Microbiology 148:2687–2695
16. Stevenson BJ, Liu JW, Ollis DL (2008) Directed evolution of yeast pyruvate decarboxylase 1 for attenuated regulation and increased stability. Biochemistry 47:3013–3025

Chapter 18

In Vitro Directed Evolution of Enzymes Expressed by *E. coli* in Microtiter Plates

Bradley J. Stevenson, Sylvia H.-C. Yip, and David L. Ollis

Abstract

A method is described for using 96-well plates to prepare libraries of *Escherichia coli* cultures for screening a library of gene variants. This approach bypasses colony-picking to allow standard molecular biology laboratories to carry out directed evolution efficiently with a 96-well plate-reader and multichannel pipettes. Initial screens are applied to cultures that are rapidly prepared by diluting transformed cells so that an average of four cells starts each culture. Subsequent screens are used to isolate individual enzyme-expressing clones that exhibit activity higher than the parental clone. The outlined method also includes guidelines for preparing a library of gene variants and for optimizing a screening method.

Key words: Molecular breeding, Catalysis, Protein expression, Error-prone PCR, Screen

1. Introduction

In directed evolution large libraries of mutant enzymes are routinely screened for improved activity or stability. The enzyme in question is often expressed in *Escherichia coli*. In some cases the activity of the target enzyme can be monitored by a genetic selection. However, simple genetic selection is not available for many enzymes, so the library is screened by an assay that is usually done with an automated plate scanner. In these experiments, the mutated genes are placed in a plasmid and transformed into *E. coli*. Now the library is usually grown up on agar plates so that individual clones can be obtained and grown up in microtiter plates prior to assaying. These libraries are typically of the order of 10^4 and without a colony picker the transfer of individual colonies to the microtiter plate is both time consuming and tedious. We offer a simple

James C. Samuelson (ed.), *Enzyme Engineering: Methods and Protocols*, Methods in Molecular Biology, vol. 978,
DOI 10.1007/978-1-62703-293-3_18, © Springer Science+Business Media New York 2013

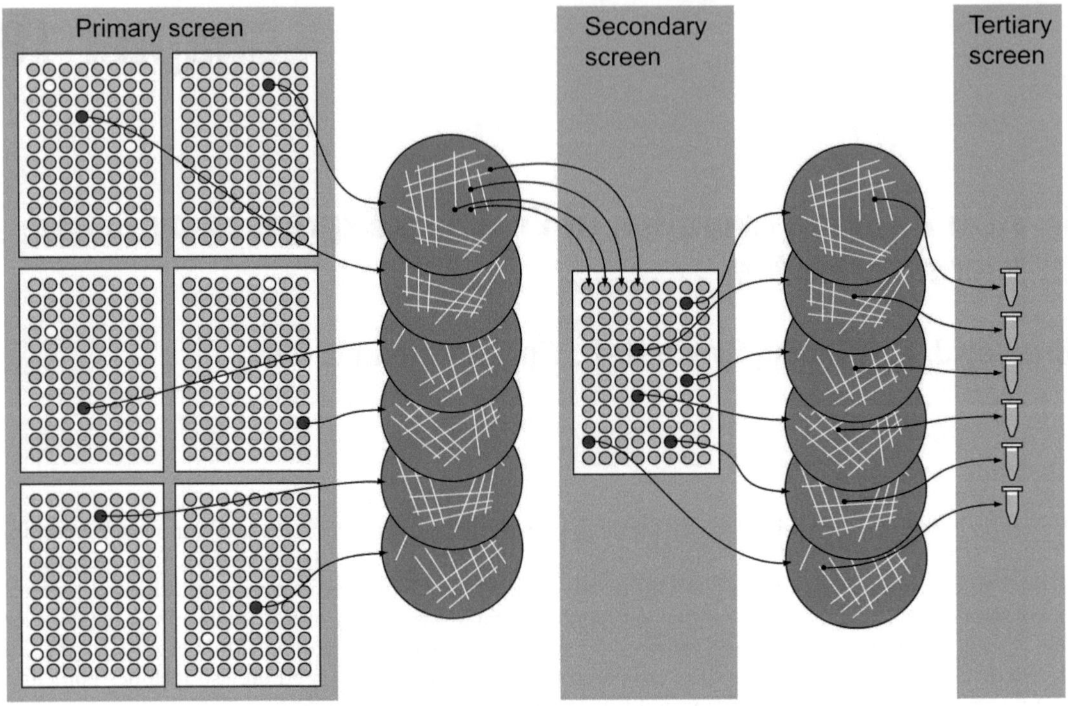

Fig. 1. General scheme for processing libraries in 96-well plates. Two isolation steps on media-agar plates are necessary, between each level of screen, to ensure that the single genotype responsible for the desired phenotype is isolated. A typical primary screen will consist of five batches of twenty 96-well plates with an average of four genotypes per well, allowing about 38,000 transformants to be screened.

time-saving alternative to this procedure that has been used in our laboratory for directed evolution of pyruvate decarboxylase (1) or glycerophosphodiesterase (2).

We describe a simple method that aims to bypass colony-picking for screening *E. coli* libraries. Transformants are prepared and the cell density measured. This library can then be diluted so that there are two to four viable cells in each well in a microtiter plate. These random culture arrays are incubated, sampled, lysed, and assayed for enzyme activity. Wells displaying high levels of activity are identified and the clones are then isolated. The method relies on additional screens to ensure that the genotype responsible for the optimal phenotype is isolated, as indicated in Fig. 1. Though colony-picking gives greater certainty when matching enzyme activity with a particular culture, this random array allows great savings in time and resources while allowing more gene mutants to be screened. This method is suitable for screening up to 50,000 transformants per generation with manual liquid-handling or more with robotic liquid-handling systems.

2. Materials

2.1. Gene Library Preparation

1. MilliQ water or equivalent, ice, and a laboratory equipped for molecular biology.
2. Reagents for PCR: DNA polymerase, reaction buffer, dNTPs, $MgCl_2$, and $MnCl_2$.
3. Equipment for PCR: Thermocycler and reaction tubes.
4. Equipment and reagents for analysis of DNA by gel electrophoresis.
5. Enzymes for molecular biology: Restriction enzymes (see Note 1), *Dpn*I (see Note 2), alkaline phosphatase (see Note 3), and T4 DNA ligase (several suppliers are available; we use New England Biolabs T4 DNA ligase).
6. Commercial kits for DNA purification: PCR reactions, agarose gel, or *E. coli* (we use Qiagen kits, but other manufacturers have similar kits available).
7. Gene library (*see* Subheadings 3.1 and 3.2).
8. Primers for gene amplification (see Note 4) and sequencing.
9. Plasmid for gene expression, such as a pET vector (see Note 5).
10. Competent non-expressing strain of *E. coli*, e.g., DH5α (see Note 6).
11. Sterile Luria Bertani (LB) media: 10 g/L Tryptone (Difco), 5 g/L yeast extract (Difco), and 10 g/L NaCl adjusted to pH 7 with 10 N NaOH and sterilized by autoclaving.
12. LB with antibiotic (LBA): We use pET vectors encoding β-lactamase and therefore use ampicillin at a final concentration of 50 mg/L to select transformants.
13. LBA agar plates: Autoclave LB with 1.5% agar, mix, cool to 50°C, add ampicillin to 50 mg/mL final concentration, and distribute among 90 mm diameter petri dishes.

2.2. Expression of Gene Library

1. Items in Subheading 2.1.
2. Competent *E. coli* for expression, such as DE3 lysogen for T7 RNA polymerase-based expression (3), e.g., BL21(DE3)^{RecA} (see Notes 5 and 6).
3. 80% (v/v) glycerol, sterilized by autoclaving.
4. LBA media with 1 mM lactose (LBAL, see Note 7).
5. Sterile 96-well plates made from clear polystyrene with round-bottomed wells and lid.
6. Multichannel pipette (50–200 μL range).
7. Petroleum jelly in 20 mL syringes (see Note 8).
8. Incubator at 37°C (see Note 9).

2.3. Assaying Enzyme Variants

1. Enzyme assay optimized for 96-well plates (see Subheading 3.4 and Note 10).
2. 96-well plate reader.
3. 96-well plate shaker (many plate readers have this function).
4. 96-well plates, flat bottom, clear for monitoring changes in absorption, opaque black for monitoring fluorescence, or white for monitoring luminescence.
5. Multichannel pipettes: 2–20 μL range and 50–200 μL range.
6. Lysis detergent: Several reagents for neutral bacterial cell lysis are available; we use B-PER (Pierce) or BugBuster (Novagen).

2.4. Isolating Optimal Variants

1. Materials listed in Subheadings 2.2 and 2.3.
2. LBA agar plates.
3. 5–10 μL sterile inoculation loops.
4. 10–15 mL culture tubes.
5. Incubator for holding and shaking culture tubes (150–200 rpm orbital).

3. Methods

3.1. Generating the Variant Gene Library

1. A simple method of generating diversity is required. This will also depend upon the requirements of the system being studied. The experimenter may want to vary the mutation rate or introduce shuffling for different generations. Several methods for generating mutant libraries are available. We have found that an effective approach is to combine error-prone PCR (4) with template switching by staggered extension process (5). Staggered extension process error-prone PCR (StEP-EPCR) generates new mutations while existing mutations are propagated in new combinations.
2. Combine the following for StEP-EPCR:
   (a) 10–100 ng of parent gene (or equimolar mixture of gene variants) amplified as plasmid in vivo.
   (b) 10 pmoles each of forward and reverse primers (see Note 4).
   (c) 10 nmoles of dNTPs.
   (d) 6.5 mM $MgCl_2$ and 0.1 mM $MnCl_2$ (for 0.5–1 kb gene, *see* Notes 11 and 12).
   (e) 5 μL of 10× polymerase buffer without Mg^{2+}.

(f) 1 μL of *Taq* DNA polymerase (see Note 12).

(g) Add pure water for a final volume of 50 μL.

At least two 50 μL reactions are required to prepare a library since the productivity of DNA polymerase is diminished by the presence of Mn^{2+}.

3. Run StEP-ePCR thermocycling with 30–50 cycles of
   (a) 10–30 s at 95°C.
   (b) 10–30 s at the annealing temperature (optimized for primers and template).
   (c) 30 s per 1 kb of PCR product at 72°C (see Note 13).

   After the final cycle, run for 2 min at 72°C to ensure that all products are complete.
4. Check the PCR product by gel electrophoresis alongside appropriate negative controls: a successful reaction will yield a single band corresponding to the gene size.
5. Isolate the PCR product (gene library) by either a PCR purification kit (if the PCR product has only one band visible by gel electrophoresis) or cutting the appropriate band after gel electrophoresis and then use a gel purification kit.

3.2. Cloning of the Gene Library

1. Digest the gene library with the chosen restriction enzymes (see Note 1) and *Dpn*I (see Note 2).
2. Digest the T7 expression plasmid (e.g., one of the pET vectors) with restriction enzymes followed by treatment with alkaline phosphatase (see Note 3).
3. Isolate the linear pET plasmid and the gene library by agarose gel electrophoresis, excise the correct bands of DNA, and purify each from the gel using a commercial kit.
4. Ligate the gene library into the expression plasmid with T4 DNA ligase following the manufacturer's guidelines for maximum ligation product. We typically use an overnight incubation at 16°C.
5. Transform the ligation product (pET-library) into an *E. coli* strain without T7 polymerase (see Note 6). Electroporation will typically provide more efficient transformation than chemically competent cells. Follow manufacturer's guidelines for electroporation and allow 1-h recovery without antibiotics.
6. After the transformation recovery period is complete, dilute the transformed culture tenfold with LBA media and mix (typically 1 mL of recovered transformants with 9 mL of LBA).
7. Take a sample of the culture and plate out 0.1% of the total volume (typically 10 μL) on an LBA agar plate. Incubate at 37°C overnight.

8. Incubate the remaining culture of transformants in a 50 mL tube with 150 rpm orbital shaking at 37°C overnight.
9. Count the number of colonies on the LBA agar plate (step 7) to determine how many transformants are present. The number of original transformants should exceed the number to be screened; 100,000 viable transformants (100 colonies) would be a suitable result from electroporation.
10. Use 2 mL of the overnight culture from step 8 to isolate plasmid DNA using a commercial kit. The result will be a library of predominantly supercoiled pET constructs ready to transform an *E. coli* T7 expression strain with high efficiency.
11. Examine the supercoiled pET-library by gel electrophoresis using an untreated sample and a sample digested with a restriction enzyme that cuts the plasmid once. Check that the DNA corresponds to a single plasmid size and determine the concentration.
12. Before proceeding to expression of the gene library, the mutation rate in the gene library should be measured (see Note 11). Select about ten colonies at random from the plate at step 7 for seeding 5 mL cultures in LBA. Isolate the plasmid DNA from each culture and sequence each gene to determine the number of mutations per gene.

3.3. Expression of the Gene Library

1. Use 1 μg of the supercoiled pET-library (Subheading 3.2, step 10) to transform a competent *E. coli* DE3 lysogen (see Note 6).
2. After the recovery period, add 80% glycerol for a final concentration of 16% as a cryoprotectant.
3. Distribute as 50 μL aliquots in 1.5 mL tubes, freeze rapidly in a dry ice/ethanol bath, and store at −80°C.
4. Thaw one aliquot to determine the density of viable (colony-forming) cells. Dilute 50-fold in LBA and spread out 50 μL on an LBA agar plate. Incubate at 37°C overnight and then count the colonies. The number of viable transformants in each frozen aliquot should exceed the number of wells to be screened.
5. On the day before running a screen, thaw an aliquot of the library for expression and dilute into LBAL (about 10 mL of media per 96-well plate) so that there are four viable transformed cells per 100 μL (see Note 14). Distribute 100 μL into each well of 96-well plates using a multichannel pipette (up to 20 plates is feasible for one researcher, depending on the assay).
6. Seal each plate with petroleum jelly around the inside edge of the lid (see Note 8).

7. Incubate at 37°C overnight (see Note 9). In our experiments, sufficient growth and enzyme expression have been possible with static cultures.
8. Verify the transformant density by counting the number of sterile wells (see Note 15). An average of four transformants starting each culture results in an average of two sterile wells per 96-well plate.

3.4. Assaying Enzyme Variants

The screening method is crucial to the success of any directed evolution project and will depend upon the enzyme being expressed and targeted properly. The expression system should be chosen carefully according to the properties of the enzyme. The screening method must deliver reproducible results with the control parental clone, and have sufficient sensitivity to detect improvements in enzyme activity (see Note 10). Before starting the screen, check the background activity (using *E. coli* transformed with the empty pET vector) to measure any activity produced by the *E. coli* expression host. Optimize the reagents and reaction timing so that it is possible to observe a linear response in assay activity with respect to the quantity of culture with native enzyme. Optimize the assay method so that cultures expressing native enzyme in each well of a 96-well plate result in a coefficient of variation (ratio of standard deviation to mean) less than 10% (6). To screen a 96-well plate of cultures with enzyme variants use the following procedure.

1. Resuspend the cells using a plate shaker (e.g., 16 Hz lateral shaking for 1 min).
2. Take a 5–20 μL sample of culture and transfer to a flat-bottom plate for assaying activity (see Note 16).
3. Add 5–20 μL of lysis detergent (see Note 17) in staggered fashion so that the time between adjacent rows receiving lysis detergent is the same (see Note 18).
4. After a defined incubation period (see Note 18), add assay buffer for a final volume of 100–200 μL in each well. Complete each row of wells in the same staggered fashion as used for the lysis buffer. This allows the lysis time to be consistent between wells and minimizes the variation in activity due to position in the plate.
5. Monitor the reactions in the same staggered fashion (if the plate reader has that capability).
6. Export data for calculations to identify wells with optimal activity. The location of sterile wells provides a useful confirmation when matching results with plates.
7. Mark the lid of the 96-well plates to indicate which wells to select for further testing (1 or 2 per plate). 96-well plates with cultures can be stored at 4°C for a few days, if required, before isolating optimal variants.

3.5. Isolating Optimal Variants

The convenience of preparing random arrays of transformants, with several genotypes likely to be present in each culture, comes with the requirement to carefully isolate a single clone responsible for any improved enzyme activity. The approach illustrated in Fig. 1 and outlined here ensures that optimal genes are isolated for further analysis.

1. Under sterile conditions, dip a sterile inoculation loop into a selected well and resuspend bacteria by rotating the loop handle between finger and thumb.
2. Transfer a sample in the loop to an LBA agar plate and streak out the culture so that isolated colonies can be grown.
3. Incubate the plates (one for each selected culture) at 37°C overnight.
4. Select four colonies at random to inoculate new 96-well plate wells with 100 μL of LBAL media in each (see Note 19).
5. Seal the 96-well plate(s) with petroleum jelly and incubate at 37°C overnight as done for the primary screen (Subheading 3.3).
6. Screen the plate(s) as in Subheading 3.4 and select the wells that contain optimal activity in this secondary screen. If all four cultures derived from colonies on the same LBA agar plate have high activity, select the one with the greatest activity (see Note 20).
7. To ensure clonal purity, take the cultures selected in the secondary screen and repeat steps 1–3 in this section. Alongside these, set up plates with cultures (or stock from −80°C) expressing the benchmark gene (best gene from the previous generation, or native gene) streaked out for isolated colonies.
8. Use one isolated colony from each LBA agar plate to inoculate 2 mL of LBAL media in a 10–15 mL culture tube and incubate overnight at 37°C with 200 rpm orbital shaking.
9. Mix 0.8 mL of each culture with 0.2 mL of 80% (v/v) glycerol in a 2 mL cryo-tube and store at −80°C (see Note 21). The remaining culture can be used for the activity assay.
10. At this point, the tertiary screen can be run as with the primary screen, but with greater replication to select the optimal cultures. We typically see one batch of twenty 96-well plates yield a few good candidates.
11. After running several batches of 96-well plates, prepare all of the "good candidates," alongside a benchmark culture, as cultures incubated together at 37°C for a final screen.
12. Aim to select about ten variants with improved characteristics from a total of about 100 plates. These can then be analyzed further and the respective plasmid samples can be purified for DNA sequencing and feeding into another round generating diversity and screening.

4. Notes

1. Choose restriction enzymes that have unique restriction sites in the plasmid and have no sites present in the enzyme gene. Check the buffer compatibility of the restriction enzymes. Ideally, they can be used with the same reaction buffer and digested in one reaction. Our laboratory often uses *Nde*I, since its restriction site can be incorporated within the start codon, and another restriction site that will yield cohesive ends that are not complementary to those of *Nde*I, e.g., *Xba*I, *Eco*RI, or others.
2. *Dpn*I recognizes GATC sites where the adenine is methylated during propagation of plasmid clones in standard *E. coli* host strains. *Dpn*I digestion is recommended to digest plasmid PCR templates to prevent carryover in subsequent transformation reactions (7). This enzyme is compatible with most other restriction enzyme buffers, so triple enzyme digests are often possible.
3. Treating the gene library vector with calf intestinal phosphatase (New England Biolabs) reduces the possibility of single-cut plasmid ligating with itself, or multiple-cut plasmid from ligating as concatemers. Plasmids purified by agarose gel electrophoresis will usually be free of single-cut product. However, if double-digestion only removes a small fragment of DNA from the plasmid (<10%) there may be contamination with single-cut vector. Phosphatase digestion prevents this contamination from producing viable plasmids that will result in transformants without variant enzyme being expressed.
4. This gene mutagenesis and cloning method will require primers that flank the target gene sequence: Points to consider for primer design:
   (a) Aim for a calculated T_m of 50–60°C with less than 2°C difference between the forward and reverse primer.
   (b) Aim for your primers to anneal to at least 15 nucleotides upstream or downstream of your gene.
   (c) Add three bases to the 5′ end of the restriction site in the primer to ensure efficient digestion of PCR products.
   (d) Avoid sequences that form stable primer dimers, heterodimers, or hairpins. In particular, avoid highly stable duplex formation that gives rise to a 3′ end that can be extended. You could also check that your primers are unique for the vector being used (particularly at the 3′ end of the primer).
5. A suitable plasmid should be chosen that can provide high expression of the target enzyme in *E. coli*. The plasmids should

provide controllable gene expression so that the variant gene library can be amplified without inducing gene expression. We used pETMCSIII (8) since it meets both these criteria. Prepare the plasmid from a construct that already has a section of DNA inserted between the restriction sites. This region should be unrelated to the function being selected for and be at least 20% of the plasmid size. This will allow efficient isolation of appropriately cut plasmid by gel purification.

6. We have had difficulty transforming BL21(DE3)^{RecA} cells with ligation product at sufficient efficiency for library screening. Instead, we transform an intermediate cloning strain (e.g., DH5α) to amplify plasmids with the gene variants before transformation into an expression strain. Other laboratory strains of *E. coli* (in addition to DH5α) that lack the T7 RNA polymerase gene can be used for this purpose. By establishing the library in a strain that does not express the target gene, the library will be amplified uniformly, without favoring gene variants that do not retard cell growth (by having an early stop codon, for example). This approach provides the gene variant library in a supercoiled form for high transformation rates of *E. coli.*

7. 100 μM lactose is sufficient to induce expression of T7 RNA polymerase and therefore the target gene within the pET vector (9). However, 1 mM lactose is necessary to ensure that all cells are importing lactose equally (10). We have found that lactose gives sufficient expression from a pET vector; however, some systems based on the *lac* operon may require isopropyl β-d-1-thiogalactopyranoside (IPTG) for effective expression.

8. It is necessary to seal the 96-well culture plates to prevent biased culture growth and a corresponding bias in assay activity: wells with greater aeration will have greater growth if the plate is not sealed. Several products are available to seal plates, but we have found that the most effective is to use petroleum jelly.

   Melt petroleum jelly in a beaker on a hot plate and aspirate the liquid into 20 mL disposable syringes. Leave the filled syringes to cool to room temperature. To seal a culture plate, use the syringe to apply a line of petroleum jelly around the inside edge of the plate lid. Press the lid down and check that a continuous seal is formed between the lid and the plate. When incubated at 37°C, the petroleum jelly will melt, but provided that the application is neat, the seal will remain without contaminating the wells.

9. A common source of variation during screening is due to temperature variation during culturing. Air currents can heat some plates, or sides of plates more than others, leading to greater culture growth and increased enzyme activity in screening assay. Therefore, make barriers to prevent direct airflow onto plates and allow an incubator to stabilize at 37°C before use.

10. Some questions to consider when developing an assay for screening:
    (a) How will enzyme activity be monitored? In the simplest case, the enzyme of interest catalyzes a reaction that results in a change that can be detected by a plate reader. Dehydrogenases offer convenient assays linked to NAD(P)H oxidation or $NAD(P)^+$ reduction for monitoring at 340 nm. Some enzyme activities can often be linked to dehydrogenase activity. Another simple option for hydrolase and kinase activity is to use a weakly buffered reaction with a pH indicator dye; bromothymol blue is ideal for physiological pH changes (11).
    (b) Is the enzyme active in the presence of a lysis detergent, as used in our method, or is it necessary to have the enzyme expressed on the surface or exported into the media?
    (c) Is the reaction suited to the end-point measurement or will the progress of the reaction need to be monitored?
11. Aim for an error rate to suit your gene size; one to two mutations per gene are recommended (12); however we have had successful results with 4.4 mutations per pdc1 gene on average (1). The total divalent cation concentration should be 6.6 mM and the appropriate Mn^{2+} concentration may need to be determined experimentally. As a guide, 0.1 or 0.4 mM Mn^{2+} resulted in an average of 2.6 or 6.4 nucleotide substitutions per 1,000 basepairs, respectively, with the yeast pdc1 gene.
12. To shuffle gene fragments without adding mutations use 6.6 mM $MgCl_2$, without $MnCl_2$. Use a proofreading polymerase such as KOD, Kapa Hi-fi, or *Pfu* instead of *Taq*. Use a shorter extension time at 72°C and use more PCR cycles (see Note 13).
13. The extension time at 72°C determines the amount of template switching during extension (recombination) and is dependent on the Mn^{2+} concentration. If aiming to recombine gene variants without adding new mutations then the absence of Mn^{2+} will allow faster extension. In this case, 10 s at 72°C will be sufficient for a proofreading DNA polymerase, but up to 100 cycles may be required (5). The shortest time at 72°C that still allows sufficient production of the PCR product will provide the most thorough recombination.
14. In our experience, an average of four transformant cells starting each culture (genotypes) resulted in a successful screen. However, if expecting only small changes in the enzyme activity, use only two or three genotypes per well. Higher genotype density will dilute the effect of improved variants, but if the assay method is sufficiently robust, with a low coefficient of variation, outstanding variants should be identified.

15. Assuming that the viable bacterial cells do not aggregate, are evenly distributed in the diluted library, and do not divide during pipetting, the initial number of viable cells placed in each well will follow the Poisson distribution:

$$P(r) = \frac{e^{-\mu}\mu^{r}}{r!}.$$

This equation describes the probability, $P(r)$, of a number of events (r, viable transformed cells) occurring if the mean number of events is μ. The only number of events that can be accurately observed is zero (a sterile well). By determining the proportion of sterile culture aliquots in a batch, it is possible to estimate the mean number of cells in each well. This is equal to the negative of the natural logarithm of the proportion of wells with no growth.

16. It may be necessary to provide additional agitation with the pipette tips so that the cells are completely resuspended. Once a row of tips have been used to transfer cultures, rinse with 80% ethanol (by pipetting up and down a few times), and eject the tips into a new tip box on a dry block incubator set to 100°C. In this way, two boxes of tips can be used alternately, while one is in use, the other is drying. The tips will remain sufficiently sterile and clean for about 100 plates.

17. In pipetting detergents for bacterial lysis: Preload the pipette tips by first depressing the pipette beyond the first stop, aspirate the detergent from the reagent trough, expel the detergent on the inside wall of the wells by depressing the pipette to the first stop only, and hold the plunger at this point before aspirating more detergent for the next row of wells. The detergent should be of sufficient volume (5–20 μL) so that it can mix effectively with the culture sample at the bottom of the well.

18. Lysis time needs to be long enough to process an entire plate. Stick to a staggered time process by using a count-up timer. We have used 15-s intervals between rows of wells. With an eight-channel pipettor, this results in a lysis time of at least 3 min before the detergent can be diluted with assay buffer. Check that the native enzyme is sufficiently active after this treatment. If the enzyme loses activity in the detergent, increased stability in the presence of the detergent will be selected for. The lysis step could be allowed to continue for longer periods to select for increased enzyme stability.

19. Aim to test at least as many colonies as there are average number of genotypes per well in the primary screen. Expression of enzyme activity may also hinder or alter colony growth, so it is important to select large and small colonies.

20. If promising activity observed in the primary screen is not picked up in the secondary screen, it will be possible to retest the original cultures. If only four colonies are picked from each plate for the secondary screen, it may also be necessary to test more colonies.
21. There is no need to freeze the glycerol stock rapidly on dry ice at this step; sufficient survival is possible if the cultures with 16% glycerol are placed directly in an ultra-freezer.

Acknowledgements

This work was supported by the Australian Research Council. We thank Nicholas Dixon for providing pETMCSIII and the *E. coli* strains.

References

1. Stevenson BJ, Liu JW, Ollis DL (2008) Directed evolution of yeast pyruvate decarboxylase 1 for attenuated regulation and increased stability. Biochemistry 47:3013–3025
2. Yip SH-C, Foo J-L, Schenk G, Gahan LR, Carr PD, Ollis DL (2011) Directed evolution combined with rational design increases activity of GpdQ toward a non-physiological substrate and alters the oligomeric structure of the enzyme. Protein Eng Des Sel 24:861–872
3. Studier FW, Rosenberg AH, Dunn JJ, Dubendorff JW (1990) Use of T7 RNA polymerase to direct expression of cloned genes. Methods Enzymol 185:60–89
4. Leung DW, Chen E, Goeddel DV (1989) A method for random mutagenesis of a defined DNA fragment using a modified polymerase chain reaction. Technique 1:11–15
5. Zhao H, Giver L, Shao Z, Affholter JA, Arnold FH (1998) Molecular evolution by staggered extension process (StEP) *in vitro* recombination. Nat Biotechnol 16:258–261
6. Salazar O, Sun L (2003) Evaluating a screen and analysis of mutant libraries. In: Arnold FH, Georgiou G (eds) Directed enzyme evolution: screening and selection methods. Humana Press, Totowa, NJ, p 383
7. Sambrook J, Russell DW (2001) Molecular cloning. A laboratory manual, 3rd edn. Cold Spring Harbor Laboratory Press, New York, p 13.19
8. Neylon C, Brown SE, Kralicek AV, Miles CS, Love CA, Dixon NE (2000) Interaction of the *Escherichia coli* replication terminator protein (Tus) with DNA: a model derived from DNA-binding studies of mutant proteins by surface plasmon resonance. Biochemistry 39:11989–11999
9. Studier FW (2005) Protein production by auto-induction in high-density shaking cultures. Protein Expr Purif 41:207–234
10. Khlebnikov A, Keasling JD (2002) Effect of lacY expression on homogeneity of induction from the Ptac and Ptrc promoter by natural and synthetic inducers. Biotechnol Prog 18:672–674
11. Stevenson BJ, Liu JW, Kuchel PW, Ollis DL (2012) Fermentative glycolysis with purified *Escherichia coli* enzymes for *in vitro* ATP production and evaluating an engineered enzyme. J Biotechnol 157:113–123
12. Moore JC, Arnold FH (1996) Directed evolution of a para-nitrobenzyl esterase for aqueous-organic solvents. Nat Biotechnol 14:458–467

Index

James C. Samuelson (ed.), *Enzyme Engineering: Methods and Protocols*, Methods in Molecular Biology, vol. 978,
DOI 10.1007/978-1-62703-293-3,

Current Topics in Biochemical Engineering

Edited by Naofumi Shiomi

Published in London, United Kingdom

IntechOpen

Supporting open minds since 2005

Current Topics in Biochemical Engineering
http://dx.doi.org/10.5772/intechopen.77355
Edited by Naofumi Shiomi

Contributors
Oscar A. Rojas-Rejon, Carlos González-Figueredo, Rene Alejandro Flores Estrella, José Manuel Pais-Chanfrau, Lorena Dominique Carrera Acosta, Paola Margarita Alvarado Cóndor, Jimmy Núñez Pérez, Milton Jimmy Cuaran Guerrero, Masayuki Azuma, Yoshihiro Ojima, Derrick E. Rancourt, Suman Nath, José Alberto Duarte-Moller, Rosa Isela Ruvalcaba, Anel Rocio Carrasco, Erasmo Orrantia-Borunda, Hilda Esperanza Esparza Ponce, Gordana Wozniak-Knopp, Stefan Vogt, Gerhard Stadlmayr, Florian Rüker, Johannes Grillari, Naofumi Shiomi

First published in London, United Kingdom, 2019 by IntechOpen
IntechOpen is the global imprint of INTECHOPEN LIMITED, registered in England and Wales, registration number: 11086078, The Shard, 25th floor, 32 London Bridge Street
London, SE19SG – United Kingdom
Printed in Croatia

British Library Cataloguing-in-Publication Data
A catalogue record for this book is available from the British Library

Additional hard and PDF copies can be obtained from orders@intechopen.com

Current Topics in Biochemical Engineering
Edited by Naofumi Shiomi
p. cm.
Print ISBN 978-1-83881-209-6
Online ISBN 978-1-83881-210-2
eBook (PDF) ISBN 978-1-83881-211-9

For EU product safety concerns:
IN TECH d.o.o., Prolaz Marije Krucifikse Kozulić 3, 51000 Rijeka, Croatia, info@intechopen.com or visit our website at intechopen.com.

Meet the editor

Dr. Naofumi Shiomi studied recombinant yeast as a researcher at the Laboratory of Production Technology of Kanena Corporation for 15 years until 1998 and earned his PhD degree in Engineering from Kyoto University. He now works as a professor at the School of Human Sciences of Kobe College in Japan, where he teaches biotechnology and life science in his "Applied Life Science" laboratory. He has studied bioremediation and biomedical science for 23 years at Kobe College and has published more than 50 papers and several books. His recent research has focused on the prevention of obesity and aging.

Contents

Preface

Gene manipulation and polymerase chain reaction technologies, which were born in the late 1870s, drastically changed the world of biotechnology. A substantial number of genes have been cloned from various species, and many genome sequences have been determined using these technologies. Moreover, the emergence of cloning technology and the discovery of embryonic stem (ES) cells drastically changed life science research. In the last several decades, the pathology of many diseases was elucidated by testing transgenic animals. The discovery of ES cells leads to the discovery of induced pluripotent stem (iPS) cells, and regenerative medicine using iPS cells is being used in several clinical trials at present. Biotechnology in life science continues to progress on a day-to-day basis.

With great expectations, the progress of biotechnology has imparted novel and important roles to biochemical engineers. Previously, the main role of biochemical engineers had been to design bioreactors that produced industrially important compounds under optimal conditions. Examples of these new roles for biochemical engineers include application of gene manipulation technology to bioremediation and production of bioenergy and novel chemicals. Moreover, the role has expanded to the medical field where tissue engineering, drug delivery, and therapy have grown drastically.

Based on this background, this book deals with current topics in biochemical engineering. The chapters of this book discuss research that has introduced artificial enzymes, kinetic models in bioprocessing, a small-scale production process, and production of energy with microbial fuel. These chapters offer novel ideas for the production of effective compounds and energy. Moreover, other research has introduced the production technology of stem cells and biomedical processes using nanoshells and extracellular vesicles, which will give important information for regenerative medicine and therapy.

I would like to thank Ms. Sara Petanjek and the publishing managers of InTech Publisher for their support and assistance throughout the writing and publication processes of this book.

Naofumi Shiomi
Kobe College, Japan

Chapter 1

Introductory Chapter: Artificial Enzyme Produced by Directed Evolution Technology

Naofumi Shiomi

1. Introduction

Until the middle of 1970s, the roles of biochemical engineers had been to design effective processes for production of industrially important proteins and to control bioreactors under optimal conditions. Many microorganisms were isolated from soils, rivers, and seawater, and their genes were randomly disrupted by mutagenic agents such as 1-methyl-3-nitro-1-nitrosoguanidine to enhance the productivity, which was the best procedure of breeding for a long time. Gene manipulation technology born in the late 1870s was a groundbreaking invention, and the technology drastically changed the world of biotechnology. Polymerase chain reaction (PCR) method and genome analysis using high-performance DNA sequencers, which were born after gene manipulation technology, enhanced the drastic progress. Recently, a huge number of genes have been cloned from isolated microorganisms, and many excellent vector plasmids and promoters have been developed.

Recently, the interest of biochemical engineers shifted from breeding using gene manipulation technology to using the discovered novel biocatalysts and the "directed evolution" technology. The idea of directed evolution was proposed by Arnold et al. They proposed that artificial evolution, which is similar to the natural evolution that slowly occurs in the nature world, can be performed at rapid speed using biotechnology. Based on this concept, many studies have been conducted and have become an important target of biochemical engineers. In the future, such artificial proteins will introduce a next wave in the world of biotechnology [1]. Based on this background, I introduce in this chapter the recent advancements in artificial enzymes. There are other important targets for biochemical engineers, and therefore, I recommend referring to other chapters of this book by other authors.

2. Construction of artificial proteins

The scheme of the screening process of an artificial protein or enzyme using directed evolution technology is shown in **Figure 1**. The screening process in a single cycle is composed of three steps: (1) construction of DNA (or RNA) variants, (2) construction of display library composed of both DNA (or RNA) variants and protein molecules, and (3) screening of target DNA variants. Directed evolution is completed by several repeats of these steps to obtain an optimal enzyme (or protein). Each step is discussed in the following sections.

IntechOpen

2.1 Computational design of artificial protein

Structural design by computational simulation provides important information to determine the optimal protein. There are many types of folds and motifs, the elements that compose the tertiary structure. Because of these elements, two proteins can have similar tertiary structures when they have more than 30% homology in their amino acid sequences, and the structural similarity is enhanced in proportion to increasing homology, according to the DBAli database. Moreover, homology models such as consensus protein design [2], phylogeny-based design [3], and design combined with simulation by molecular dynamics [4] have been recently developed. Therefore, the tertiary structure of an unknown protein can be predicted using a known homologous protein.

In the twentieth century, precise prediction of the tertiary structure of proteins had been difficult due to insufficient operation speed of computers. However, the operation speed of computers has drastically improved in the last 10 years. Using a recently developed supercomputer such as Kei computer [5], complicated structural analysis of proteins can be completed in a considerably shorter time. In addition to the progress in computer technology, high-performance DNA sequencers have also entered the next generation. HiSeq2000 DNA sequencer, which equips many flow cells having a huge number of template DNA, can determine giga bases of DNA sequences at one run. Data on genomes of many organisms have been added to data libraries at high speed owing to breakthrough in the DNA sequencing technology. Data on protein structure also have been enriched. The Protein Data Bank (PDB) is a data bank which includes experimental data such as X-ray crystal structural analysis and NMR data, which are important for prediction of 3D structures of proteins [6]. The data in PDB has rapidly increased in the past several years. Accordingly, computational analysis can now yield sufficiently correct 3D structures of proteins and is one of the most prominent tools to design artificial proteins.

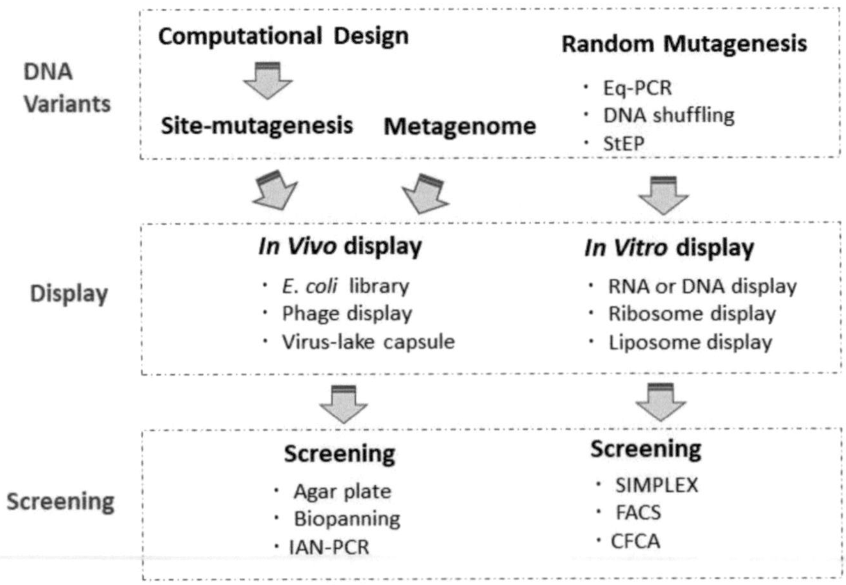

Figure 1.
Scheme of the screening process for artificial protein or enzyme.

2.2 Site mutagenesis and random mutagenesis

Mutation at the protein parts determined by computational structure simulation is generally performed by site mutagenesis or saturation mutagenesis method (site-direct mutagenesis at hotspots) [7]. For instance, activity of α-amylase was enhanced 16.7-fold using site-directed mutagenesis [7], and thermostability of transglutaminase was enhanced twelvefold at 60°C by saturation mutagenesis [8]. Truncation was also used for the mutation based on computationally designed results [9]. Enzymes include unnecessary domains, and characteristics of proteins are often improved by truncation. For example, the enzymes truncated at the N-terminus and C-terminus selected using a randomly truncated library showed high thermostability [10] and activity [11]. The most important advantage of site-direct mutagenesis is that the improved protein can be rapidly selected, because library size required for screening is much smaller than that in random mutagenesis.

Random mutagenesis is also a powerful tool to obtain novel enzymes [9], and several procedures for random mutagenesis such as error-prone PCR (ep-PCR), DNA shuffling, and staggered extension process (StEP) were proposed. In ep-PCR, random mutagenesis is introduced by PCR using error-prone polymerase under high concentration of Mn^{2+} and/or nonuninform dNTP concentration [12]. For example, activity of the subtilis E variant obtained by the ep-PCR method was 256 times higher in 60% DMF solution [13]. In DNA shuffling, many DNA variants containing mutations at different sites are digested with deoxyribonuclease, and PCR is performed using these fragments as template DNAs [14]. StEP is an improved method of DNA shuffling [15]. In this method, random mutation is introduced by repeated short DNA extension times in PCR with many DNA variants as template DNA. Random mutagenesis is a powerful tool to create unknown proteins.

2.3 Novel proteins based on mutagenesis analysis

Soils and seawater are abundant resources of microorganisms. According to the study on the sequences of DNAs of 16S rRNA of soil microorganisms [16], the ratio of microorganism which cannot grow under normal culture conditions to total microorganisms was 99%, suggesting that prominent biocatalysis may be present among microorganisms which cannot be cultured in media.

Metagenome is all genomes contained in the soil or water sample, and metagenomics which deals with metagenome is the other powerful tool to obtain proteins that we cannot even imagine [17, 18]. The procedure to obtain metagenome is as follows: DNAs of all microorganisms obtained from a target portion are purified without culture, digested to appropriated length with a restriction enzyme, and metagenomic library is constructed by combining the digested DNAs with adequate vectors. Metagenomic library of microorganisms living in special and harsh environment is especially useful because the enzymes have extremely prominent activity even though most of them are hard to culture under normal condition. Screening is recently performed to the environment such as hypersaline soda lake sediments [19] and hot environment [20, 21], and their metagenome data is accumulating day by day.

2.4 Display and screening to obtain important artificial proteins

2.4.1 In vitro display

Construction of display libraries and screening are the most important step in directed evolution technology [22]. In vitro and in vivo displays were proposed

[23, 24], and in vitro displays use a cell-free system. Cell-free systems contain components for protein synthesis such as ribosome, energy regeneration substrates, amino acids, and cofactors. Crude extract purified from wheat germ or insect cells [25, 26] and PURE system [27], a mixture of components separately purified from *Escherichia coli* cells, can be used as cell-free system. Protein can form normal tertiary structure by the addition of adequate chaperones.

In vitro display is a protein which is connected to a DNA or RNA variant coding the protein, and various types of this display were proposed [7]. In RNA display, RNA variants combined with puromycin via linker molecules at 3′ terminal of mRNA are used. Puromycin stops the protein synthesis by combining to C-terminus of protein; the protein combined with the mRNA is produced. In ribosome display, RNA variants of which stop codon is removed are used. The ribosome cannot be demolished because of loss of the stop codon; the complex composed of mRNA, protein, and ribosome is produced. In DNA display, the biotin-labeled DNA variant which codes streptavidin gene at the terminal of the gene is used, and protein is synthesized in W/O emulsion. The protein tagged by streptavidin combines with biotin-labeled DNA. Liposome display, a single vesicle liposome containing the cell-free synthesis system, is used. Advantage of liposome display is that membrane protein can be displayed at surface of lipid bilayer.

Target proteins must be screened from displays. In case of screening based on coupling constant, selection of protein is performed based on strength of binding with a target molecule immobilized on a plate or micro-beads. In the case of screening based on enzyme activity, the SIMPLEX method using microplate [28] can be used for the selection of DNA, RNA, and ribosome displays. Display using emulsion [29] and liposome display [30] is more useful for high-throughput screening because FACS or confocal fluorescence coincidence analysis (CFCA) [31] can be used. The advantage of in vitro display using a cell-free system is that construction of library and automation at microscale are easy.

2.4.2 In vivo displays

In vivo system using *E. coli* cells is also used for screening of proteins. The most general procedure is construction of a gene library of *E. coli* transformants: Plasmids containing DNA variants are transformed into *E. coli* cells. Phage display and display using virus-like molecules were also proposed. Phage display, which was developed to display V-region of antibody, can display the target protein at the surface of the phage by the fusion with coat protein of bacteriophage. Relatively large proteins can be displayed by the fusion with N-terminus of g3p protein of M13 phage or C-terminus of g10 protein of T7 phage [32]. Recently, several groups developed display using nucleocapsids as artificial virus [33, 34]. Non-viral cage of lumazine synthase (AaLS) was spontaneously formed, and mRNA could be contained in the capsule.

Selection method of target protein in in vivo displays is as follows: In case of the selection based on bonding strength, protein can be selected with phage display [32]. Biopanning is generally used for the selection from phage display: Phage displays connecting with immobilized molecules are collected and are transformed into *E. coli* cells. In the case of the selection based on enzyme activity, agar plate containing screening medium, which is the general method for screening of microorganisms, is used in the selection of *E. coli* transformant library. Besides this method, a method such as IAN-PCR method, which is based on amino acid sequence, is proposed. The procedure using *E. coli* library is not adapted to select from a huge size of library because the screening speed is very low, but this screening method is adapted to select from small-scale library such as site-mutated library combined with computer simulation or metagenome library.

3. Targets of directed evolution technology

Directed evolution is a trend in twenty-first century, and interesting study subjects have been proposed [35]. One of the interesting study subjects is, of cause, development of super-proteins which are not present in the nature world and can catalyze novel reaction, and the study on Kemp elimination reaction is its pioneer. Natural enzyme catalyzing the Kemp elimination reaction had not been discovered. Röthlisberger et al. produced the eight novel enzymes catalyzing Kemp elimination reaction by using computational design and site mutagenesis [36]. Similarly, Hilvert et al. conducted the site-directed mutagenesis and screened by using droplet-based microfluidic screening platform. The obtained artificial aldolases showed high activity although the original protein showed few activities [37]. Tryptophan synthetase and cytochrome P450 variants obtained by directed evolution could catalyze novel reactions [38]. In addition to directed evolution, metagenomics is a powerful tool to discover such unknown enzymes [39, 40]. For instance, more than unknown 300 nitrilases were discovered from the metagenome library, although nitrilase discovered during many years of study was only less than 20 [41].

Directed evolution technology can be also applied to improvement of metabolic pathways [42]. For example, 1-propanol is expected as a biofuel, but adequate producing strains have not been discovered. Thus, Atsumi et al. produced *E. coli* recombinant by expressing a series of genes for 1-propanol production, and improvement of citramalate synthase gene was conducted by using error-prone PCR and DNA shuffling. The productivities of 1-propanol and 1-butanol in the improved strain were enhanced [43]. Arnold et al. improved alcohol dehydrogenase and ketol-acid reducto-isomerase to enhance the productivity of 1-propanol by directed evolution. The obtained enzymes could use NADH instead of NADPH as a coenzyme [44]. Otherwise, directed evolution technology is applied to regulatory genes for improvement of metabolic pathway. Promoter [45], operon connection [46], and enhancer sequences [47] were improved by the method.

The other interesting study subject is to understand the evolution of life. One approach to the subject is to know which genome length is required for a micro-organism. Gibson et al. constructed artificial genome DNA which only includes chemically synthesized and introduced the genome into microorganisms in which the genome was previously removed [48]. Another research group tried to construct minimal bacterial genome; the constructed genome JCVI-syn3.0 was only 531 kb (473 genes) [49], and the obtained microorganism could grow and showed several characteristics of the microorganism.

The other approach to understand evolution of life is to generate artificial cells using a cell-free system. Szostak et al. randomly synthesized RNAs (1015 of 90 mer) and created a ribozyme having sufficient RNA ligase activity using error-prone PCR with only 10 rounds of repeats [50]. Noireaux et al. constructed cell-sized synthetic vesicle (artificial cells) containing components for translation and transduction [51]. Yomo et al. also produced novel artificial cells, which can progress artificial evolution of RNAs by themselves [52]. Induction of unnatural compounds into cells is another approach. Unnatural basic pairs and more than 100 unnatural amino acids were synthesized, and they were site-specifically introduced into proteins [53]. Accordingly, evolutional RNA engineering may impart validity to the hypothesis of RNA word in the future.

4. Conclusion

Enzymes are useful for industrial chemical reactions, and the role of biochemical engineers was to study the industrial use of enzymes. However, most of the enzymes that occur in nature do not have sufficient activity for industrial use, and the bioreactors were not as successful as expected. Directed evolution technology is proposed to overcome these problems. Artificial enzymes have been drastically produced by exploiting the direct evolution technology. The technology may be advanced by using editing tools such as TALEN and CRISPR/Cas9. Artificial microorganisms whose genomes are directly improved using CRISPR/Cas9 may be more suitable than recombinant *E. coli* for use as bioreactors. As such, biochemical engineers will play more important roles in the development of bioreactors in the future.

Author details

Naofumi Shiomi
School of Human Sciences, Kobe College, Japan

*Address all correspondence to: n-shiomi@mail.kobe-c.ac.jp

References

[1] Bornscheuer UT, Huisman GW, Kazlauskas RJ, Lutz S, Moore JC, Robins K. Engineering the third wave of biocatalysis. Nature. 2012;**485**:185-194. DOI: 10.1038/nature11117

[2] Porebski BT, Buckle1 AM. Consensus protein design. Protein Engineering, Design & Selection. 2016;**29**:245-251. DOI: 10.1093/protein/gzw015

[3] Akanuma S, Iwami S, Yokoi T, Nakamura N, Watanabe H, Yokobori S, et al. Phylogeny-based design of a B-subunit of DNA gyrase and its ATPase domain using a small set of homologous amino acid sequences. Journal of Molecular Biology. 2011;**412**:212-225. DOI: 10.1016/j.jmb.2011.07.042

[4] Wijma HJ, Floor RJ, Bjelic S, Marrink SJ, Baker D, Janssen DB. Enantioselective enzymes by computational design and in silico screening. Angewandte Chemie. 2015;**54**:3726-3730. DOI: 10.1002/anie.201411415

[5] Greene T. 10 of the world's fastest supercomputers. IDG News Service. 2018

[6] RCSB PDB. Available from: https://www.rcsb.org/

[7] Yang H, Li J, Shin HD, Du G, Liu L, Chen J. Molecular engineering of industrial enzymes: Recent advances and future prospects. Applied Microbiology and Biotechnology. 2014;**98**:23-29. DOI: 10.1007/s00253-013-5370-3

[8] Buettner K, Hertel TC, Pietzsch M. Increased thermostability of microbial transglutaminase by combination of several hot spots evolved by random and saturation mutagenesis. Amino Acids. 2012;**42**:987-996. DOI: 10.1007/s00726-011-1015-y

[9] H1 Y, Liu L, Shin HD, Chen RR, Li J, Du G, et al. Structure-based engineering of histidine residues in the catalytic domain of α-amylase from *Bacillus subtilis* for improved protein stability and catalytic efficiency under acidic conditions. Journal of Biotechnology. 2013;**164**:59-66. DOI: 10.1016/j.jbiotec.2012.12.007

[10] Kim Y-M, Shimizu R, Nakai H, Mori H, Okuyama M, Kang M-S, et al. Truncation of N- and C-terminal regions of *Streptococcus* mutans dextranase enhances catalytic activity. Applied Microbiology and Biotechnology. 2011;**91**:329-339. DOI: 10.1007/s00253-011-3201-y

[11] Wang Y, Yuan H, Wang J, Yu Z. Truncation of the cellulose binding domain improved thermal stability of endo-beta-1,4-glucanase from *Bacillus subtilis* JA18. Bioresource Technology. 2009;**100**:345-349. DOI: 10.1016/j.biortech.2008.06.001

[12] Shao W, Ma K, Le Y, Wang H, Sha C. Development and use of a novel random mutagenesis method: In situ error-prone PCR (is-epPCR). In: Reeves A, editor. In Vitro Mutagenesis. NY: Humana Press; 2016. pp. 497-506

[13] You L, Arnold FH. Directed evolution of subtilisin E in *Bacillus subtilis* to enhance total activity in aqueous dimethylformamide. Protein Engineering. 1994;**9**:77-83

[14] Stemmer WP. DNA shuffling by random fragmentation and reassembly: In vitro recombination for molecular evolution. Proceedings of the National Academy of Sciences of the United States of America. 1994;**91**:10747-10751

[15] Zhao H, Giver L, Shao Z, Affholter JA, Arnold FH. Molecular evolution by staggered extension process (StEP)

in vitro recombination. Nature Biotechnology. 1998;**16**:258-261

[16] Hugenholtz P, Goebel BM, Pace NR. Impact of culture-independent studies on the emerging phylogenetic view of bacterial diversity. Journal of Bacteriology. 1998;**180**:4765-4774

[17] Uchiyama T, Miyazaki K. Functional metagenomics for enzyme discovery: Challenges to efficient screening. Current Opinion in Biotechnology. 2009;**20**: 616-622. DOI: 20:616-622. 10.1016/j.copbio.2009.09.01

[18] Simon C, Daniel R. Metagenomic analyses: Past and future trends. Applied and Environmental Microbiology. 2011;**77**:1153-1161. DOI: 10.1128/AEM.02345-10

[19] Vavourakis CD, Andrei A-S, Mehrshad M, Ghai R, Sorokin DY, Muyzer G. A metagenomics roadmap to the uncultured genome diversity in hypersaline soda lake sediments. Microbiome. 2018;**6**:168. DOI: 0.1186/s40168-018-0548-7

[20] Wohlgemuth R, Littlechild J, Monti D, Schnorr K, et al. Discovering novel hydrolases from hot environments. Biotechnology Advances. 2018;**36**:2077-2100. DOI: 10.1016/j.biotechadv.2018.09.004

[21] Kaushal G, Kumar J, Sangwan RS, Singh SP. Metagenomic analysis of geothermal water reservoir sites exploring carbohydrate-related thermozymes. International Journal of Biological Macromolecules. 2018;**119**:882-895. DOI: 10.1016/j.ijbiomac.2018.07.196

[22] Newton MS, Arcus VL, Gerth ML, Patrick WM. Enzyme evolution: Innovation is easy, optimization is complicated. Current Opinion in Structural Biology. 2018;**48**:110-116. DOI: 10.1016/j.sbi.2017.11.007

[23] Lu Y. Cell-free synthetic biology: Engineering in an open world. Synthetic and Systems Biotechnology. 2017;**2**:23-27. DOI: 10.1016/j.synbio.2017.02.003

[24] Jijakli K, Khraiwesh B, Fu W, Luo L, Alzahmi A, et al. The in vitro selection world. Methods. 2016;**106**:3-13. DOI: 10.1016/j.ymeth.2016.06.003

[25] Takai K, Sawasaki T, Endo Y. Practical cell-free protein synthesis system using purified wheat embryos. Nature Protocols. 2010;**5**:227-238. DOI: 10.1038/nprot.2009.207

[26] Ezure T, Suzuki T, Shikata M, Ito M, Ando E. A cell-free protein synthesis system from insect cells. Methods in Molecular Biology. 2010;**607**:31-42. DOI: 10.1007/978-1-60327-331-2_4

[27] Kazuta Y, Matsuura T, Ichihashi N, Yomo T. Synthesis of milligram quantities of proteins using a reconstituted in vitro protein synthesis system. Journal of Bioscience and Bioengineering. 2014;**118**:554-557. DOI: 10.1016/j.jbiosc.2014.04.019

[28] Ohuchi S, Nakano H, Yamane T. In vitro method for the generation of protein libraries using PCR amplification of a single DNA molecule and coupled transcription/translation. Nucleic Acids Research. 1998;**26**:4339-4346

[29] Griffiths AD, Tawfik DS. Directed evolution of an extremely fast phosphotriesterase by in vitro compartmentalization. The EMBO Journal. 2003;**22**:24-35. DOI: 10.1093/emboj/cdg014

[30] Fujii S, Matsuura T, Sunami T, Kazuta Y, Yomo T. In vitro evolution of α-hemolysin using a liposome display. Proceedings of the National Academy of Sciences of the United States of America. 2013;**110**:16796-16801. DOI: 10.1073/pnas.1314585110

[31] Winkler T, Kettling U, Koltermann A, Eigen M. Confocal fluorescence coincidence analysis: An approach to ultrahigh-throughput screening. Proceedings of the National Academy of Sciences of the United States of America. 1999;**96**:1375-1378

[32] Bazan J, Całkosiński I, Gamian A. Phage display—A powerful technique for immunotherapy. Human Vaccines & Immunotherapeutics. 2012;**8**:1817-1828. DOI: 10.4161/hv.21703

[33] Butterfield GL, Lajoie MJ, Gustafson HH, Sellers DL, Nattermann U, et al. Evolution of a designed protein assembly encapsulating its own RNA genome. Nature. 2017;**552**:415-420. DOI: 10.1038/nature25157

[34] Terasaka N, Azuma Y, Hilvert D. Laboratory evolution of virus-like nucleocapsids from nonviral protein cages. Proceedings of the National Academy of Sciences of the United States of America. 2018;**115**:5432-5437. DOI: 10.1073/pnas.1800527115

[35] Martínez R, Schwaneberg U. A roadmap to directed enzyme evolution and screening systems for biotechnological applications. Biological Research. 2013;**46**:395-405. DOI: 10.4067/S0716-97602013000400011

[36] Röthlisberger D, Khersonsky O, Wollacott AM, Jiang L, DeChancie J, et al. Kemp elimination catalysts by computational enzyme design. Nature. 2008;**453**:190-195. DOI: 10.1038/nature06879

[37] Obexer R, Godina A, Garrabou X, Mittl PR, Baker D, Griffiths AD, et al. Emergence of a catalytic tetrad during evolution of a highly active artificial aldolase. Nature Chemistry. 2017;**9**:50-56. DOI: 10.1038/nchem.2596

[38] Romney DK, Murciano-Calles J, Wehrmüller JE, Arnold FH. Unlocking reactivity of TrpB: A general biocatalytic platform for synthesis of tryptophan analogues. Journal of the American Chemical Society. 2017;**139**:10769-10776. DOI: 10.1021/jacs.7b05007

[39] Madhavan A, Sindhu R, Parameswaran B, Sukumaran RK, Pandey A. Metagenome analysis: A powerful tool for enzyme bioprospecting. Applied Biochemistry and Biotechnology. 2017;**183**:636-651. DOI: 10.1007/s12010-017-2568-3

[40] Steele HL, Jaeger KE, Daniel R, Streit WR. Advances in recovery of novel biocatalysts from metagenomes. Journal of Molecular Microbiology and Biotechnology. 2009;**16**:25-37. DOI: 10.1159/000142892

[41] Robertson DE, Chaplin JA, DeSantis G, Podar M, Madden M, et al. Exploring nitrilase sequence space for enantioselective catalysis. Applied and Environmental Microbiology. 2004;**70**:2429-2436. DOI: 10.1128/AEM.70.4.2429-2436.2004

[42] Pröschel M, Detsch R, Boccaccini AR, Sonnewald U. Engineering of metabolic pathways by artificial enzyme channels. Frontiers in Bioengineering and Biotechnology. 2015;**3**:168. DOI: 10.3389/fbioe.2015.00168

[43] Atsumi S, Liao JC. Directed evolution of *Methanococcus jannaschii* citramalate synthase for biosynthesis of 1-propanol and 1-butanol by *Escherichia coli*. Applied and Environmental Microbiology. 2008;**74**:7802-7808. DOI: 10.1128/AEM.02046-08

[44] Brinkmann-Chen S, Flock T, Cahn JK, Snow CD, Brustad EM, McIntosh JA, et al. General approach to reversing ketol-acid reductoisomerase cofactor dependence from NADPH to NADH. Proceedings of the National Academy of Sciences of the United States of America. 2013;**110**:10946-10951. DOI: 10.1073/pnas.1306073110

[45] Alper H, Fischer C, Nevoigt E, Stephanopoulos G. Tuning genetic control through promoter engineering. Proceedings of the National Academy of Sciences of the United States of America. 2005;**102**:12678-12683. DOI: 10.1073/pnas.0504604102

[46] Pfleger BF, Pitera DJ, Smolke CD, Keasling JD. Combinatorial engineering of intergenic regions in operons tunes expression of multiple genes. Nature Biotechnology. 2006;**24**:1027-1032. DOI: 10.1038/nbt1226

[47] Biyani M, Biyani M, Nemoto N, Husiki Y. Evolutionary molecular engineering to efficiently direct in vitro protein synthesis. In: Biyani M, editor. Cell-Free Protein Synthesis. Rejeca: InTech; 2012. pp. 551-562

[48] Gibson DG, Glass JI, Lartigue C, Noskov VN, Chuang RY, Algire MA, et al. Creation of a bacterial cell controlled by a chemically synthesized genome. Science. 2010;**329**:52-56. DOI: 10.1126/science.1190719

[49] Hutchison CA, Chuang RY, Noskov VN, Assad-Garcia N, Deerinck TJ, et al. Design and synthesis of a minimal bacterial genome. Science. 2016;**351**:aad6253. DOI: 10.1126/science.aad6253

[50] Bartel DP, Szostak JW. Isolation of new ribozymes from a large pool of random sequences. Science. 1993;**261**:1411-1418. DOI: 10.1126/science.7690155

[51] Noireaux V, Maeda YT, Libchaber A. Development of an artificial cell, from self-organization to computation and self-reproduction. Proceedings of the National Academy of Sciences of the United States of America. 2011;**108**:3473-3480. DOI: 10.1073/pnas.1017075108

[52] Ichihashi N, Usui K, Kazuta Y, Sunami T, Matsuura T, Yomo T. Darwinian evolution in a translation-coupled RNA replication system within a cell-like compartment. Nature Communications. 2013;**4**:2494. DOI: 10.1038/ncomms3494

[53] Uzawa T, Tada S, Ito Y. Expansion of the aptamer library from a "natural soup" to an "unnatural soup". Chemical Communications. 2013;**49**:1786-1795. DOI: 10.1039/c2cc36348h

Chapter 2

Fermentation: Metabolism, Kinetic Models, and Bioprocessing

Carlos González-Figueredo,
René Alejandro Flores-Estrella and Oscar A. Rojas-Rejón

Abstract

Biochemical and metabolic interpretation of microbial growth is an important topic in bioreactor design. We intend to address valuable information about the relation of critical operation variables and the simulation of bioprocesses with unstructured and structured kinetic models. Process parameters such as nutrient supply, pH, dissolved oxygen, and metabolic end-products directly impact the physiology and metabolism of microorganisms. Changes in the membrane as well as cell viability are of interest since protein expression and maturation in prokaryota are directly related to membrane integrity. This chapter intends to deliver an insight of different alternatives in kinetic modeling.

Keywords: metabolism, unstructured kinetic models, black box models, gray box models

1. Introduction

Bacteria are the dominant form of life spread across the whole planet. Their biochemistry machinery is well adapted to scarcity conditions; also, they can biosynthesize complex molecules in various environmental conditions. For this reason, the growth and proliferation of bacteria in controlled environments represent an interest of biochemical engineers, microbiologists, and cell-growth enthusiasts since they allow bioprocess simulation and control scheme design. Substrate transformations into cell biomass, organic molecules, therapeutic proteins, biofuels, enzymes, and food additives are of attention since application to actual fields and laboratory experiments are very difficult to scale-up to industrial level with strict and complete control of key variables determined as an ideal process [1]. It is known that the complexity in a mathematical model may increase with the inclusion of environmental conditions such as multisubstrate consumption and product formation, pH change during fermentation, variable temperature, rheological changes in culture media, multiphasic environmental variability, and nonideality of mixing and stirring [2]. The kinetic model had been evolved from simple exponential growth to complex mathematical expressions to predict heterogeneity in single cells, describe multiple reactions, explain internal control mechanisms, and even predict genetic variability between bacterial populations [3]. However, despite the efforts to represent the progress of biological reactions in microbial cultures, the actual application of the model in real production processes is impractical due to a significant amount of information fed to the model [4].

Many of the kinetic growth models base their structure on and take information from empirical observations through experimental data. The white box models (WBMs) use information from mass balances in a single stoichiometric equation where inputs, outputs, and the conversion from substrates to products are followed [5]. Despite effectiveness and advanced reaction representation in WBMs, the representation of the reaction advance degree, some information on metabolic flux analysis (MFA) can be obtained. Models based on detailed MFA can be used to define optimal operation conditions based on biochemical pathways. It has been established that kinetic models of biological reactions are more complicated than "common" chemical reaction models. Microbial growth models require specialized knowledge of rapid changes of environmental conditions, stoichiometric individual reactions, and the appearance of new steady states in different culture stages [6]. In many cases, mechanistic models, based on first principles, are ineffective because of metabolic complexity of microorganisms.

In this sense, complex microbial consortium behavior and culture media with different types of substrates are difficult to model. Nonmechanistic models, or black box models (BBMs), or a combination between mechanistic and nonmechanistic models, or gray box models (GBMs), are more suitable to describe them. Kinetic parameter fitting for WBMs requires experimental measurements of multiple variables, and frequently, model validation may be impractical. BBMs and GBMs constitute alternatives which describe the general dynamic behavior of bioreactors, without requiring many experimental measurements of the system. These models do not offer mechanistic information about metabolic phenomenology present in the system, but they can optimize and control without it. Then, models can be classified based on the mathematical formulation of the system (**Figure 1**). These are classified into mechanistic, empirical, and fermentation models. A mechanistic model is based on deterministic principles. On the other hand, empirical models represent input-output relations without the knowledge of a mechanism. Fermentation process models are usually represented with a combination of both, mechanistic and empirical models.

An important characteristic of modeling is the assumption of homogeneous or heterogeneous conditions. In this sense, a homogeneous system is related to a single continuous phase. In most cases, bioreactors are described as single liquid phases. However, if the biofilm is included in the study, a solid or semisolid phase needs to be considered in the model. On the other hand, heterogeneous systems are related to the description of two or more continuous phases and the interactions between them. Complex heterogeneous systems can be described as multiple phases: liquid,

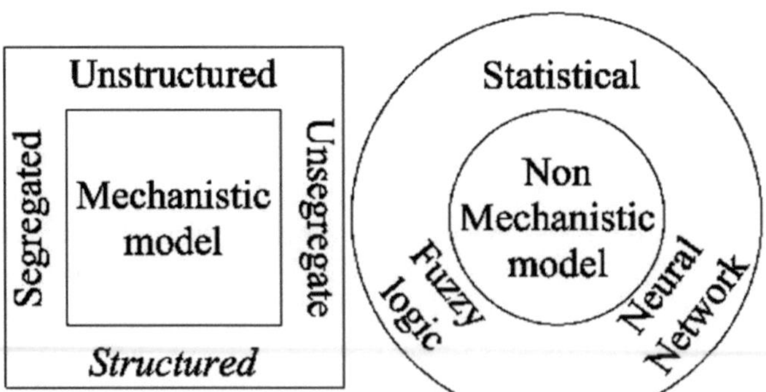

Figure 1.
Classification of models as mechanistic and nonmechanistic.

solid or semisolid, and gaseous phases (e.g. solid-state fermentation). Within this classification, parameters in a model can be classified as distributed and nondistributed (lumped). Distributed parameter models assume that operation parameters vary as a function of space. One, two, or three dimensions are considered in the description of key variables as a function of parameter distribution. As a result, the system is described by a set of partial differential equations (PDEs). On the other hand, a lumped model is necessary, and the system can be described by a set of ordinary differential equations (ODEs), since these parameters do not vary as a function of space.

In this chapter, we provide an overview of mechanistic and empirical models for cell population in fermentation processes.

2. Simple and unstructured kinetic growth models

Unstructured kinetic models (UKMs) represent, in a simple global point of view, the metabolic behavior of the biomass cell production. Mainly, mathematical descriptions for microbial growth kinetics in fermentation processes are based on semiempirical observations. From simple experimental data, we can obtain information to represent cellular growth with unstructured kinetic models.

2.1 Unstructured kinetic models for simple systems

To get the most efficient description of a kinetic model, it is essential to be clear about the application purpose. The application determines the complexity level and structure of the model. The correlation among cell growth, substrate consumption and inhibition [7], or description of the substrate profiles within the reactor during expression of extracellular proteins is the central goal of the model process [8]. The description of key variables is the contribution of the model [9]. These representations are expressed as equations in a simple mathematical model. The UKMs, which are unstructured, unsegregated, are based on the monitoring of cell and nutrient concentration and describe the fermentation process as an average of the species under ideal conditions. Also, it describes the cell and its components as a single species in solution. UKMs consider the apparent rate obtained by metabolic processes, which are carried out by microorganisms. These models are based on conservation equations for cell mass, nutrients, metabolites, and species generation/consumption rates. Most of the UKMs can be divided into three terms: rate expressions for cell growth, rate expressions for nutrient uptake, and rate expressions for metabolite production.

In the case of exponential growth phase, which is the simplest representation of microbial growth, nutrient concentration profiles and decrease rate in several cases are not almost considered.

$$r_X = \frac{dX}{dt} = (\mu - k_D) \cdot X \tag{1}$$

$$r_i = \frac{\alpha}{Y_i}\frac{dX}{dt} \tag{2}$$

where r is the reaction rate, X represents biomass, μ is specific growth rate, k_D is the death rate, α is the stoichiometric factor, and *Yi* is the yield.

The simplest example of multiple reaction models includes substrate consumption for cell maintenance and true yield coefficients (g DCW/g DW) [5]. One of the most used UKMs is Monod's model [10]. This is one of the simplest models to deal with microbial growth, physiology, and biochemistry. The Monod equation describes the proportional relationship between the specific growth rate and low substrate concentrations (Eq. (3)).

$$\mu = \frac{\mu_{MAX}[S]}{K_S + [S]} \tag{3}$$

where μ_{MAX} is the maximum specific growth rate, [S] is the substrate concentration, and K_S is the saturation constant.

The disadvantage of the model is that the individual entity, regulatory complex, adaptive response to environmental changes, and capacity of cell organelles to generate various products in inherent metabolism cannot be considered. The simplest mathematical models used to estimate microbial growth and substrate consumption are still used for monoclonal antibody production by Chinese hamster ovary (CHO) cells [11, 12]. UKMs can predict specific growth rate in simple systems by calculation of mass balances with independent variables.

2.2 Unstructured kinetic models for a more complicated system

The Monod equation is not able to predict the substrate inhibition effect. Thus, several models including such effects have been developed. For example, Andrew's kinetic equation includes an inhibition function to relate substrate concentration and specific growth rate [13].

$$\mu = \frac{\mu_{MAX}}{1 + K_S/[S] + [S]/K_i} \tag{4}$$

where K_i is the substrate inhibition parameter.

Under the assumption of steady state in continuous operation, substrate concentration is low, and the term $[S]/K_i$ is neglected. Under these conditions, specific growth rate of Andrew's kinetic equation follows Monod equation [13]. Another inhibition function is Aiba's equation for alcoholic fermentation.

$$\mu = \frac{\mu_{MAX}[S]}{K_S + [S]} e^{(-[P]/K_i)} \tag{5}$$

where [P] is the product concentration.

Under the assumption of low product concentration, the term $[P]/K_i \approx 0$, resulting in a simplification to Monod equation [13].

The Monod model assumes that the fermentation culture media has only one limiting substrate. More than one limiting substrate is present and impacts specific growth rate. Thus, the following model considering multiple substrates is proposed [14].

$$\mu = \left(1 + \sum_i^n \frac{[S_{e,i}]}{[S_{e,i}] + K_{e,i}}\right) \left[\prod_j^n \frac{\mu_{MAX,j}[S_j]}{[S_j] + K_{S,j}}\right] \tag{6}$$

where subscript i is the number of each substrate species and e represents the essential substrate.

The limited but accurate information provided by UKMs may help to represent global reactions effectively. In addition to substrate consumption and microbial growth, fermentations present catabolic inhibition. Therefore, several research groups propose complete UKMs, which include the empirical observation such as variables regarding cells, substrates, and products. Hans and Levenspiel [15] proposed a kinetic model that assumes the existence of inhibitor critical concentrations.

$$\mu = \mu_{MAX} \left[\prod_{i=1}^{h} \left(1 - \frac{[I_i]}{[I_i^*]} \right)^{n_i} \right] \left(\frac{[S]}{[S] + K_S \left[\prod_{i=1}^{h} \left(1 - \frac{[I_i]}{[I_i^*]} \right)^{m_i} \right]} \right) \tag{7}$$

where [I] is the inhibitor species concentration.

The inhibition function proposed by Levenspiel [16] takes into account the inhibition of ethanol production of alcoholic fermentation modeling, where subscript i corresponds to the substrate or product concentrations. Linear ($n, m = 1$), nonlinear ($n, m > 1$), and fractional ($n, m = 0.5$) applications of these models are possible for fermentation bioreactors (Eq. (7)). An extension of the model was proposed by Luong [7]. He assumes a common mechanism to describe substrate inhibition. Inhibitory factors acting simultaneously could be represented by the following equation:

$$\mu = \mu_{MAX} \left(\frac{[S]}{[S] + K_S} \right) \left(1 - \frac{[P]}{[P_{MAX}]} \right)^n \tag{8}$$

where [P] is the product species concentration and n is the index of cooperativity between inhibitors.

These models can also explain multiple reactions and include biochemical information of metabolites in the global net effect, making them experimentally accurate. This characteristic is useful for structured and segregated modeling [17]. These models can also describe mixed metabolism [18] and hetero-fermentations [19]. The duality of *Saccharomyces cerevisiae* metabolism, aerobic and anaerobic metabolism, is the best example of multiple reactions. The aerobic growth of the yeast yields biomass by favoring metabolic pathways designed for anabolism and cell division. This metabolism is oxidative in amphibolic reactions. However, at low oxygen concentrations, the yeast metabolism changes from being purely respiratory to partially fermentative. The fermentative pathway mainly leads to ethanol production as a final electron acceptor. Thus, there is a limited growth with high ethanol yields in fermentation culture media. Both metabolisms can occur during the growth of *S. cerevisiae* in a wide range of simple carbohydrate fermentations. At high substrate concentrations, there are limitations in respiratory pathways, which lead to an overflow to ethanol production with enhanced fermentative pathways. The simple WBM with overall reactions could not explain in detail the dualism of both fermentative and respiratory metabolisms. Thus, there are two stoichiometric reactions proposed to explain oxidative and fermentative metabolisms [18].

$$\gamma_1 X + \beta_{11} CO_2 - S - \alpha_{12} O_2 = 0 \tag{9}$$

$$\gamma_2 X + \beta_{21} CO_2 + \beta_{22} P - S = 0 \tag{10}$$

where α, β, and γ are stoichiometric coefficients.

This system considers nearly ideal Monod kinetics, no by-product formation, linear specific oxygen consumption rate, and correlation with substrate uptake. If the primary carbon source is glucose (instead of ethanol), glucose can be used aerobically and anaerobically. Ethanol can be used as a carbon source only aerobically. Then, different sets of linear algebraic equations can be derived concerning carbon, oxygen, and hydrogen balance.

The respiratory quotient (RQ) is often used as an indicator of fermentative processes. When RQ is close to one, there is no fermentative metabolism, whereas if RQ is above one, the fermentative metabolism occurs.

$$RQ = \left|\frac{r_{CO_2}}{r_{O_2}}\right| = \begin{cases} >1, \textit{fermentative metabolism} \\ \approx 1, \textit{non fermentative metabolism} \end{cases} \tag{11}$$

The mechanistic characteristics of an unstructured, unsegregated kinetic model contribute to the knowledge of the complex metabolism of *S. cerevisiae*. Despite giving relevant information of simple metabolic processes with multiple reactions, UKMs cannot give information about complete intracellular oxidative metabolism. An example of the application of these models is explained in subsequent sections.

3. Structured growth kinetics

There are several classifications of mechanistic and statistical models of cell population for bioprocess applications. Two terms are essential for mathematical description of cell populations: segregated and structured models. A structured model is related to cell material description using multiple chemical components. A segregate model is related to the description of individual cells in a heterogeneous population. Additionally, it is possible to combine a structured approach with a segregated approach. Structured kinetic models are introduced in this section.

3.1 Simple structured kinetic models

Structured kinetic models (SKMs) describe changes in cell population. The liquid phase (abiotic phase) usually contains nutrients for cell growth and some extracellular metabolites. The microorganisms suspended in the liquid phase behave as multicomponent systems. SKMs consider the internal structure of cells (e.g. mitochondria), and the description of cell growth and its metabolism is used to assume a more accurate growth rate. The information used is a starting point to generate schemes that represent more accurately the growth of microorganisms and their cellular components. The complexity of the information variables and parameters increases in SKMs with the mathematical representation of cellular growth.

SKMs are generally classified into morphologically structured models, chemically structured models, genetically structured models, and metabolically structured models [20].

Morphologically structured models consider the kinetics of nutrient consumption and product formation. These models consider different cell types as living species in terms of the role that they play in the overall reaction. Chemically structured models consider the effects of chemical species in fermentation kinetics; all viable cells are functionally similar, and all the fermentation rates and transport phenomena parameters are accounted for. Genetically structured models assume molecular mechanism knowledge. The model includes the rate of expression of an operator-regulated gene and kinetic equations for the transcription, translation, and folding processes. Metabolically structured models provide a better understanding of process regulation mechanisms such as feedback regulation. This model is based on the main metabolic pathways and in most cases is included in MFA. In the presence of metabolite concentration changes, the network structure represents the reaction and metabolite concentration as a matrix array. Then, SKMs can be classified as dynamic and structural [21]. Dynamic models are described as a set of ordinary differential equations (ODEs). Structural models, which are simplified from ODEs, are represented by a set of algebraic equations through two main approaches: MFA and elementary mode analysis (EMA).

4. Nonmechanistic models

The structured and unstructured kinetic models in the previous sections describe, with a high degree of accuracy, the dynamic behavior of microbial growth in bioreactors. These models, associated with material and energy balances, also help to understand the phenomena associated with microbial metabolism, giving clues to the process design and control.

Black box models (BBMs) usually fall into two main categories: statistical models (SMs) and artificial intelligence tools (AITs). SMs use experimental design, response surface analysis, and exploratory data analysis, whereas AITs consider tools such as data mining, artificial networks, and fuzzy logic [22]. Also, several methodologies to combine mechanistic approaches with nonmechanistic modeling strategies have been developed. The hybrid models, which are known as gray box models (GBMs), inherit the advantages of BBMs such as data analysis and can achieve semi-mechanistic description to each metabolic phenomenon. GBMs offer greater estimation accuracy, calibration ease, better extrapolation properties, and more detailed information on the phenomenology of the system [23]. The advantages of GBMs in the application of bioreactor modeling are direct control and optimization. In this section, we will describe some of these nonmechanistic modeling tools and some of their applications, such as the design of soft sensors.

4.1 Neural network models

Artificial neural networks (ANNs) are mathematical models that are devised from the need to characterize biological neural processes. As the system of ANNs imitates the way which is used to interact with each other in brain neuron, ANNs are simple and strong processes to interconnect the elements that transmit and process information through electrical impulses. In ANNs, these simple process elements are also known as neurons, and depending on the complexity of the connection schemes, they can develop the ability to describe the nonlinear behavior of many dynamic systems [24]. ANNs are computational models that aim to achieve mathematical formalizations of the brain structure and functions, which are constantly reformed by learning through experience and extracting knowledge from the same experience. In ANNs, the hierarchical structure similar to that in brain is established, where neurons connect with each other and transmit the response to other neurons. Once the ANN's structure is defined, it is necessary to develop memory form experience (experimental data). In order to introduce this experience, the ANN training algorithm performs a weight (ω) fitting process associated with each neuron, such that the actions introduced (input signals) converge to the reactions produced (output signal) [24]. Although ANNs do not provide a physical interpretation of the phenomena that take place in the system, these models can approximate the dynamic behavior of the system, making them suitable universal approximators [22]. ANNs are defined based on three basic characteristics: their architecture, activation functions, and training algorithm. The architecture deals with the type of interconnections between their processing units or neurons, while the activation function corresponds to the dynamic characteristics of the neuron transfer functions. The training algorithm refers to the parameter fitting procedure, which provides the learning ability.

Feedforward networks (FNN) represent the simplest network configuration capable of describing the nonlinear behavior of bioreactors. In FFNs, neurons of each layer propagate their information to all neurons in subsequent layers. In each neuron, the input information corresponds to the weighted sum of all the outputs of the previous layer, and the weighting factors, weights and thresholds, are internally fitted for better system description [24].

Another type of ANNs frequently used is recurrent neural networks (RNNs). The structure of RNNs differ from FFNs, in the sense that some of the last layer neuron output signals are fed back as inputs to any previous layer. RNNs could converge to stable system solutions and include the effects of response delays. These characteristics make these models especially useful in the modeling of continuous bioreactors [25].

4.2 Gray box models

BBMs are based on the analysis of data generated to detect correlations, and basic functionalities between the variables and the WBMs are constructed from first principles. A hybrid model category between WBMs and BBMs is GBMs; these models implement a set of tools that combine some of the characteristics of both. Some of these characteristics include properties of process and control design, without losing the ability to explain the phenomena present in the system. Defining a parallel or serial data flow structure allows the integration of both, mechanistic and nonmechanistic information (e.g., **Figure 2**).

Parallel arrays are mainly used when there is a well-defined mechanistic model of the process and are suitable to improve its estimation performance. It is especially useful in cases where dynamic aspects of the system can be decoupled. **Figure 2A** shows a conceptual diagram of parallel interaction where the circle represents WBMs and the square BBMs, and the circle inside the square corresponds to a hybrid model. In the case of serial arrays, BBMs describe just specific terms of the WBMs, such as growth kinetics or transport parameters. **Figure 2B** represents the hybrid model where the BBMs (square) are substituted into the WBMs (circle). Stosch et al. presented a detailed panorama of this model [23].

In the design of bioreactors and their associated controllers, one of the difficulties is the determination of the kinetic model that adequately describes the growth rate of the respective microorganisms. The selection of a kinetic model leads to restrictive models for fixed operating conditions with little extrapolation possibility.

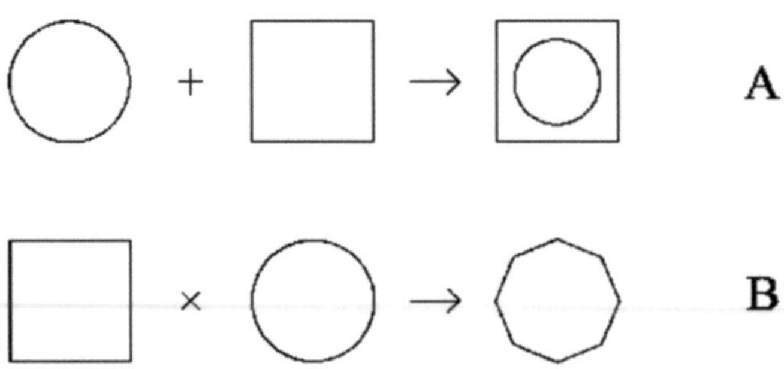

Figure 2.
Classification of models as (A) parallel flow nonmechanistic model and (B) serial flow nonmechanistic model.

GBMs take advantage of the well-known mechanistic information through mass and energy balances to describe specific dynamics of the system and expand its parametric scenario of applicability.

In the determination of bacterial growth kinetic models, GBMs have some advantages compared to BMMs; FFNs are able to describe the system state values, but have important deviations in their estimates, due to their high sensitivity to noise. GBMs combine WBMs with neural network structures, either by feeding the outputs of the ANNs to the state space model, or by backpropagating the estimation error from output layers. Therefore, GBMs offer better forecasting properties and strengthen their performance in the presence of noise [26].

4.3 State observers and soft sensors

Monitoring culture media requires the measurement of variables such as biomass concentration, substrates, dissolved oxygen, carbon dioxide, ammonia, and temperature, among others. These values are used for growth kinetics determination and bioreactor design. However, on multiple occasions, implementation of specific sensors is complicated, and there may be limitations in sensing frequency for variables such as biomass concentration. The implementation of indirect measurement methodologies, such as signal filtering, observer design, and ANNs, allows the estimation of some of these variables, and even the estimation of complex variables such as overall microbial growth rate and the heat flux produced by the system.

In control system theory, a major implementation of state observers is a common complementary strategy. These observers estimate some state variables that cannot be easily measured, either by the absence of suitable sensors, or because of low sampling frequency and high delay times. The main types of observers used for these purposes are those based on the Luenberger scheme, finite-dimensional observers, Bayesian estimators such as Kalman filters, interval observers, observers for fault detection, and even models of artificial intelligence such as ANNs and hybrid models [27].

State observer design requires that the estimated variables are detectable and observable. These states are observable if for a set of specific initial conditions, the internal states of the system are inferred from the knowledge of their outputs. Once its observability is determined, the observer can be designed; the desired type of observer is selected from categories mentioned above. Afterwards, tests of the estimator are carried out by comparing the real values against observer estimates, and in the case of important discrepancies between these values, the observer is adjusted or a different one is selected.

In the case of stirred tank reactors (batch, continuous, or semi-continuous), we may often assume that homogeneous conditions are available, so the system models obtained consist of aggregate parameter systems (ODEs). However, frequently in the case of tubular reactors or solid substrate fermentation systems, homogeneity assumptions are not adequate, so it is necessary to construct using distributed parameter models (PDEs). In the latter case, the design and performance of the main observers, such as Luenberger or Bayesian type, are usually limited, and therefore, it is usual to resort to another type of observer. In such a case, the observers based on a discretized system are substituted for traditional observers [28].

Soft sensors or virtual sensors are used as state observers in specific application. These soft sensors combine several physical measurements with dynamic characteristics to calculate other variables that are not measured.

Soft sensors can not only provide variable information to characterize a system but also facilitate the design of the control schemes.

In bioreactor design, soft sensors can be used to estimate unavailable variables such as biomass. Traditionally, biomass has been traditionally determined by use of a variety of methodologies such as optical density, dry weight, and microbial counts, among others. These techniques present several problems, the most important being the lack of continuous online measurements. To overcome this problem, various strategies have been applied, such as the implementation of low-cost sensors combined with signal processing strategies. For instance, the RGB sensor is used for biomass measurement in microalgae production reactors [29]. This type of sensors uses the intensity of the red, green, and blue (RGB) colors, which correlates with the biomass concentration using dry weight and/or colony formation unit (CFU) information, using the Beer-Lambert law principles. The correlation is described through linear fitting [30]. Additionally, it is possible to compensate background noise by use of ANNs even in the case of nonlinear correlation [31].

Soft sensors can also be applied to nonexplicit system states. These observers can estimate lumped system variables, such as growth rate. As the simplest factor, temperature is commonly used, since it allows estimating system concentrations, due to intrinsic dependence between reaction rates and reaction enthalpy. The heat of reaction, either consumed or dissipated by the system, is one of the implicit system states used for reaction rate determination. The same strategy may also be used to determine microbial growth rates [32].

Microbial growth rates are inherently variable due to their metabolic nature and operation conditions. For example, as fluctuation in substrate concentration occurs in fed-batch bioreactors, the condition of osmotic pressure within cells is modified through the plasma membrane, which may change cellular energetics and the viability of cell division. A suitable strategy for these cases is the design of a substrate consumption rate observer. This kind of observer helps to design a robust control strategy against important fluctuations in maintaining constant substrate concentrations.

The use of observers or soft sensors is an interesting alternative to elucidate approximate values of system states, whether these are explicit or implicit, in cases where online continuous physical measurement is not available. These approximations can be used to design process control schemes that ensure proper functioning.

5. Application examples

There are many practical applications of structured or unstructured kinetic models. In the acetone-butanol-ethanol (ABE) production, the models for the bioprocess have evolved from simple stoichiometric equations to sophisticated and elaborate kinetic models based on metabolic pathways [33, 34], genome-scale metabolic flux modeling [35], system-level modeling [34], and metabolic network [21]. Gordeeva classified mathematical modeling of specific growth rate (dependent or independent on substrate concentration), specific rate of substrate consumption, and specific rate of product formation in batch fermentations [36]. In this study, the states in fermentation are described by a system of three ODEs [36]. Cui reported unstructured lactate formation by enzymatic hydrolysis of sugarcane bagasse, and the model is based on Logistic equations, Luedeking-Piret equations and Luedeking-Piret-like equations [37]. Similarly, Sharma reported an unstructured model to describe growth, substrate utilization, and lactate production by *Lactobacillus plantarum* [38]. On the other hand, the common mathematical descriptions of the fermentation process are based on UKMs. For example, the fermentation of sweet sorghum stalk juice by immobilized *Saccharomyces cerevisiae* is explained by the kinetic parameters of Hinshelwood's model [39]. Another example using the UKM model is the basic logistic model incorporated with the Luedeking-Piret model (hybrid model) to describe the production of bioethanol from banana and pineapple wastes [40].

Cephalosporium acremonium (ATCC 36225) is one example of the utilization of SKMs where morphological differentiation and catabolite repression are the main aspects of the model approach [41]. SKMs can also effectively represent diauxic growth as well as the monitoring of an intracellular reactant in acetic acid production by *Bacillus licheniformis* [42]. Sansonetti reported a biochemically structured model for ethanol production from ricotta cheese whey by *Kluyveromyces marxianus* [43]. Wang studied a segregated kinetic model in fed-batch culture to represent simultaneous saccharification and co-fermentation (SSCF) for bioethanol production from lignocellulosic raw materials at high substrate concentrations [44]. Another interesting process is the solid-state fermentation. In most proposed models, a set of PDEs is used to describe how intraparticles are diffused or how the growth can be affected by intraparticle diffusion of oxygen, enzymes, hydrolysis products, and other nutrients and the role in the fermentation of other phenomena such as particle shrinkage and spatial microbial biomass distribution [45]. Computational fluid dynamics (CFD) provides information concerning the mixing modeling and design of bioreactors [46]. Another example of CFD is cephalosporin production by *Acremonium chrysogenum;* it was found that the oxygen transfer rate (OTR) directly affects fermentation performance with different impeller combinations [47]. Applications of CFD to fermentation modeling include effects of stress on cell morphology and mass transfer from the bulk solution to the organisms [46]. Biochemical models should be coupled to the CFD models in order to give a closed link between biochemistry and fluid dynamics of the system [33]. Haringa assesses the effect of substrate heterogeneity on the metabolic response of *P. chrysogenum* in industrial bioreactors via coupling of a 9-pool metabolic model with Euler-Lagrange CFD simulations toward rational scale-down and design optimization [48].

Another way to construct mathematical models of microbial growth is the use of FFNs, which describe the behavior of different configurations of bioreactors. An example of this type of applications is the modeling of the production of bioethanol obtained from sugar beets [49]. Here, a three-layer FFN is used to describe the dynamic behavior of the reactor. The first neuron layer consists of system inputs, which correspond to substrate concentration, substrate type, and fermentation time. The second layer corresponds to hidden neurons that process the information through their activation function. Finally, the third layer matches the output of the system that corresponds to the viable cell count of yeasts and the concentration of ethanol produced. On the other hand, GBMs and their hybrid models are not only used to characterize fermentation kinetics but can also describe general behaviors of bioprocesses. For example, in fed batch cultures of *Chlorella pyrenoidosa*, a hybrid scheme of ANN with mass balance mechanistic models describes the general behavior of the states of the system, reducing considerably the variability of their predictions, and achieving versatility in application [50]. These types of GBMs are useful in cases of high complexity due to metabolic dynamics of microorganisms [51]. GBMs or hybrid models are not only combinations of first principles with ANNs, but there may also be hybrid models obtained through the combination of statistical models with ANNs. This type of models usually has special applicability in the optimization of operating conditions of bioreactors (e.g., fed batch fermentation of *Ralstonia eutropha* for poly-β-hydroxybutyrate production) [52].

Soft sensors are also useful in control design. For example, sliding mode observers can describe the behavior of sulfate reduction rate which results from *Desulfovibrio alaskensis* fermentation [53]. These observers use turbidimetric and colorimetric titration information, and formulate based on sliding modes and sigmoidal functions, but their performance depends strongly on the nature of the system and its monitoring schemes.

Abrupt leaps in substrate concentration can be detected and prevented by the strategy of adaptive or optimal control by coupling with an observation scheme such as ANNs. For example, in L-glutamate production with *Corynebacterium glutamicum* fermentation, physical sensor applications are limited because of high costs and system complexity. However, it is possible to use simpler measurements such as oxygen concentrations, temperature, pH, and carbon dioxide production to train models of ANNs that can approximate the dynamic behavior of glucose concentration [54].

6. Conclusion

In bioprocesses, representation through simple or complex models, fermentation must consider process variables and analytes in mathematical models to achieve optimization, to develop simulations, and to calculate the output of critical variables in bioprocesses. Kinetic models allow predicting the behavior of biochemical reactions. This useful information is critical to techno-economic analysis. The incorporation of simple or complex models could represent phenomena more precisely and thus enhance our comprehension. In the design of bioreactors, a mathematical model is necessary to allow selecting the optimal operating conditions. There is a wide variety of types of models ranging from simple statistical descriptions to artificial intelligence tools. Appropriate model selection depends on the specific application: unstructured models can describe the global behavior, while unstructured models can describe specific phenomena such as metabolic pathways.

Another alternative in the modeling of bioreactors is the black or gray box models, which can be used for bioreactor design, without describing in detail the phenomenology present in the system, which is mainly focused on the global behavior of the system. An important part of the modeling, design, and control of bioreactors is the selection of appropriate sensors. It is often difficult to find suitable sensors for the process, so soft sensors are an interesting alternative to solve this problem.

Once a model describing the dynamical behavior of the bioreactor reaches the available condition, the control scheme can be designed. The goal may be different in each scenario: in the case of variables such as pH, this objective is usually regulation, but in variables such as concentrations and temperature, tracking is usually the goal. In any of these cases, slow and smooth dynamics inherent in these processes usually allow PID controllers to bring system states to the set point efficiently.

Acknowledgements

The authors want to thank ITESO Fund for Research Support 2017-2018.

Conflict of interest

The authors declare not to have a conflict of interest.

Author details

Carlos González-Figueredo, René Alejandro Flores-Estrella
and Oscar A. Rojas-Rejón*
Department of Technological and Industrial Processes, Western Institute of Technology and Higher Education (ITESO), San Pedro Tlaquepaque, Jalisco, México

*Address all correspondence to: orojas@iteso.mx

References

[1] Stanescu D, Chen-Charpentier B. Random coefficient differential equation models for monod kinetics. Dynamical Systems. 2009:719-728 http://aimsciences.org/journals/pdfs.jsp?paperID=4686&mode=full

[2] Van Impe JF, Poschet F, Geeraerd AH, Vereecken KM. Towards a novel class of predictive microbial growth models. International Journal of Food Microbiology. 2005;**100**:97-105. DOI: 10.1016/j.ijfoodmicro.2004.10.007

[3] Esser DS, Leveau JHJ, Meyer KM. Modeling microbial growth and dynamics. Applied Microbiology and Biotechnology. 2015;**99**:8831-8846. DOI: 10.1007/s00253-015-6877-6

[4] Almquist J, Cvijovic M, Hatzimanikatis V, Nielsen J, Jirstrand M. Kinetic models in industrial biotechnology—Improving cell factory performance. Metabolic Engineering. 2014;**24**:38-60. DOI: 10.1016/j.ymben.2014.03.007

[5] Villadsen J, Nielsen J, Lidén G. Bioreaction Engineering Principles. US: Springer; 2011. DOI: 10.1007/978-1-4419-9688-6

[6] Ferrell JE, Ha SH. Ultrasensitivity part I: Michaelian responses and zero-order ultrasensitivity. Trends in Biochemical Sciences. 2014;**39**:496-503. DOI: 10.1016/j.tibs.2014.08.003

[7] Luong JH. Generalization of monod kinetics for analysis of growth data with substrate inhibition. Biotechnology and Bioengineering. 1987;**29**:242-248. DOI: 10.1002/bit.260290215

[8] Kiefhaber T, Rudolph R, Kohler HH, Buchner J. Protein aggregation in vitro and in vivo: A quantitative model of the kinetic competition between folding and aggregation. Biotechnology (N. Y). 1991;**9**:825-829. DOI: 10.1038/nbt0991-825

[9] Deindoerfer FH. Fermentation kinetics and model processes. Advances in Applied Microbiology. 1960;**2**:321-334. DOI: 10.1016/S0065-2164(08)70134-5

[10] Monod J. The growth of bacterial cultures. Annual Review of Microbiology. 1949;**3**:371-394. DOI: 10.1146/annurev.mi.03.100149.002103

[11] Feisther VA, Ulson De Souza AA, Trigueros DEG, De Mello JMM, De Oliveira D, Guelli Ulson De Souza SMA. Biodegradation kinetics of benzene, toluene and xylene compounds: Microbial growth and evaluation of models. Bioprocess and Biosystems Engineering. 2015;**38**:1233-1241. DOI: 10.1007/s00449-015-1364-0

[12] López-Meza J, Araíz-Hernández D, Carrillo-Cocom LM, López-Pacheco F, Rocha-Pizaña M del R, Alvarez MM. Using simple models to describe the kinetics of growth, glucose consumption, and monoclonal antibody formation in naive and infliximab producer CHO cells. Cytotechnology. 2016;**68**:1287-1300. DOI: 10.1007/s10616-015-9889-2

[13] Andrews JF. A mathematical model for the continuous culture of microorganisms utilizing inhibitory substrates. Biotechnology and Bioengineering. 1968;**10**:707-723. DOI: 10.1002/bit.260100602

[14] Tsao GT, Hanson TP. Extended Monod equation for batch cultures with multiple exponential phases. Biotechnology and Bioengineering. 1975;**17**:1591-1598. DOI: 10.1002/bit.260171104

[15] Han K, Levenspiel O. Extended monod kinetics for substrate, product,

and cell inhibition. Biotechnology and Bioengineering. 1988;**32**:430-447. DOI: 10.1002/bit.260320404

[16] Levenspiel O. The monod equation: A revisit and a generalization to product inhibition situations. Biotechnology and Bioengineering. 1980;**22**:1671-1687. DOI: 10.1002/bit.260220810

[17] Nielsen J, Villadsen J. Modelling of microbial kinetics. Chemical Engineering Science. 1992;**47**: 4225-4270. DOI: 10.1016/0009-2509 (92)85104-J

[18] Sonnleitner B, Käppeli O. Growth of Saccharomyces cerevisiae is controlled by its limited respiratory capacity: Formulation and verification of a hypothesis. Biotechnology and Bioengineering. 1986;**28**:927-937. DOI: 10.1002/bit.260280620

[19] Tobajas M, Mohedano AF, Casas JA, Rodríguez JJ. Unstructured kinetic model for reuterin and 1,3-propanediol production by *Lactobacillus reuteri* from glycerol/glucose cofermentation. Journal of Chemical Technology and Biotechnology. 2009;**84**:675-680. DOI: 10.1002/jctb.2098

[20] Bapat PM, Bhartiya S, Venkatesh KV, Wangikar PP. Structured kinetic model to represent the utilization of multiple substrates in complex media during rifamycin B fermentation. Biotechnology and Bioengineering. 2006; **93**:779-790. DOI: 10.1002/bit.20767

[21] Millat T, Winzer K. Mathematical modelling of clostridial acetone-butanol-ethanol fermentation. Applied Microbiology and Biotechnology. 2017; **101**:2251-2271. DOI: 10.1007/ s00253-017-8137-4

[22] Sablani S, Datta A, Rahman M, Mujumdar A, Cronin K, Dou J, et al., editors. Handbook of Food and Bioprocess Modeling Techniques. Boca Raton: CRC Press; 2006. DOI: 10.15713/ ins.mmj.3

[23] Von Stosch M, Oliveira R, Peres J, Feyo S, Azevedo D. Hybrid semi-parametric modeling in process systems engineering: Past, present and future. Computers and Chemical Engineering. 2014;**60**:86-101

[24] Haykin S. Neural Network and Learning Machines. Ca: Pearson; 2008. pp. 978-0131471399

[25] Sivakumaran N, Radhakrishnan TK, Babu JSC. Identification and Control of Bioreactor using Recurrent Networks. Instrumentation Science and Technology. 2006;**34**:635-651. DOI: 10.1080/10739140600963871

[26] Acuña G, Cubillos F, Thibault J, Latrille E. Comparison of Methods for Training Grey-Box Neural Network Models. Process Engineering. 2003

[27] Mohd Ali J, Ha Hoang N, Hussain MA, Dochain D. Review and classification of recent observers applied in chemical process systems. Computers and Chemical Engineering. 2015;**76**: 27-41. DOI: 10.1016/j. compchemeng.2015.01.019

[28] Aguilar-Garnica E, García-Sandoval JP, González-Figueredo C. A robust monitoring tool for distributed parameter plug flow reactors. Computers and Chemical Engineering. 2011;**35**:510-518. DOI: 10.1016/j. compchemeng.2010.06.001

[29] Benavides M, Mailier J, Hantson AL, Muñoz G, Vargas A, Van Impe J, et al. Design and test of a low-cost RGB sensor for online measurement of microalgae concentration within a photo-bioreactor. Sensors (Switzerland). 2015;**15**:4766-4780. DOI: 10.3390/s150304766

[30] González-Figueredo C, Aguilar-Garnica E, Guzmán M. Implementación

de un colorímetro webcam para el monitoreo en línea de un sistema de reacción. In: XXXII Encuentro Nac. y 1er Congr. Int. La AMIDIQ; Academia Mexicana de Investigación y Docencia en Ingeniería Química; Riviera Maya, Quintana Roo, México. 2011. pp. 1731-1735

[31] González-Figueredo C, Haro-Córdova F, González-Guerrero FC. Monitoreo de un sistema de agitación utilizando webcam. In: XXXIII Encuentro Nac. y II Congr. Int. AMIDIQ; Academia Mexicana de Investigación y Docencia en Ingeniería Química; San José del Cabo, BCS, México. 2012. pp. 1610-1614

[32] Paulsson D, Gustavsson R, Mandenius CF. A soft sensor for bioprocess control based on sequential filtering of metabolic heat signals. Sensors (Switzerland). 2014;**14**: 17864-17882. DOI: 10.3390/s141017864

[33] Mayank R, Ranjan A, Moholkar VS. Mathematical models of ABE fermentation: Review and analysis. Critical Reviews in Biotechnology. 2013; **33**:419-447. DOI: 10.3109/07388551.2012.726208

[34] Chen L, Seung-Oh S, Ting Lu. System-level modeling of acetone-butanol-ethanol fermentation. FEMS Microbiology Letters. 2016;**363**:1-20. DOI: 10.1093/femsle/fnw074

[35] Senger RS, Yen JY, Fong SS. A review of genome-scale metabolic flux modeling of anaerobiosis in biotechnology. Current Opinion in Chemical Engineering. 2014;**6**:33-42. DOI: 10.1016/j.coche.2014.08.003

[36] Gordeeva YL, Rudakovskaya EG, Gordeeva EL, Borodkin AG. Mathematical modeling of biotechnological process of lactic acid production by batch fermentation: A review. Theoretical Foundations of Chemical Engineering. 2017;**51**:282-298. DOI: 10.1134/S0040579517030058

[37] Cui Y, Liu R, Xu L, Zheng W, Sun W. Fermentation kinetics of enzymatic hydrolysis bagasse solutions for producing L-lactic acid. Sugar Tech. 2018;**20**:364-370. DOI: 10.1007/s12355-018-0592-4

[38] Sharma V, Mishra HN. Unstructured kinetic modeling of growth and lactic acid production by Lactobacillus plantarum NCDC 414 during fermentation of vegetable juices. LWT-Food Science and Technology. 2014;**59**:1123-1128. DOI: 10.1016/j.lwt.2014.05.039

[39] Jin H, Liu R, He Y. Kinetic models and effects of initial total soluble sugar concentrations on batch ethanol fermentation of sweet sorghum stalk juice by *Saccharomyces cerevisiae* strain. Energy Sources, Part A: Recovery, Utilization, and Environmental Effects. 2015;**37**:1282-1290. DOI: 10.1080/15567036.2011.616871

[40] Teoh YP, Ooi ZX. Evaluation of unstructured kinetic models for the production of bioethanol from banana and pineapple wastes. BioResources. 2016;**11**:4295-4305. DOI: 10.15376/biores.11.2.4295-4305

[41] Chu WB, Constantinides A. Modeling, optimization, and computer control of the cephalosporin C fermentation process. Biotechnology and Bioengineering. 1988;**32**:277-288

[42] Larsen R, Kjaergaard L. A structured model for microbial growth and product formation. European Journal of Applied Microbiology and Biotechnology. 1978;**5**: 177-188. DOI: 10.1007/BF00579337

[43] Sansonetti S, Hobley TJ, Calabrò V, Villadsen J, Sin G. A biochemically structured model for ethanol fermentation by *Kluyveromyces marxianus*: A batch fermentation and

kinetic study. Bioresource Technology. 2011;**102**:7513-7520. DOI: 10.1016/j.biortech.2011.05.014

[44] Wang R, Koppram R, Olsson L, Franzén CJ. Kinetic modeling of multi-feed simultaneous saccharification and co-fermentation of pretreated birch to ethanol. Bioresource Technology. 2014; **172**:303-311. DOI: 10.1016/j.biortech.2014.09.028

[45] Rahardjo YSP, Tramper J, Rinzema A. Modeling conversion and transport phenomena in solid-state fermentation: A review and perspectives. Biotechnology Advances. 2006;**24**: 161-179. DOI: 10.1016/j.biotechadv.2005.09.002

[46] Formenti LR, Nørregaard A, Bolic A, Hernandez DQ, Hagemann T, Heins AL, et al. Challenges in industrial fermentation technology research. Biotechnology Journal. 2014;**9**:727-738. DOI: 10.1002/biot.201300236

[47] Duan S, Yuan G, Zhao Y, Ni W, Luo H, Shi Z, et al. Simulation of computational fluid dynamics and comparison of cephalosporin C fermentation performance with different impeller combinations. Korean Journal of Chemical Engineering. 2013;**30**:1097-1104. DOI: 10.1007/s11814-013-0010-2

[48] Haringa C, Tang W, Wang G, Deshmukh AT, van Winden WA, Chu J, et al. Computational fluid dynamics simulation of an industrial *P. chrysogenum* fermentation with a coupled 9-pool metabolic model: Towards rational scale-down and design optimization. Chemical Engineering Science. 2018;**175**:12-24. DOI: 10.1016/j.ces.2017.09.020

[49] Grahovac J, Jokić A, Dodić J, Vučurović D, Dodić S. Modelling and prediction of bioethanol production from intermediates and byproduct of sugar beet processing using neural networks. Renewable Energy. 2016;**85**: 953-958. DOI: 10.1016/j.renene.2015.07.054

[50] Wu Z, Shi X. Optimization for high-density cultivation of heterotrophic *Chlorella* based on a hybrid neural network model. Letters in Applied Microbiology. 2007;**44**:13-18. DOI: 10.1111/j.1472-765X.2006.02038.x

[51] Teixeira A, Cunha AE, Clemente JJ, Moreira JL, Cruz HJ, Alves PM, et al. Modelling and optimization of a recombinant BHK-21 cultivation process using hybrid grey-box systems. Journal of Biotechnology. 2005;**118**:290-303. DOI: 10.1016/j.jbiotec.2005.04.024

[52] Patnaik PR. Neural and hybrid optimizations of the fed-batch synthesis of poly-β-hydroxybutyrate by *Ralstonia eutropha* in a nonideal bioreactor. Bioremediation Journal. 2008;**12**: 117-130. DOI: 10.1080/10889860802261687

[53] Gómez-Acata RV, Neria-González MI, Aguilar-López R. Robust software sensor design for the state estimation in a sulfate-reducing bioreactor. Theoretical Foundations of Chemical Engineering. 2016;**50**:67-75. DOI: 10.1134/S0040579516010036

[54] Ding J, Jia L, Mpofu E, Gao M, Ren X. An on-line adaptive glucose feeding system incorporating patterns recognition for glucose concentration control in glutamate fermentations. Biotechnology and Bioprocess Engineering. 2016;**21**:758-766. DOI: 10.1007/s12257-016-0394-z

Chapter 3

Small-Scale Process for the Production of Kefiran through Culture Optimization by Use of Central Composite Design from Whey and Kefir Granules

José Manuel Pais-Chanfrau, Lorena D. Carrera Acosta, Paola M. Alvarado Cóndor, Jimmy Núñez Pérez and Milton J. Cuaran Guerrero

Abstract

Cheese is one of the most demanded dairy products worldwide. However, during the conversion of milk to cheese, about 10 liters of milk are employed and about 9 liters of whey are generated for each 1 kg of cheese produced. The whey has traditionally been used for animal feed and as starting material for obtaining whey proteins. Furthermore, whey has the significant values of BOD and COD, becoming the most important contaminant in the dairy industry. For this reason, further growth of cheese sector is being limited by the surplus of whey as a by-product of the production of the cheeses. One of the many possibilities offered by the whey is its use as a starting material to produce many biotech products with a higher added value. The kefiran is a degradable biopolymer and is formed by galactose and glucose units, in almost similar proportions, which have been found with numerous benefits for human health. It is produced by a consortium of acid-lactic bacteria and yeasts, which coexist within the kefir granules, which are able to grow and multiply using the lactose present in the whey. The objective of the present study is to establish a small-scale process that allows the obtaining of kefiran.

Keywords: kefiran, central composite design, statistical optimization, *Lactobacillus kefiranofaciens*, kefiran production process

1. Introduction

Although it is still below the per capita consumption level recommended by FAO, milk production in Ecuador has been growing for some years now [1]. The stimulus of milk consumption and its derivatives was a state policy in recent years [2]. Within Ecuador, the Andean region marches at the head of production levels, accounting for about 77% of the country's production [3].

Of the 5.5 million liters of milk produced daily in Ecuador, 48% is sent to the industry, 35% of this goes to the industrial production of cheese [4], and about 12% of the total milk goes to the artisan cheese production. Therefore, around 47% of the total daily production of fresh milk is destined to produce different types of cheeses.

To produce 100 kg of cheese, about 1000 liters of fresh milk are used and about 900 liters of liquid milk whey (WL) are generated [5]. The WL is a yellowish liquid, whose characteristics vary with the region, the time of year, and the forage used in cattle feed. The majority composition is formed by 4.5–5% (m/v) of lactose, 0.6–0.8% (m/v) of whey proteins, 0.5% (m/v) fat, and 8–10% (m/v) (dry basis) mineral salts [6]. Its own composition gives the WL high values of COD and BOD, reporting values of 60–80 and 30–50 kg/m^3, for the COD and BOD_5, respectively [7, 8]. This by-product of the dairy industry, therefore, classifies as the most polluting effluent of this industry and constitutes the main impediment, for the further growth of the dairy industry [9].

There are numerous reviews on the use of whey as a raw material to obtain products with a higher added value than cheese or other dairy products [5, 10–12].

One of the products that can be obtained from WL by fermentation is the exopolysaccharides (EPS), which can be formed by sugars of equal or lower sweetening level than commercial sugar, but which can provide other beneficial effects for human health, classifying many of them as prebiotic substances [13].

Kefiran is an EPS, soluble in water, edible, and biodegradable, which is part of the structure of the kefir granule [14–17], which forms a symbiotic consortium of lactic acid bacteria (LAB), fungi, and yeasts, which can transform fresh milk in a fermented and effervescent milk drink, with a certain ethanol content, called kefir, very popular in central-eastern Europe and in Asia [18–21].

The role of kefiran seems to protect the microbiota inside the kefir granules [22]. Its synthesis is attributed to the LAB *Lactobacillus kefiranofaciens*, but it has been proven that its levels depend on the interactions that this bacterium establishes with others and the presence of some yeasts of the consortium present in the kefir granule [23, 24].

Kefiran, like kefir, has been attributed to several beneficial properties for human health, such as promotion of the growth of *Bifidobacterium bifidum* and modulating its genetic expression [25], its healing capacity [26, 27], hypotensive [28], hypercholesterolemia lowering [29, 30], anti-inflammatory agent [31], body weight regulator [32], immunoenhancer [33], and it has some anticancer effects [18, 34, 35]. It is for this reason that interest has recently grown to produce this EPS [24, 36, 37].

The production of kefiran goes through a fermentative phase where the polysaccharide is synthesized by *L. kefiranofaciens*, using a suitable carbon source such as lactose, present in the WL [38]. There are numerous reports suggesting that the temperature, the amount of inoculum, and the lactose concentration determine the amount and productivity of this phase [23, 24, 39]. However, the diversity of origins of the kefir granules and the logical variation of the amounts and types of microorganisms present in the granule mean that the optimal conditions may vary in each case and region.

2. Statistical optimization of kefiran production from whey and kefir granules by using response surface methodology

Response surface methodology (RSM) has been extensively used to find optimal fermentation conditions for the production of metabolites and chemicals from different microorganisms [40, 41].

Certain response (i.e., product concentration or productivity) depending for many well defines cultivation variables (like pH, temperature, substrate concentration, etc.) can be related via experimentation through a mathematical model with certain response like concentration or productivity of a metabolite produce by microorganism. Such a model can be used, after its verification, to "navigate" inside defined region of variables, to find the combination of such a variable able to maximize or minimize the response function [42–45].

Some designs of experimental (DOE) arrangements have been employed in RSM. One of the most popular is the central composite design (CCD), which the experimental points surrounded certain central point in equidistance from them [46].

In CCD, the extreme values of the response are searching inside a square in scripted in the sphere formed by the experimental points. In the CCD, in its minimal expression, only the central point needs to be replicated, assumed that the statistical variation of the rest of all experimental points is the same that in the central one. Additional point, however, can enhance the accuracy of the model. With this kind of DOE, a quadratic model of response could be obtained. A second degree of a polynomial can be used to find the extreme values (minimum or maximum) inside the experimental surface. Before that, however, it can be demonstrated that the quadratic model of response is suitable to navigate inside his values to find the optima. Analysis of variance (ANOVA) of the mathematical model and the analysis of his residuals can be served for this purpose.

Several professional statistical software can be used to perform the planning and analysis of the mathematical model. Design-Expert 11, from the Stat-Easy, Inn. (Minneapolis, USA) was used through this work.

In previous work, whey powder (WP) and kefir grains have been used to find the suitable conditions for the WP concentration and temperature to maximize kefiran production [37] or maximize certain desirability function formed by equally weighed kefiran as a prebiotic, and the cell concentration of LAB and yeast in a functional drink [47].

In those works, however, some inconveniences associated to cost of WP as a starting raw material has been detected when the preliminarily economic analysis was performed, suggesting the direct use of a cheaper sweet or acidic whey instead of the WP for similar purposes.

The way of optimization passes for a well-designed experiment where the response (concentration of kefiran in the fermentation broth) is correlated with relevant independent variables, like concentration of liquid sweet whey (WL) in the culture medium and temperature. The optimization algorithm used is shown in **Figure 1**.

For its simplicity, a central composite design was selected, bringing as a central point 25°C and 42% (m/m), for temperature and WL, respectively. A temperature of 25°C brings a maximum value of kefiran concentration in previous work with the same origin kefir grains ("Yogurt-Kombucha-Tibicos en Ecuador," Quito, Pichincha, Ecuador, www.kefir.ec), but employing whey powder as a lactose source [37].

Kefiran was determined according to phenol-sulfuric acid method [48] which employed the glucose as a reference sugar similar as it had been determined elsewhere [24, 37].

Due to a complexity of the consortia of microorganism presented in the kefir grain, a culture medium is supplemented either with glucose or sucrose, adjusted the brix of the medium at a constant value of 14%, as was suggested elsewhere [43, 44], and it was performed in previous work [37].

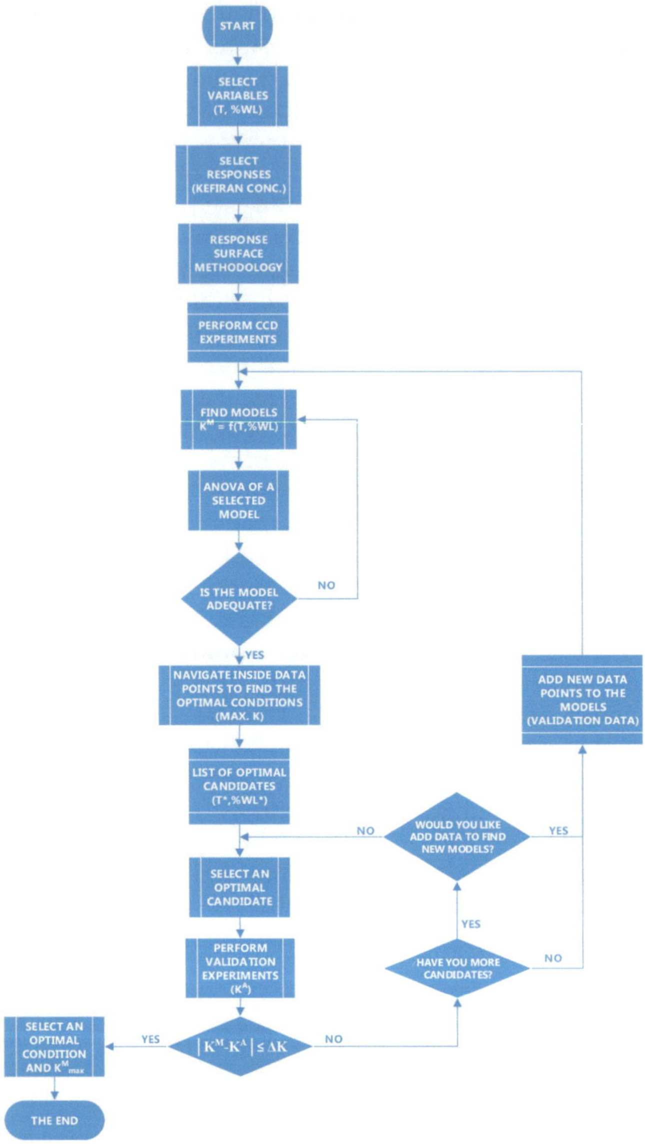

Figure 1.
Flow-chart to find the maximum of kefiran production by using CCD in RSM.

The starting CCD experiments for two independent variables (A: Temp and B: %WL) at a minimum of 13 points experiments are recommended [46], five of which corresponds to the central point and the rest of eight surrounding the central point. Due to the existence of position in a shaker, however, two experiments per point surrounding four experiments for a central experiments points were performed (**Figure 2**).

Table 1 shows the coded and actual values of the independent variables and the corresponding value for the response variable.

As a response range was a ratio of max to min of 12.4082, the response could be transformed. In this case, a power transformation was selected, which brings a model for coded variables of:

$$(Kefiran + 10.0)^{0.28} = 4.33 + 0.2231\,A + 0.0665\,B + 0.2322\,AB \\ -0.2233\,A^2 - 0.2107\,B^2 + 0.3001\,A^2B - 0.7354\,AB^2 - 0.2848\,A^2B^2 \quad (1)$$

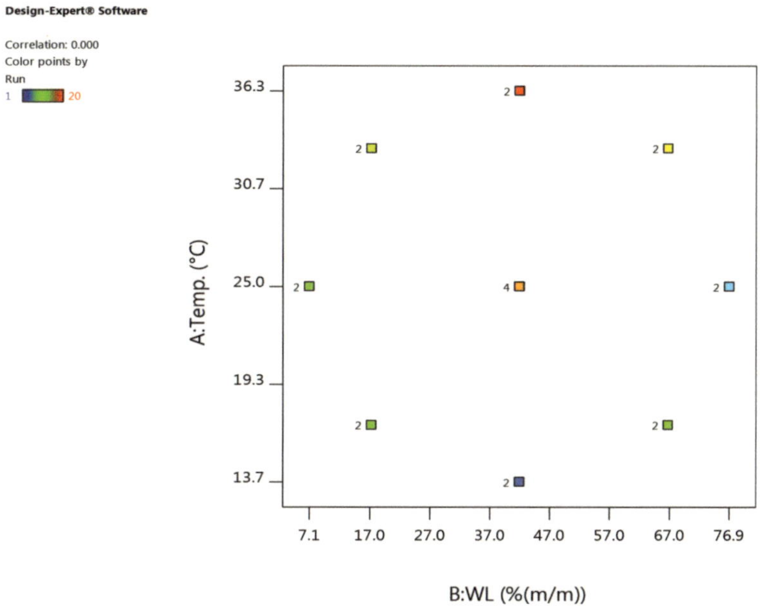

Figure 2.
Original surface data for independent variables.

Std	Code variables		Actual variables		Response
	A: Temp.	B: WL	A: T, °C	B: WL, % (m/m)	Kefiran, g Glu/ml
1	−1.000	−1.000	17.0	17.3	130.0
2	1.000	−1.000	33.0	17.3	17.8
3	−1.000	1.000	17.0	66.7	171.0
4	1.000	1.000	33.0	66.7	105.5
5	0.000	−1.414	25.0	7.1	109.0
6	0.000	1.414	25.0	76.9	129.7
7	0.000	0.000	25.0	42.0	180.7
8	0.000	0.000	25.0	42.0	182.4
9	0.000	0.000	25.0	42.0	173.3
10	0.000	0.000	25.0	42.0	171.4
11	−1.414	0.000	13.7	42.0	84.7
12	−1.414	0.000	13.7	42.0	82.4
13	−1.000	1.000	17.0	66.7	161.1
14	−1.000	−1.000	17.0	17.3	128.8
15	0.000	1.414	25.0	76.9	132.5
16	0.000	−1.414	25.0	7.1	108.6
17	1.000	1.000	33.0	66.7	87.6
18	1.000	−1.000	33.0	17.3	14.7
19	1.414	0.000	36.3	42.0	169.7
20	1.414	0.000	36.3	42.0	145.8

Table 1.
Coded and actual values of independent variables and its corresponding values of response.

ANOVA for reduced quadratic model for the transformed response of kefiran is shown in **Table 2**.

The model *F-value* of 190.66 implies that the model is significant. There is only a 0.01% chance that an F-value that is large could occur due to noise.

P-values less than 0.0500 indicate model terms are significant. In this case A, B, AB, A^2, B^2, A^2B, AB^2, A^2B^2 are significant model terms.

The accuracy of the model can be judged by the plot of the actual responses *versus* predicted responses (**Figure 3**), and the normal plot of residuals (**Figure 4**).

The *predicted* R^2 of 0.9729 is in reasonable agreement with the *adjusted* R^2 of 0.9876; i.e., the difference is less than 0.2.

Adeq. precision measures the signal-to-noise ratio. A ratio greater than 4 is desirable. Your ratio of 45.727 indicates an adequate signal. This model can be used to navigate the design space.

The *fit statistics* of the model is:

Std. dev.	**0.0597**	**R^2**	**0.9928**
Mean	3.86	Adjusted R^2	0.9876
C.V. %	1.54	Predicted R^2	0.9729
		Adeq. precision	45.7268

After checking the model for kefiran, an optimization procedure to find the maximum of kefiran was performed. A wide range of region for temperature and WL concentration was found, and suggested a combination of 90 points of temperature and WL concentration for the maximum of kefiran concentration.

Taking one of this point is to perform the validation experiments and confirm or refuse the model for the kefiran. This point corresponds with a combination of temperature and WL concentration of 30.5°C and 47.8% (m/m), respectively (**Figure 5**).

Three experiments at the selected condition were performed, reaching values of kefiran concentration of 181.1, 146.6, and 167.9 g Glu/ml. As can be seen, only the first are located inside the range suggested by the model, whilst two are outside. These results cannot confirm the validity of the model of kefiran.

Source	Sum of squares	df	Mean square	F-value	P-value	
Model	5.43	8	0.6786	190.66	<0.0001	Significant
A-T	0.3981	1	0.3981	111.86	<0.0001	
B-WL	0.0353	1	0.0353	9.93	0.0092	
AB	0.4312	1	0.4312	121.16	<0.0001	
A^2	0.3990	1	0.3990	112.11	<0.0001	
B^2	0.3553	1	0.3553	99.84	<0.0001	
A^2B	0.3602	1	0.3602	101.22	<0.0001	
AB^2	2.16	1	2.16	607.79	<0.0001	
A^2B^2	0.3245	1	0.3245	91.19	<0.0001	
Pure error	0.0391	11	0.0036			
Cor total	5.47	19				

Table 2.
ANOVA for the kefiran model.

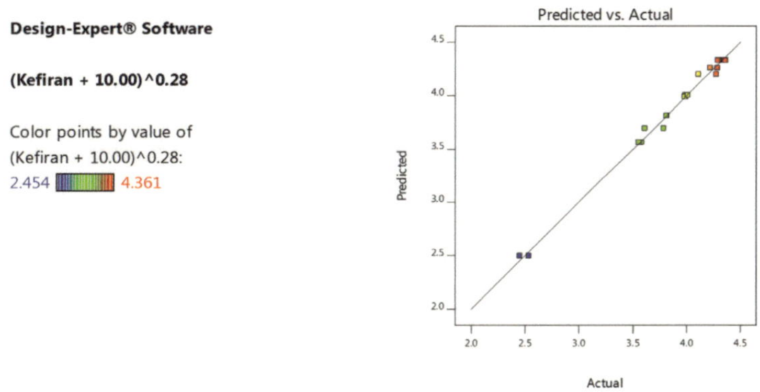

Figure 3.
Actual versus *model responses of the model for kefiran.*

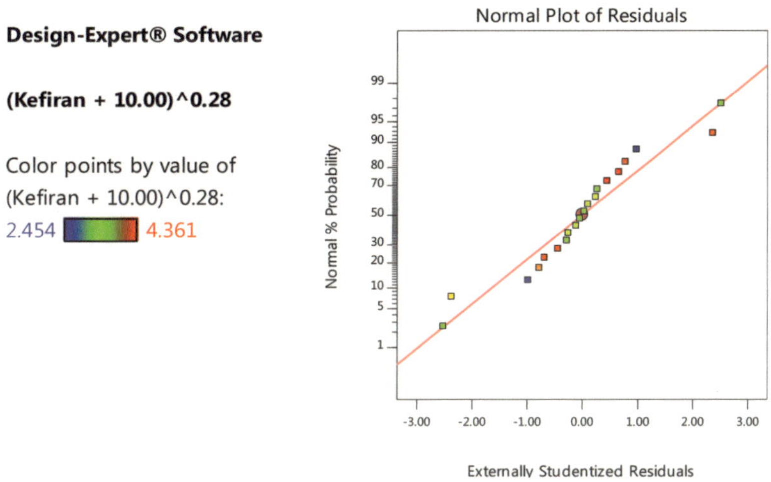

Figure 4.
Normal plot of residual in the kefiran model.

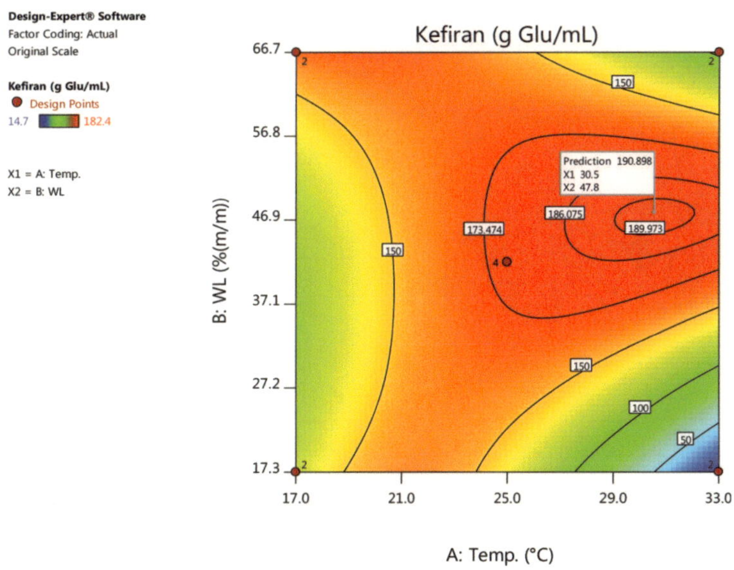

Figure 5.
Contour plot of the point selected for the validation experiments.

To check the usefullness of the kefiran model (Eq.1), one of the suggest combination of temperature (30.5°C) and sweet liquid whey (47.8% (m/m)) that should maximize the kefiran concentration was used to perform three validation experiments. Those experiments could not validate the model (**Table 3**).

According to the flow-chart shown in **Figure 1**, at this point could be evaluated another of 90 data points suggest in the optimization procedure to maximize kefiran for this model and perform with him the additional validation experiments or, alternatively, can be add the values of validation experiments to the originally data and find a new, more robust, suitable cuadratic model for determination of kefiran. The second alternative was followed, and a new set of experimental points was obtained (**Figure 6**).

The coded and actual quadratic model for kefiran were, respectively:

$$\begin{aligned} Kefiran = {} & 176.75 - 8.56\,A + 7.88\,B + 10.91\,AB - 28.15\,A^2 \\ & -28.50\,B^2 + 21.35\,A^2B - 54.50\,AB^2 + 17.40A^3 - 18.24\,A^2B^2 \end{aligned} \quad (2)$$

$$\begin{aligned} Kefiran = {} & -1017.04387 + 132.1778\,T + 12.3776\,WL - 1.6439\,T \bullet WL \\ & - 4.3795\,T^2 - 0.059484\,WL^2 + 0.052743\,T^2 \bullet WL \\ & + 0.012188\,T \bullet WL^2 + 0.033978\,T^3 - 0.000467\,T^2 \bullet WL^2 \end{aligned} \quad (3)$$

ANOVA, the model for kefiran concentration (**Table 4**), its fit statistics (**Table 5**), and the plot of predicted *vs* actual kefiran concentration values (**Figure 7**) suggests that the model could be used to find inside the space of independent variables the maximum value of kefiran.

Response	Predicted mean	Predicted median*	Std. dev.	n	95% PI low	Data mean†	95% PI high
Kefiran	190.9	190.7	9.7	3	175.5	**164.8**	206.9

**For transformed responses, the predicted mean and median may differ on the original scale.*
†For transformed responses, the data mean is calculated on the transformed scale.

Table 3.
Confirmation experiments for the first kefiran model.

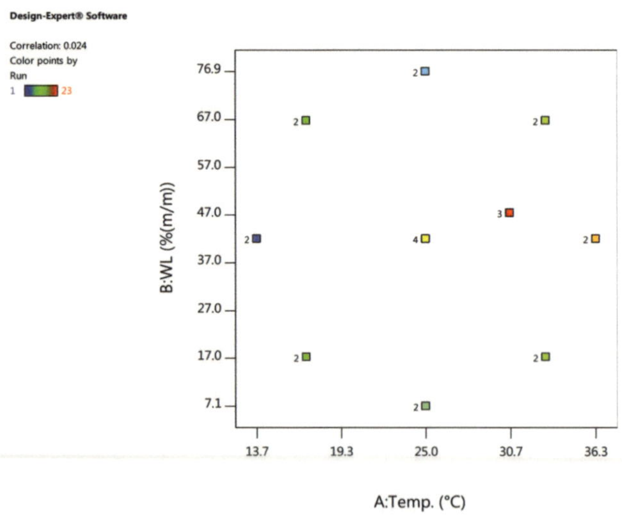

Figure 6.
New design space data of independent variables.

The *model F-value* of 60.14 implies the model is significant. There is only a 0.01% chance that an F-value this large could occur due to noise. *P-values* less than 0.0500 indicate model terms are significant. In this case, B, AB, A^2, B^2, A^2B, AB^2, A^3, A^2B^2 are significant model terms. The model term A is insignificant, but it counts to support hierarchy of the model (**Table 4**).

The fit statistics of the quadratic model for the kefiran concentration is shown in **Table 5**. The *Predicted R^2* of 0.9307 is in reasonable agreement with the *Adjusted R^2*

Source	Sum of squares	df	Mean square	F-value	P-value	
Model	50004.24	9	5556.03	60.14	< 0.0001	significant
A-T	39.19	1	39.19	0.4242	0.5262	
B-WL	497.29	1	497.29	5.38	0.0372	
AB	952.66	1	952.66	10.31	0.0068	
A^2	6339.38	1	6339.38	68.62	< 0.0001	
B^2	6498.00	1	6498.00	70.34	< 0.0001	
A^2B	1823.85	1	1823.85	19.74	0.0007	
AB^2	4845.51	1	4845.51	52.45	< 0.0001	
A^3	584.72	1	584.72	6.33	0.0258	
A^2B^2	1330.43	1	1330.43	14.40	0.0022	
Pure error	1200.93	13	92.38			
Cor total	51205.17	22				

Table 4.
ANOVA of the quadratic model for kefiran concentration.

Std. dev.	**9.61**	**R^2**	**0.9765**
Mean	129.67	Adjusted R^2	0.9603
C.V. %	7.41	Predicted R^2	0.9307
		Adeq. precision	25.3566

Table 5.
Fit statistics of the quadratic model for kefiran concentration.

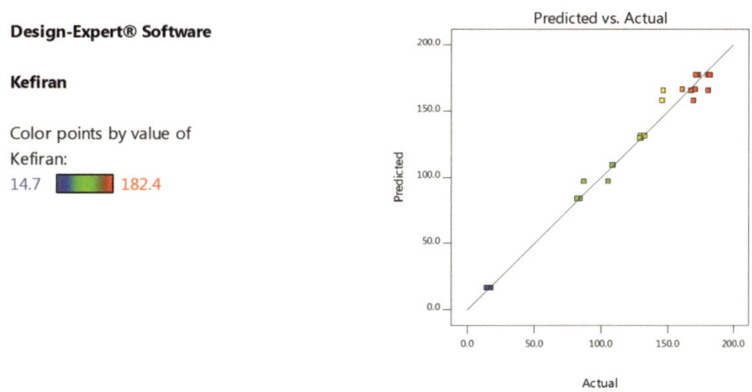

Figure 7.
Predicted versus *actual values of kefiran concentration.*

of 0.9603; i.e., the difference is less than 0.2 *and Adequate precision* measures the signal-to-noise ratio. A ratio greater than 4 is desirable. The ratio of 25.357 indicates an adequate signal. This model can be used to navigate the design space.

Name	Goal	Lower limit	Upper limit	Lower weight	Upper weight	Importance
A:T	Is in range	17	33	1	1	3
B:WL	Is in range	17.3	66.7	1	1	3
Kefiran	Maximize	14.7	182.4	1	1	5

Table 6.
Constrains to find the maximum of kefiran concentration.

Number	T, °C	WL, %(m/m)	Kefiran, mg Glu/ml	Desirability
1	23.864	45.932	178.090	0.974
2	19.962	66.700	174.901	0.955

Table 7.
Solution candidates to values for independent variables to reach maximum for the kefiran concentration.

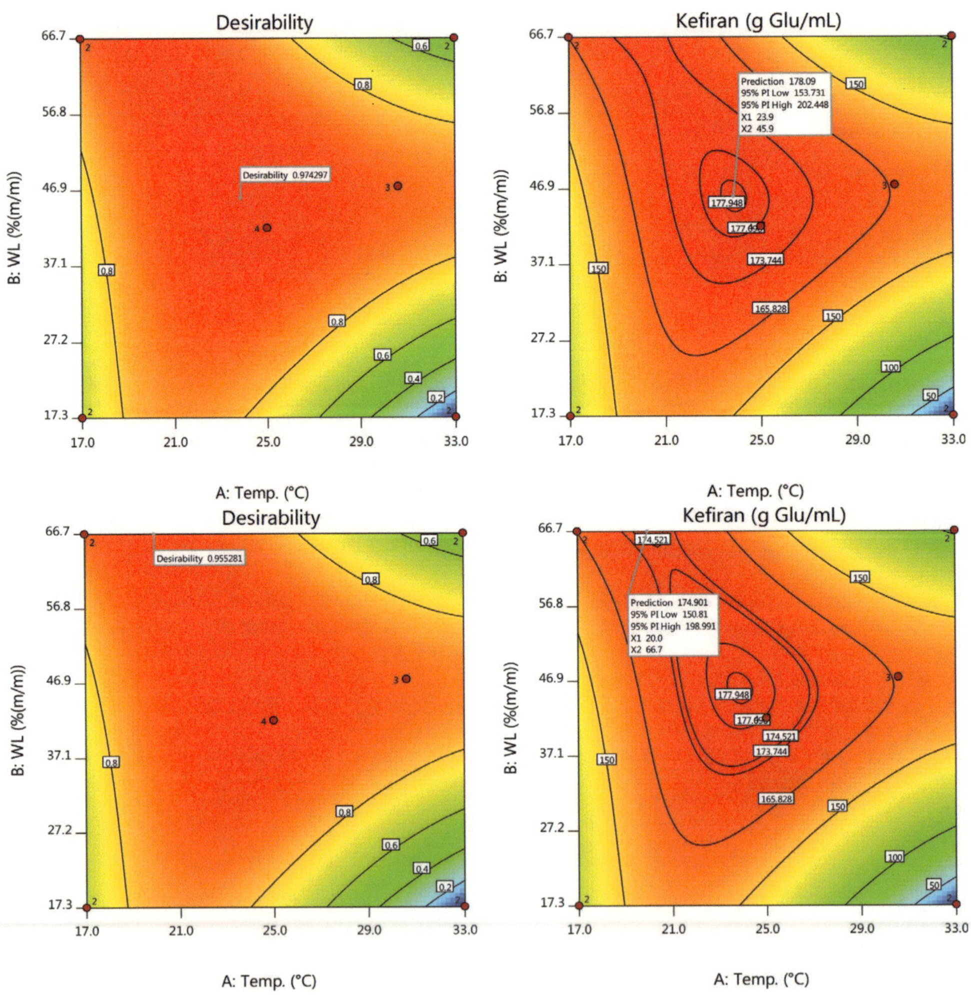

Figure 8.
Contour plots of solution values for independent variables and candidates to reach a maximum for kefiran concentration.

Response	Predicted mean	Std. dev.	n	SE Pred	95% PI low	Data mean	95% PI high
Kefiran	178.1	9.6	3	8.1	160.7	177.5	195.5

Table 8.
Value of the validation experiments for two-side confidence of 95%.

A numerical optimization to find the maximum of kefiran concentration inside the space of independent variables with constraints shown in **Table 6** was performed.

And two, equally in possibilities, solution candidates to reach a maximum for the kefiran concentration were obtained (**Table** 7).

The first suggested solution (23.9°C and 45.9% (m/m) for temperature and whey liquid, respectively) was selected to perform the three validation experiments.

The validation experiments for the kefiran concentration were 175.5, 184.5, and 172.6 mg Glu/ml, reaching a mean of 177.5 ± 6.2 mg Glu/ml, which is inside the probability interval of the model. In this manner, the validation experiments confirm the adequateness of the kefiran model (Table 8).

Further experiments to establish the small-scale production of kefiran were conducted starting with a fermentation stage with the same media, inoculum concentration, and a temperature and concentration of whey at the optimal conditions find here.

3. Small-scale production of kefiran

The soluble exopolysaccharide kefiran, synthetized by *Lactobacillus kefiranofaciens* as a part of kefir granules, could be isolated and purified from the culture broth after the lactose-fermentation from whey [37, 49]. To achieve this, several isolated steps can be executed like filtration, homogenization, thermal inactivation of glycolytic enzymes, polysaccharide precipitation, separation, and washing the precipitate (**Figure 9**).

Due to kefiran representing 50% (dry basis) of a granule [50], after fermentation, the broth could be homogenized to use all the granules to isolate the kefiran. However, together with kefiran, certain contaminants and glycolytic enzymes that eventually could degrade the kefiran itself will be aggregated. For this reason, it could be convenient to ask the question if it is necessary to employ the disrupted whole fermentation broth including in it the kefir granules to isolate kefiran, or only use the supernatant of the culture after separate the kefir granules from the broth of culture.

Another question to ask is, if the kefiran grains is separated from the supernatant is the option choice, how many times the kefiran granules can be reused? Will it be convenient to reuse many times and replace the inoculum at all or to refresh inoculum partially with a fresh grain?

At this point, at least until three times was reused, the same kefiran granules with the washing the granules with abundant sterilized distillate water between batches, with no significant difference in the final kefiran concentration (results not shown). Further experiments, however, should be performed to know how many times this process could be repeated.

Either two alternatives can be implemented instead, but only if the homogenization of whole fermentation broth brings a significant increment in the kefiran concentration in respect to use the supernatant. Additionally, in the case that only the culture supernatant was used, less time and energy would be used, and the kefir

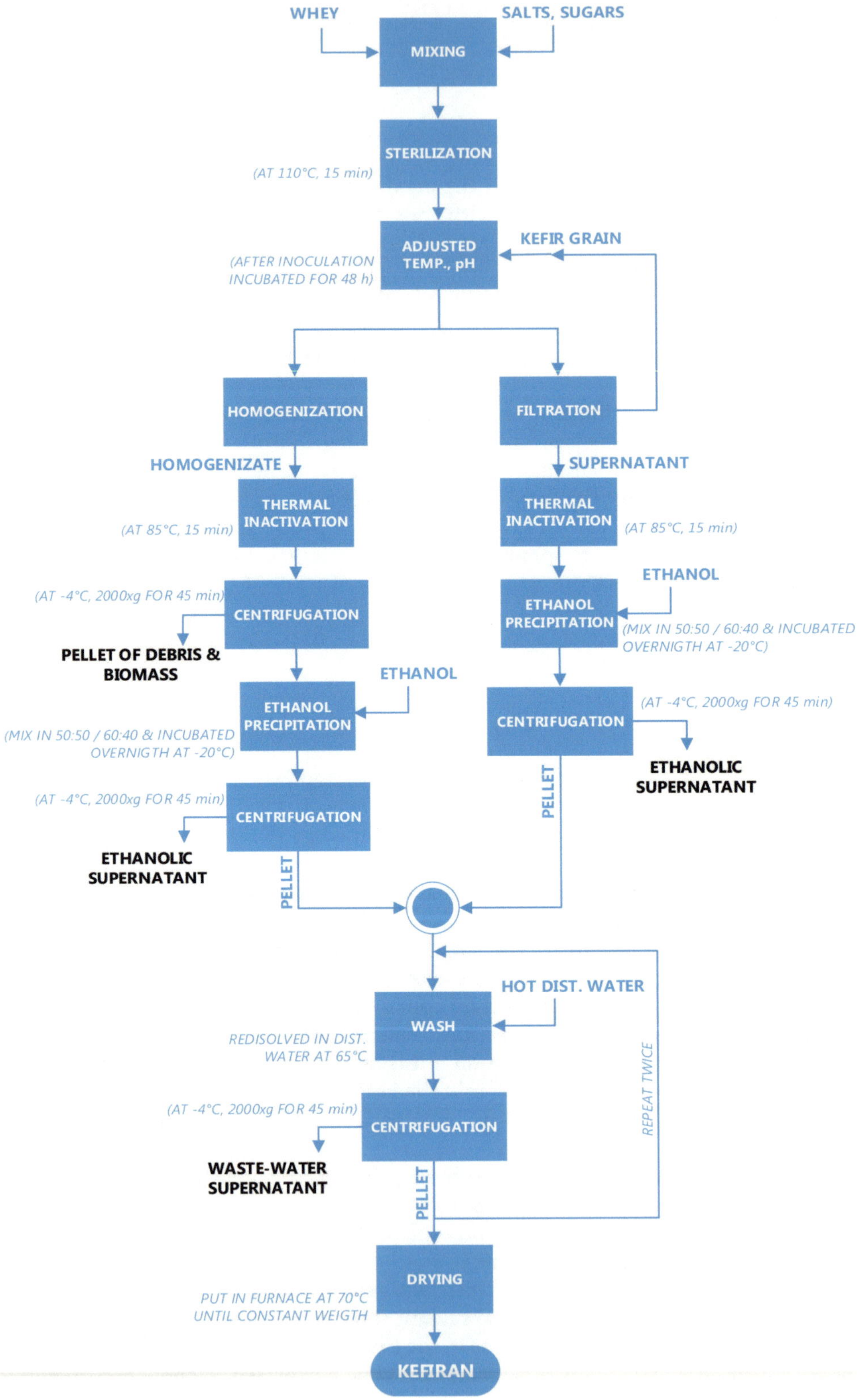

Figure 9.
Lab-scale procedure for the kefiran isolation from the fermentation of whey with kefir granules.

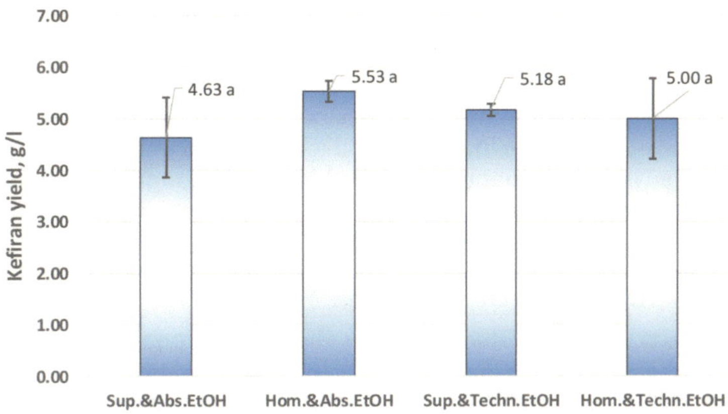

Figure 10.
Experiments to compare supernatant versus *homogenate, and ethanol absolute (anhydrous)* versus *technical in the yield of kefiran. Legend: Sup—supernatant, Hom–homogenate, Abs—EtOH-absolute ethanol, and Techn. EtOH—technical ethanol (96% (v/v)).*

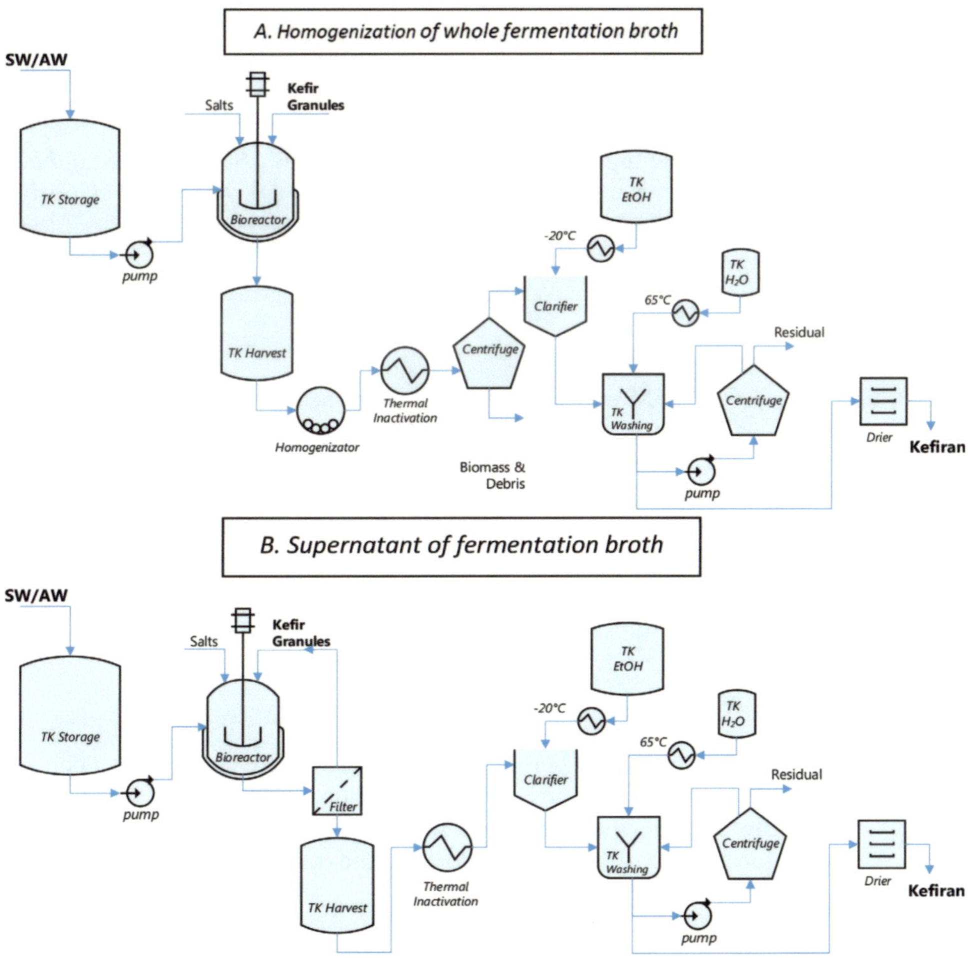

Figure 11.
Two possible flow diagrams for the kefiran production, by using: (A) the homogenization of whole fermentation broth and (B) the supernatant of fermentation broth. Legend: SW/AW: sweet or acidic whey, EtOH: ethanol, and TK: tank.

grains can be reused inside the process and wash residuals should be less contaminated that the similar is obtained for the homogenization variant.

Another question is associated to the use of solvent precipitation. Several reports referred the use of absolute (anhydrous) ethanol [24, 38], with a higher price respect his counterpart, the technical ethyl alcohol at 96% (v/v). If the technical ethyl alcohol at 96% (v/v) could be employed instead the more expensive absolute (anhydrous) one during the solvent precipitation of kefiran, a significant impact over the manufacturing cost of kefiran could be obtained.

At this point, three independent experiments were conducted to compare homogenization *versus* supernatant alternatives, and the utilization of absolute (anhydrous) *versus* technical ethyl alcohol as the precipitation agent. When the precipitation experiments were performed using absolute alcohol (anhydrous) a proportion of 50:50 with a material was employed; and when technical alcohol at 96% was used instead the anhydrous ethanol, the technical alcohol: material was mixed in a proportion of 60:40. Each experiment was started with the same fresh whey and inoculum and was performed as shown in **Figure 9**.

The alternatives shown in **Figure 10** do not have significant differences ($p > 95\%$) between them, suggesting that the more suitable and economic way for production of kefiran is to use the supernatant of fermenting broth and technical ethanol as a precipitation solvent (**Figure 11B**), instead of the homogenization of whole broth and using absolute ethanol as a solvent (**Figure 11A**).

Kefiran isolation process was carried out from three 100 ml batches each, yielding 5.2 ± 0.1 g/l of kefiran in the culture supernatant.

These results are slightly higher than the 3.1 ± 1.3 [37] or 1.91 g/l of kefiran [24] reported recently, and higher to the values between 1.5–3.7 g/l reported previously [38, 51].

4. Conclusions

The usefulness of the RSM was demonstrated in the search for a condition where the concentration of kefiran is maximized from liquid sweet whey, significantly cheaper than whey powder. It was also possible to satisfactorily replace the "absolute" (anhydrous) ethyl alcohol with the technical-grade ethyl alcohol at 96%(v/v). Both facts represent a remarkable financial saving in the production of the kefiran, which would allow to begin the studies to scale-up the production technology of this attractive EPS.

Acknowledgements

We wish to express our gratitude to the authorities of the **Universidad Técnica del Norte** (UTN, Ibarra, Ecuador) for their unconditional support, and also to **Dr. Luis E. Trujillo Toledo** from the University of the Armed Forces (ESPE, Quito, Ecuador) for his contributions made during the investigation and in the revision of this manuscript.

Author details

José Manuel Pais-Chanfrau*, Lorena D. Carrera Acosta, Paola M. Alvarado Cóndor, Jimmy Núñez Pérez and Milton J. Cuaran Guerrero
North-Technical University (Universidad Técnica del Norte, UTN), FICAYA, Ibarra, Imbabura, Ecuador

*Address all correspondence to: jmpais@utn.edu.ec

References

[1] Requelme N, Bonifaz N. Caracterización de sistemas de producción lechera de ecuador. In: La Granja, Revista De Ciencias De la Vida. Vol. 15(1). Ecuador: Univ Politécnica Sales; 2012. pp. 55-68

[2] Orozco M. Un tercio de la producción láctea se dedica al queso [Internet]. Líderes. 2015. Available from: https://www.revistalideres.ec/lideres/ecuador-produccion-lactea-queso.html

[3] Real L. Industria láctea con mejores condiciones de producción. Gestion Digital. 2013;**226**:36-39

[4] Anonymous. La AGSO firmó convenio para producción de leche. El Telegrafo [Internet]. 2017. Available from: https://www.eltelegrafo.com.ec/noticias/economia/4/la-agso-firmo-convenio-para-produccion-de-leche

[5] Pais-Chanfrau J, Núñez-Pérez J, Lara-Fiallos M, Rivera-Intriago L, Abril V, Cuaran-Guerrero M, et al. Milk whey-from a problematic byproduct to a source of valuable products for health and industry: An overview from biotechnology. La Prensa Medica Argentina. 2017;**103**(4):1-11. DOI: 10.4172/lpma.1000254

[6] Tsakali E, Petrotos K, Allessandro AD. A review on whey composition and the methods used for its utilization for food and pharmaceutical products. In: 6th Int Conf Simul Model Food Bio-Industry; FOODSIM; 2010. p. 8

[7] González Siso MI. The biotechnological utilization of cheese whey: A review. Bioresource Technology. 1996;**57**:1-11

[8] Kotoupas A, Rigas F, Chalaris M. Computer-aided process design, economic evaluation and environmental impact assessment for treatment of cheese whey wastewater. Desalination. 2007;**213**(1–3):238-252

[9] Marwaha SS, Kennedy JF. Whey—Pollution problem and potential utilization. International Journal of Food Science and Technology. 1988;**23**(4):323-336

[10] Spalatelu C. Biotechnological valorisation of cheese whey. Innovative Romanian Food Biotechnology. 2012;**10**(March):1-8

[11] Grosu L, Fernandez B, Grigoraş CG, Patriciu OI, Grig ICA, Nicuţă D, et al. Valorization of whey from dairy industry for agricultural use as fertiliser: Effects on plant germination and growth. Environmental Engineering and Management Journal. 2012;**11**(12):2203-2210

[12] Parra R. Lactosuero: Importancia en la industria de alimentos. Rev. Fac. Nac. Agron. Medellín. 2009;**62**:4967-4982. DOI: 10.15446/rfnam

[13] Ruas-Madiedo P, Hugenholtz J, Zoon P. An overview of the functionality of exopolysaccharides produced by lactic acid bacteria. International Dairy Journal. 2002:163-171

[14] la Riviére JWM, Kooiman P, Schmidt K. Kefiran, a novel polysaccharide produced in the kefir grain by *Lactobacillus brevis*. Archiv für Mikrobiologie. 1967;**59**:269-278. DOI: 10.1007/BF00406340

[15] Kooiman P. The chemical structure of kefiran, the water-soluble polysaccharide of the kefir grain. Carbohydrate Research. 1968;**7**:200-211

[16] Frengova GI, Simova ED, Beshkova DM, Simov ZI. Exopolysaccharides produced by lactic acid bacteria of kefir

grains. Zeitschrift fur Naturforschung. Section C, Journal of Biosciences. 2002; **57**(9–10):805-810

[17] Wang Y, Ahmed Z, Feng W, Li C, Song S. Physicochemical properties of exopolysaccharide produced by *Lactobacillus kefiranofaciens* ZW3 isolated from Tibet kefir. International Journal of Biological Macromolecules. 2008;**43**(3):283-288

[18] Sharifi M, Moridnia A, Mortazavi D, Salehi M, Bagheri M, Sheikhi A. Kefir: A powerful probiotics with anticancer properties. Medical Oncology. 2017;**34**: 183

[19] Arslan S. A review: Chemical, microbiological and nutritional characteristics of kefir. CyTA Journal of Food [Internet]. 2015;**13**:340-345

[20] Leite AM de O, Miguel MAL, Peixoto RS, Rosado AS, Silva JT, Paschoalin VMF. Microbiological, technological and therapeutic properties of kefir: A natural probiotic beverage. Brazilian Journal of Microbiology. 2013; **44**:341-349

[21] Nielsen B, Gürakan GC, Ünlü G. Kefir: A multifaceted fermented dairy product. Probiotics and Antimicrobial Proteins. 2014;**6**:123-135

[22] Badel S, Bernardi T, Michaud P. New perspectives for *Lactobacilli* exopolysaccharides. Biotechnology Advances. 2011;**29**(1):54-66

[23] Pop C, Apostu S, Salanță L, Rotar AM, Sindic M, Mabon N, et al. Influence of different growth conditions on the kefir grains production, used in the Kefiran synthesis. Bulletin UASVM Food Science and Technology. 2014; **71**(2):2344-2344

[24] Dailin DJ, Elsayed EA, Othman NZ, Malek R, Phin HS, Aziz R, et al. Bioprocess development for kefiran production by *Lactobacillus kefiranofaciens* in semi industrial scale bioreactor. Saudi Journal of Biological Sciences. 2016;**23**(4):495-502

[25] Serafini F, Turroni F, Ruas-Madiedo P, Lugli GA, Milani C, Duranti S, et al. Kefir fermented milk and kefiran promote growth of *Bifidobacterium bifidum* PRL2010 and modulate its gene expression. International Journal of Food Microbiology. 2014;**178**:50-59

[26] Piermaria JA, de la Canal ML, Abraham AG. Gelling properties of kefiran, a food-grade polysaccharide obtained from kefir grain. Food Hydrocolloids. 2008;**22**(8):1520-1527

[27] Kazazi H, Khodaiyan F, Rezaei K, Pishvaei M, Mohammadifar MA, Moieni S. Rheology and microstructure of kefiran and whey protein mixed gels. Journal of Food Science and Technology. 2017;**54**:1168-1174

[28] Furuno T, Nakanishi M. Kefiran suppresses antigen-induced mast cell activation. Biological & Pharmaceutical Bulletin [Internet]. 2012;**35**(2):178-183

[29] Wang Y, Xu N, Xi A, Ahmed Z, Zhang B, Bai X. Effects of *Lactobacillus plantarum* MA2 isolated from Tibet kefir on lipid metabolism and intestinal microflora of rats fed on high-cholesterol diet. Applied Microbiology and Biotechnology. 2009;**84**(2):341-347

[30] Maeda H, Zhu X, Omura K, Suzuki S, Kitamura S. Effects of an exopolysaccharide (kefiran) on lipids, blood pressure, blood glucose, and constipation. BioFactors. 2004;**22**(1–4): 197-200

[31] Rodrigues KL, Carvalho JCT, Schneedorf JM. Anti-inflammatory properties of kefir and its polysaccharide extract. Inflammopharmacology [Internet]. 2005;**13**(5–6):485-492

[32] Aebi H, Ahmed Z, Wang Y, Ahmad A, Khan ST, Nisa M, et al. Hypocholesterolaemic effects of milk-kefir and soyamilk-kefir in cholesterol-fed hamsters. The British Journal of Nutrition [Internet]. 2013; **28**(1):939-946

[33] Medrano M, Racedo SM, Rolny IS, Abraham AG, Pérez PF. Oral administration of kefiran induces changes in the balance of immune cells in a murine model. Journal of Agricultural and Food Chemistry. 2011; **59**(10):5299-5304

[34] Shiomi M, Sasaki K, Murofushi M, Aibara K. Antitumor activity in mice of orally administered polysaccharide from kefir grain. Japanese Journal of Medical Science & Biology. 1982;**35**:75-80

[35] Murofushi M, Shiomi M, Aibara K. Effect of orally administered polysaccharide from kefir grain on delayed-type hypersensitivity and tumor growth in mice. Japanese Journal of Medical Science & Biology. 1983;**36**: 49-53

[36] Blandón LM, Noseda MD, Islan GA, Castro GR, de Melo Pereira GV, Thomaz-Soccol V, et al. Optimization of culture conditions for kefiran production in whey: The structural and biocidal properties of the resulting polysaccharide. Bioactive Carbohydrates and Dietary Fibre. 2018; **16**:14-21

[37] Pais-Chanfrau JM, Toledo LET, Cóndor PMA, Guerrero MJC, Pérez JN, Intriago LMR. Response surface methodology to optimize a bioprocess for kefiran production. Prensa Médica Argentina. 2018;**104**(2):1-5. DOI: 10.4172/0032-745X.1000285

[38] Taniguchi M, Nomura M, Itaya T, Tanaka T. Kefiran production by lactobacillus kefiranofaciens under the culture conditions established by mimicking the existence and activities of yeast in kefir grains. Food Science and Technology Research. 2001;7(4): 333-337

[39] Yokoi H, Watanabe T. Optimum culture conditions for production of kefiran by *Lactobacillus sp.* KPB-167B isolated from kefir grains. Journal of Fermentation and Bioengineering. 1992; **74**(5):327-329

[40] Myers R, Montgomery D, Anderson-Cook C. Response surface methodology: Process and product optimization using designed experiments. 4th ed. Chichester: Wiley; 2016. 854 p

[41] Yolmeh M, Jafari SM. Applications of response surface methodology in the food industry processes. Food and Bioprocess Technology. 2017;**10**:413-433

[42] Ghasemlou M, Khodaiyan F, Jahanbin K, Gharibzahedi SMT, Taheri S. Structural investigation and response surface optimisation for improvement of kefiran production yield from a low-cost culture medium. Food Chemistry. 2012;**133**(2):383-389

[43] Sabokbar N, Khodaiyan F, Moosavi-Nasab M. Optimization of processing conditions to improve antioxidant activities of apple juice and whey based novel beverage fermented by kefir grains. Journal of Food Science and Technology. 2015;**52**(6):3422-3432

[44] Sabokbar N, Moosavi-Nasab M, Khodaiyan F. Preparation and characterization of an apple juice and whey based novel beverage fermented using kefir grains. Food Science and Biotechnology. 2015;**24**(6):2095-2104

[45] Magalhães KT, Dias DR, de Melo Pereira GV, Oliveira JM, Domingues L, Teixeira JA, et al. Chemical composition and sensory analysis of cheese whey-based beverages using kefir grains as

starter culture. International Journal of Food Science and Technology. 2011; **46**(4):871-878

[46] Bezerra MA, Santelli RE, Oliveira EP, Villar LS, Escaleira LA. Response surface methodology (RSM) as a tool for optimization in analytical chemistry. Talanta. 2008;**76**:965-977

[47] Alvarado-Cóndor PM. Desarrollo de un bioproceso para la obtención de una bebida funcional a partir de lactosuero en polvo y gránulos de kéfir [thesis]. Universidad Técnica del Norte (UTN); 2018

[48] DuBois M, Gilles KA, Hamilton JK, Rebers PA, Smith F. Colorimetric method for determination of sugars and related substances. Analytical Chemistry. 1956;**28**(3):350-356

[49] Hui HAN, Bao-Ping JI, Bo LI. Study on extraction and isolation of Kefiran from kefir. Food Science. 2006;**27**: 130-133

[50] Exarhopoulos S, Raphaelides SN, Kontominas MG. Flow behavior studies of kefiran systems. Food Hydrocolloids. 2018;**79**:282-290

[51] Cheirsilp B, Shimizu H, Shioya S. Kinetic modeling of kefiran production in mixed culture of *Lactobacillus kefiranofaciens* and *Saccharomyces cerevisiae*. Process Biochemistry. 2007; **42**(4):570-579

Chapter 4

Catalyst Development of Microbial Fuel Cells for Renewable-Energy Production

Masayuki Azuma and Yoshihiro Ojima

Abstract

In this chapter, we focus on microbial fuel cells (MFCs) that convert the energy from organic matters into electrical energy using microorganisms. MFCs are greatly expected to be used as a relatively low-cost and safe device for generating renewable energy using waste biomass as a raw material. At present, however, it has not reached the desired practical application because of the low-power generation; hence, improvements on fuel cell efficiency, such as electrode materials, are still being examined. Here, we focus on the microorganisms that can be used as catalysts and play a central role in improving the efficiency of the fuel cells. Several kinds of microbial catalysts are used in MFCs. For example, *Shewanella oneidensis* has been well studied, and as known, since *S. oneidensis* transports the electrons generated within the cell to the surface layer, it does not require a mediator to pass the electrons from the cells to the electrode. Furthermore, *Escherichia coli* and *Saccharomyces cerevisiae*, a model organism for MFCs, are also used. The improvements of such microbial catalysts have also been proceeding actively. Here, we elaborated on the principle of MFCs as well as the current situation and latest research on the catalyst development.

Keywords: microbial fuel cell, MFC, catalyst, renewable energy, bioelectrochemical device, microorganism, biomass

1. Introduction

The fossil fuel depletion and inevitable global warming have become worldwide problems; thus, significant efforts have been made to generate and utilize renewable energy to alleviate these crises. Methods for obtaining energy compounds from biomass, such as ethanol, methane, and hydrogen, have been developed using environment-friendly technology, and some of these technologies have been put to practical use. It is important to establish the technologies that are able to obtain energy in various forms according to the environments and circumstances of each region. Apart from the above technologies, biofuel cells utilizing microorganisms and enzymes, which can generate renewable electrical energy from organic matters contained in biomass, begin to attract attention as a means to obtain sustainable energy. It has not been put into practical use yet, but without the problem of by-products, electricity can be directly obtained from the devices, whereby multiple operations, such as product distillation (e.g., ethanol), are not necessary. Moreover,

IntechOpen

if biomass waste is used as the fuel, no food competition will occur. Therefore, using this method, energy can be obtained sustainably (**Figure 1**).

There are various types of biomass, e.g., sustainably harvested wood, waste paper, food waste, sewage sludge, and various wastewaters. Taking wood-based biomass as a fuel example, when everything is burnt using available technology for thermal power generation, there will be nothing left, and we will lose some other useful compounds contained in it. On the other hand, in biofuel cells, although electricity is generated from the sugar obtained from the biomass, other components in the wood, such as lignin, can be used for purposes other than power generation. Generally, the energy density of the biomass used as a fuel for MFCs is high. For example, glucose and xylose, found in various plant biomass, can produce up to 20 or 24 electrons per molecule, provided that they are completely oxidized to carbon dioxide. It is possible to generate 4430 Wh power per kg of glucose according to the calculation described later. For reference, a typical lithium-ion battery has a weight energy density of about 200 Wh per kg. This comparison means that glucose and xylose are two biofuel sources of interest, especially as electron donors. Therefore, MFCs using glucose and/or xylose as their fuel have great potential as a means of obtaining high energy.

In biofuel cells, biological reactions are used for the oxidation reaction of biomass, and they are divided into two based on the type of catalyst used: (1) enzymes and (2) microorganisms. When enzymes are used, the most widely studied mechanism is the two-electron oxidation system by glucose oxidase (GOx) or glucose dehydrogenase (GDH) [1]. Since purified enzymes are generally used, the reaction rate is faster than using microorganisms. However, the number of electrons obtained by one enzymatic reaction is smaller than that of a microorganism. For example, when GDH is used as the catalyst, glucose is oxidized to gluconic acid, and at most, only two electrons are obtained from one glucose molecule. Therefore, if only one enzyme is used, high-energy production per glucose consumed cannot be much expected. Further, the addition of cofactors, such as nicotinamide adenine dinucleotide and pyrroloquinoline quinone (PQQ), is necessary for the enzymatic reaction to enhance the energy production. Furthermore, the cost of enzyme purification is also high; hence, enzymes are better utilized in sensor applications than energy production. By contrast, as mentioned above, one completely oxidized glucose molecule gives 24 electrons when using microorganisms as the catalyst. It

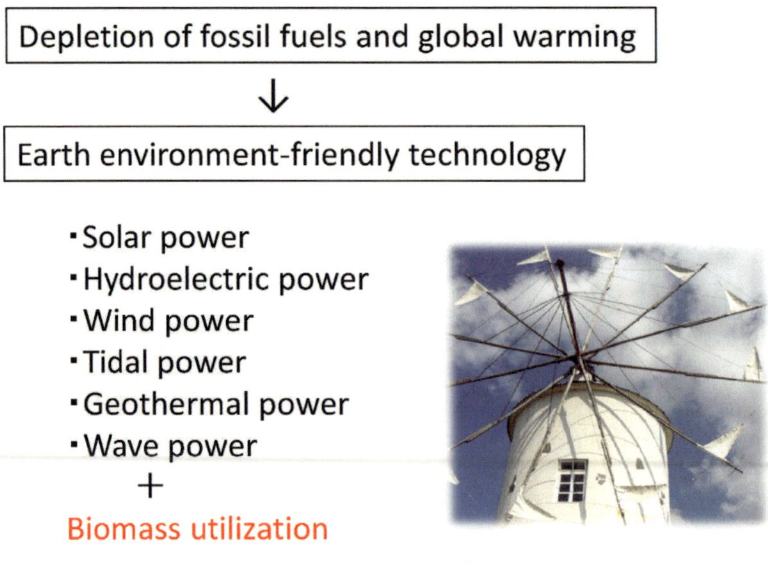

Figure 1.
Environmentally friendly energy.

shows a possibility of obtaining more electrons per glucose consumed. Moreover, the addition of cofactors is not necessary, unlike when enzymes are used. These are some substantial advantages of MFCs. However, the production of low power in MFCs is still a problem because of typical processes in living organisms, such as the uptake of glucose into cells, metabolism repression, and extraction of electrons from the inside of cells (**Figure 2**). Many researchers are working to solve such problems, and those results are summarized in recent review articles [2, 3].

Looking back at the historical background, research on the MFCs has been conducted for a long time, whereby the first idea of using microorganisms to produce electricity was conceived and reported by Potter in 1911 [4]. *Escherichia coli* and *Saccharomyces cerevisiae* were used as the catalyst, and platinum was used as the electrode. Further, in 1931, Cohen showed that a number of small fuel cells connected in series produced 2 mA of electricity at over 35 V [5]. Early MFCs used an artificial mediator, e.g., thionine, methyl viologen, and humic acid, to carry electrons from the microbial cells to the electrodes [6, 7]. The oxidized mediators came into contact with the microbial cells, and were reduced by accepting electrons, and they were then separated from the microbial cells. They diffused and came into contact with the surface of the electrode to release the electrons and were reoxidized thereafter. Overall, the addition of artificial mediators promotes the flow of electrons. Next, in the 1990s, several bacteria were found capable of acquiring electrons from the electrode via a self-mediator without the addition of an artificial mediator. Moreover, they used electrons for their growth; for example, a ferric-iron-reducing bacterium *Shewanella putrefaciens* grew on lactate by obtaining electrons from the electrode [8], and similar reports were found on *Shewanella oneidensis* and *Geobacter sulfurreducens* [9, 10]. Although the flow direction of electrons between cells and electrodes is opposite from the MFC explanation just before, these findings led to the development of mediator-free MFCs [11–13]. In the 2000s, the mechanism at the cell surface whereby bacteria directly came into contact with the electrodes and carried the electrons was reported [14, 15]. Since then, MFC research, including the analysis of the electron transport mechanism at the cell surface, has been actively conducted. In fact, the number of publications related to the MFCs grew significantly since 2010 and reached nearly 1000 in 2016 and 2017 according to ISI WEB OF SCIENCE [2]. In particular, with the discovery of new fuel cell components other than the microbial catalysts, the performance of MFCs could be increased further.

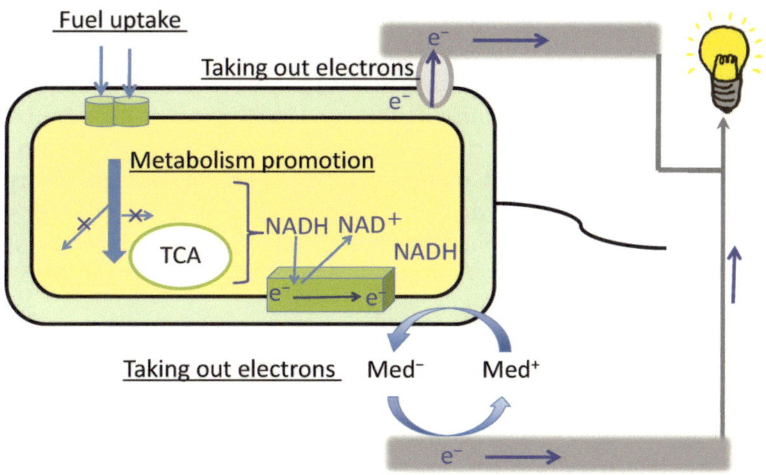

Figure 2.
Electron generation and extraction in the microbial fuel cell system. Med: mediator.

The performance of MFCs is evaluated based on some indicators. The electrical energy (Wh) used to express the capacity of dry batteries is also an important indicator, but only a few papers have reported it so far. In most cases, the performance is expressed as the maximum power per anode electrode area (power density per area) or the maximum power per cell volume (power density per volume). The latter is a straightforward index and important for practical use. For example, a relatively high-performance small-scale fuel cell (2.5 mL) using a complex (mixed) microorganism system was reported in 2007 with a power density of 1550 W/m^3 [16]. Other fuel cells performing beyond 500 W/m^3 were also reported [17–22], but many of them are still at a microliter or milliliter scale. Owing to the low proton diffusion speed and high internal resistance, the maximum power per volume tends to be small for a large-scale fuel cell. MFCs with a volume more than 1 L were also being studied in the laboratory, but the maximum power per volume was still low at the level of several W/m^3 to tens of W/m^3 [23, 24]. Scaling-up is also another issue of MFCs, and further improvements are still being conducted.

Practical applications of MFCs are still problematic because of the high cost and low-power generation. Despite this situation, research on the implementation of MFCs has been carried out. For example, an artificial stomach called Gastrobot (aka Chew-Chew train) using *E. coli* and sugar as the catalyst and fuel, respectively, EcoBot-III (a self-sustainable robot with its own circulatory system, such as ingestion, digestion, and excretion), and several environmental sensors using the MFCs for powering [2]. Such implementation studies are also important to understand the desired performance level for MFCs. Meanwhile, in order to bring MFCs closer to practical use, a combination of power generation and other effects is one of the promising methods. For example, MFCs that are installed at a wastewater treatment facility are expected not only to reduce the generated sludge amount from the treatment but also to cover a part of the electricity load used by the plant. Recently, the minimum performance of MFCs required for reaching energy neutrality in a wastewater treatment facility has been calculated [25], and the realization has been highly expected. Such research on MFCs installed for wastewater treatment has been actively carried out so far, and the experiments were examined at a pilot-scale plant of more than 10 L [26–28]. Accordingly, the practical use of MFCs is expected soon. Besides this, although it deviates from energy production, the use of MFCs as a sensor has also been extensively studied. In order to perform on-site real-time monitoring, it is important to recognize the toxic compounds rapidly. Several analytical techniques based on electrochemistry have been developed for this purpose, but in many cases, they lack practicality for environmental measurements. The MFC-based biosensor is one promising candidate, and it has already been shown that not only toxic heavy metals but also toxic organic compounds can be detected [29]. The sensitivity adjustment suitable for the detection of specific contaminants is left as a challenge, but there is still a great expectation for its application.

Here, we will explain the mechanism of electron generation in microorganisms, introduce the principle of MFCs, describe the microbial catalysts used for various MFCs mentioned above, and discuss the recent topics on microbial catalysts.

2. General principles of MFCs and various microbial catalysts

2.1 Mechanism of electron generation in microbial cells

MFCs utilize the decomposition energy of organic matters by the organisms to produce ATP, known as the energy currency, based on the energy obtained from

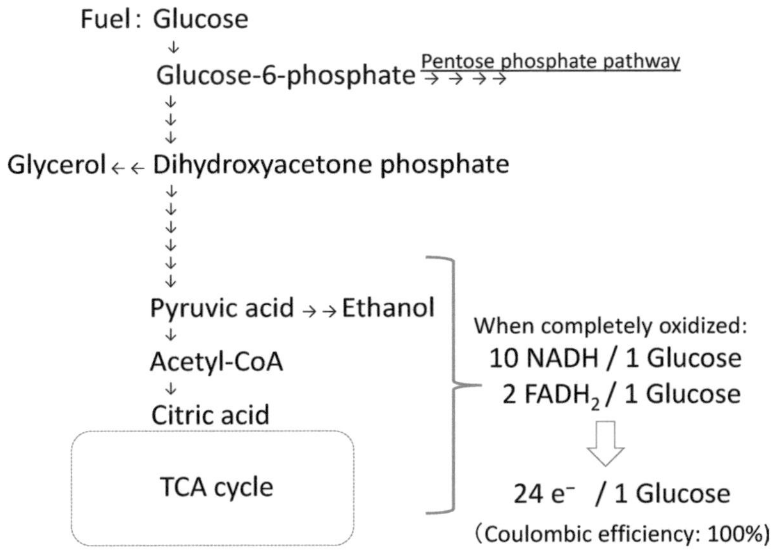

Figure 3.
Glucose metabolism (Saccharomyces cerevisiae).

this process. Taking glucose decomposition in *Saccharomyces cerevisiae* as an example (**Figure 3**), glucose taken into the microbial cells by cell membrane enzymes is oxidized and decomposed to pyruvic acid by various enzymes via an intracellular glycolysis system. Next, pyruvic acid becomes carbon dioxide and water when it is completely oxidized via the TCA cycle. The electrons generated are then collected in the mitochondrial inner membrane in eukaryotes, and in prokaryotes, they were accumulated in the cell membrane via NADH and FADH2 (both of which provide two electrons to the membrane electron flow; **Figure 3**). In addition, the flow of electrons in these membranes is maintained through multiple protein complexes localized in the membrane. Quinone compounds and cytochrome proteins are also included in the complex. ATP is synthesized by the membrane enzyme, ATP synthase, using the proton concentration gradient, which is generated inside and outside the membrane because of the flow of electrons. In MFCs, it is thought that it takes the electrons directly from NADH or the flow of electrons generated in the membrane by the decomposition of organic substances. A part of the electrons generated within the microbial cell is carried to an electrode outside the cell via an electron carrier called a mediator. When one molecule of glucose is completely oxidized into carbon dioxide and water in the cell, 10 molecules of NADH and 2 molecules of FADH2 are also generated. In total, 24 electrons are obtained from 1 glucose molecule. If this principle were to apply to fuel cells, the Coulomb efficiency, which is an index of energy efficiency, would become 100%. Therefore, in order to obtain electrons more efficiently from the cells via such metabolism, it is important to adjust the redox balance within the microbial cells in the MFCs.

2.2 Calculation of the energy obtained from glucose

In the case of using glucose as the fuel source, the reaction occurring in the anode tank is represented by Eq. (1), and the reduction reaction occurring in the cathode tank is represented by Eq. (2).

$$C_6H_{12}O_6 + 6H_2O \rightarrow 6CO_2 + 24H^+ + 24e^- \quad (1)$$

$$6O_2 + 24H^+ + 24e^- \rightarrow 12H_2O. \quad (2)$$

The oxidation-reduction potential of Eq. (1) is −0.42 V, whereas the oxidation-reduction potential of Eq. (2) is 0.82 V. Therefore, the total potential difference of the MFC reaction (Eq. (1) + Eq. (2)) as represented by Eq. (3) is 1.24 V. Theoretically, the voltage exceeds 1 V, but in most cases, it has never reached that value.

$$C_6H_{12}O_6 + 6O_2 + 6H_2O \rightarrow 6CO_2 + 12H_2O. \quad (3)$$

Assuming that 24 electrons are obtained from 1 glucose molecule and that they can be recovered in 1 h, the quantity of electricity (Ah) obtained from the glucose (1 kg) can be calculated using the Faraday constant (96,485 C/mol) as shown in Eq. 4. As a result, the electrical energy of 4430 Wh can be achieved if the potential is 1.24 V; accordingly, this value is the same as the value mentioned in the Introduction section.

$$(24 \times 96{,}485 \times 1000)/(3600 \times 180) = 3574 \text{ Ah}. \quad (4)$$

2.3 Basic components of dual-chambered MFCs using a mediator

A dual-chambered fuel cell consisting of an anode tank and a cathode tank is the simplest and has been used for a long time for MFCs. In many cases, they are separated by a cation exchange membrane (CEM) to create a potential difference between the two tanks (**Figure 4**). CEM prevents mixing of each content and allows the protons generated in the anode to migrate to the cathode. In addition, CEM selection, especially based on its proton transfer efficiency, is important

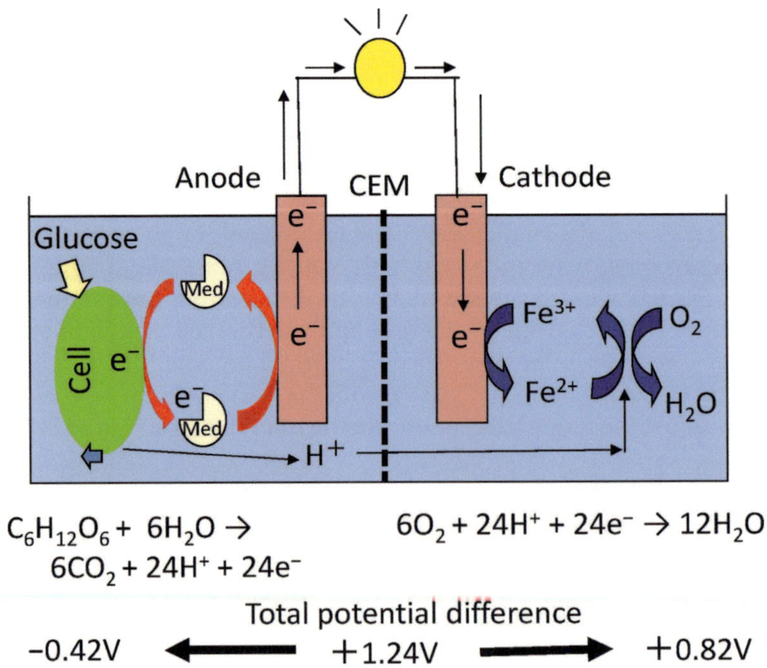

Figure 4.
General dual-type MFC. Med: mediator, CEM: cation exchange membrane.

because it significantly regulates the movement of the protons responsible for the pH reduction at the anode affecting the activity of microorganisms and the delivery of electrons to the oxygen at the cathode. Also, some factors to consider, such as durability and cost, are important for selecting CEM. At present, Nafion is popular for many MFCs [30, 31].

Numerous research studies are being conducted to evaluate the influence of the electrode materials on the performance and cost of the MFCs. Carbon materials, which are noncorrosive, have been widely used because of their high electrical conductivity and chemical stability, e.g., carbon rod, carbon fiber, carbon felt, and carbon cloth [3]. Biocompatibility, specific surface area, electrical conductivity, and cost are important factors for its selection. Since its discovery in 2004, graphene has been attracting much attention for its use as an electrode because of its high specific surface area, electrical conductivity, and biocompatibility [32]. In fact, graphene has been already used in lithium-ion batteries, and the development of graphene-modified materials to increase the power density has progressed actively [33, 34]. Moreover, since biofilm formation by microorganisms on the electrodes affects the performance of MFCs, the preference of electrode materials tends to shift from two-dimensional to three-dimensional surfaces, where a larger surface area is obtained; thus, the contact with microorganisms increases. Furthermore, metals are also used as the electrodes. The conductivities are higher than those of carbon materials, but they are prone to corrosion in the anode solution. Therefore, metals are problematic to use, except for stainless and titanium. To improve such problem, materials in which metal is incorporated into graphite have been made [3].

A phosphate buffer or bicarbonate buffer solution is often used for the anode electrode solution to achieve high performance [16, 35]. The pH of the solution affects not only the activity of microorganisms but also the transfer of hydrogen ions used from the anode to the cathode when the electrons are transferred to oxygen at the cathode. The solutions contain microorganisms as the catalyst, organic matter as the fuel, and mediator as the electron carrier. In addition, there are reports that the performance of MFCs was improved by adding NaCl to increase the ionic strength [36].

Regarding the fuel, many substrates have already been studied [37]. For example, acetic acid, lactic acid, glycerol, glucose, xylose, sucrose, starch, yeast extract, malt extract, various real wastewaters, and synthetic wastewater were used depending on the purpose of each research. Generally, the fermentable substrate of microorganisms is used to generate electricity more efficiently. There is a trend where glucose is used when using *S. cerevisiae*, lactic acid when using *S. oneidensis*, and acetate when using *G. sulfurreducens* in the experiments. On the basis of a calculation, when lactic acid, acetic acid, and glycerol are completely oxidized, there are 14, 8, and 14 electrons obtained, respectively. The number of electrons obtained from each substrate depends on the metabolic pathway.

Regarding the mediator, although some microorganisms can carry electrons directly to the electrode as described later, in many cases, the electrons cannot be carried, or the performance is low even if carried, so an artificial mediator that can pass through the cell membrane is added to the anode solution. The typical compounds for artificial mediators are methylene blue, neutral red, 2-hydroxy-1,4-naphthoquinone (HNQ), thionine, benzyl viologen, 2,6-dichlorophenolindophenol, and various phenazines [38]. It was reported that the hydrogenase donates electrons to the neutral red [39], but the process was not yet clearly proven as to how these mediators deprive electrons of the cell. It is thought that, depending on the type of mediator, the electrons may be taken directly from NADH or obtained from the electron transfer system of the cell membrane. On the other hand, there is also a difficult aspect of using a mediator. In order to increase the electron transfer efficiency by the mediator, it requires a high concentration, but because of its high toxicity, it has a strong influence on the cells; therefore, the level of use is necessary to be controlled.

Finally, the cathode solution is explained as follows. The electrons generated at the anode are carried to the cathode, where the reduction reaction takes place. When oxygen, the most common electron acceptor, is used as an oxidizing agent, aeration is necessary because oxygen has low solubility (about 8 mg/L DO). There are cases where oxygen generation by the photosynthesis of algae is used for oxygen supply [40, 41]. In the reaction at the cathode, H_2O is produced by oxygen, whereby the electrons were carried from the anode via an external circuit and protons were carried from the anode solution via CEM. There is also a report that the addition of hydrogen peroxide leads to an improvement in power generation [42]. Besides oxygen, there are various electron acceptors [43]; for example, an oxidizing agent such as iron ferricyanide is also used for the cathode. In many cases, the ferricyanide has a high mass transfer efficiency and a high cathode potential so that a high output can be obtained. The combination of carbon electrodes and ferricyanides to achieve power 50–80% higher than the combination of Pt/carbon electrodes and oxygen was reported [44]. In the case of using ferricyanide, once the trivalent iron ion receives the electrons, it becomes divalent, and when it delivers the electrons to oxygen, it reverts to the trivalent state. However, the latter reaction is less likely to occur owing to the low solubility of oxygen. Ferricyanide is an excellent electron acceptor, but owing to its toxicity, its use is generally limited to the laboratory. Other than oxygen and ferricyanide, there are also many candidates, for example, nitrate, persulfate, permanganate, and manganese dioxide. It is also possible to use the nitrate contained in the wastewater because its redox potential is close to that of oxygen, and then, the nitrate is reduced to nitrogen gas at the cathode [43].

2.4 Other types of MFCs

MFCs are typically divided into a dual-chambered cell described above and a single-chambered cell (**Figure 5**). In the latter, a membrane-type positive electrode with oxygen permeability called an air cathode is used [45]. The electrode is coated with the platinum catalyst, and H_2O is produced from the oxygen permeated from the atmosphere, the proton in solution, and the electron from the anode. However,

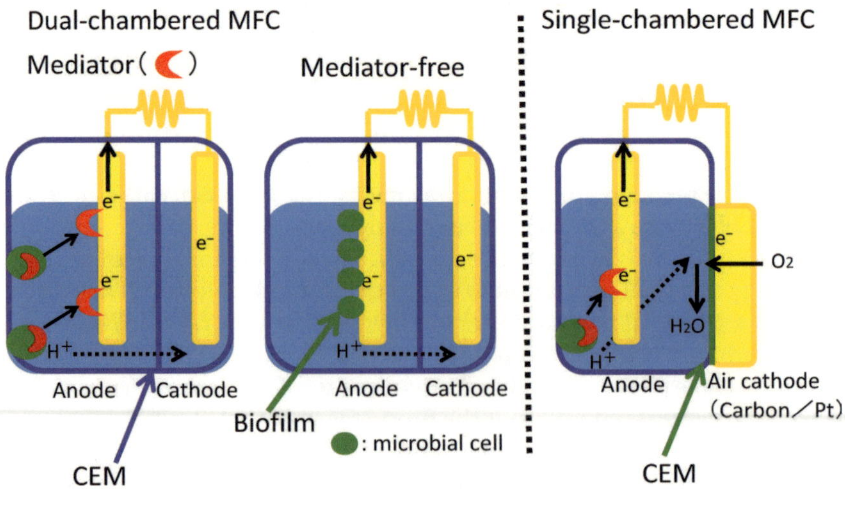

Figure 5.
Other types of MFCs. Med: mediator, CEM: cation exchange membrane.

if the permeated excess oxygen diffuses and is delivered to the microorganisms at the anode, the electrons generated in the microorganisms are then transferred to the oxygen, and the energy recovery rate decreases. Therefore, a CEM is used between the anode solution and the cathode in order to prevent a decrease in energy recovery. This single-chambered type has already been widely used currently.

Moreover, there is a mediator-free type that does not require an artificial electron compound [46]. The microbial strains used here can synthesize mediators themselves and/or have an electron transfer function on the cell surface. In the former, the self-synthesized mediators are flavin compounds, hydroquinone, and phenazine that are able to transfer the electrons to the electrode. In the latter, membrane-bound proteins such as pili, c-type cytochromes, and filaments are known as cell surface structures that can directly transmit electrons. The biofilm formation on the electrodes, namely, biocompatibility of the electrodes, is also important for power generation via such direct electron transfer. Therefore, research on electrodes promoting the formation of the biofilm is actively being conducted.

In addition, with the use of MFCs in wastewater treatment, contamination on the CEM results in reduced power generation; hence, membrane-free MFCs have also been studied [46].

2.5 Various microbial catalysts

Various microorganisms have been studied for a long time since the first experiments on *S. cerevisiae* and *E. coli* [4]. The classification of these catalysts is largely based on the purity and complexity of the cultured microbial systems. Many different microorganisms are used in the pure system [37, 38, 47, 48]. *S. cerevisiae* is a safe microorganism used in foods and can grow even in the presence of a high concentration of sugar, sulfate, and ammonium nitrogen. MFCs show high performance when using *S. cerevisiae* and glucose as a catalyst and a fuel, respectively [43]. *E. coli* can also ferment sugar well and is used for the study of MFCs using glucose as a fuel. Although it can generate electricity without a mediator, in the present situation, the power generated is low, so an artificial mediator is added in order to achieve better performance. Besides the two examples, there are also *Pseudomonas aeruginosa*, *Enterococcus faecalis*, *Rhodoferax ferrireducens*, *Geothrix fermentans*, *Shewanella* species, *Geobacter* species, *Clostridium* species, and sulfate-reducing bacteria. The possibility of utilizing extremophilic microorganisms is also being studied [49], and to add a new perspective to power generation by MFCs, the utilization of photosynthetic bacteria at the anode is also examined [40]. One of the advantages of these MFCs is the elimination of carbon dioxide released into the atmosphere. Meanwhile, in complex systems, the use of various wastewater and waste sludge has been reported [25, 37, 46, 50]. Many studies on bacterial communities under the control of MFCs have been conducted using those aforementioned resources. It is thought that the bacteria belonging to the phylum Proteobacteria were involved in power generation [51, 52]. However, owing to the complexity of bacterial interactions, their contribution to power generation within these communities is not well understood yet.

In such a research situation, there are relatively many examples of research on *S. oneidensis* and *G. sulfurreducens*, and the details of their power generation mechanisms are being clarified. *S. oneidensis* can produce self-synthesized mediators, like flavin compounds. The strain has not only such a mediator but also an extracellular electron transport system involved in power generation. This system, present from the inner membrane to the outer membrane, plays a role in carrying the electrons to the extracellular receptors (i.e., the electrodes in this case) by contacting them directly. In particular, cell-surface-localized cytochromes (MtrC and OmcA) are important components for the electron transfer [53]. On the other hand, *G. sulfurreducens* has electrically conductive pili, called nanowires, which can transfer electrons to extracellular electron acceptors on the cell surface [54]. *S. oneidensis* also has an electrically conductive structure similar to the pili, but its structure is different, whereby the membrane structure containing the cytochrome protein described above was raised [54]. In any case, it has been confirmed that electrons can be delivered via such protrusions. It is expected that new developments will be made in the future, such as introducing the genes related to these mechanisms into other species, especially model organisms, such as *E. coli* and *S. cerevisiae*.

3. Recent topics of microbial catalyst and future directions

3.1 Modification of microorganisms

The utilization of chemical and biotechnological techniques is important to modify the microbial cells as biocatalysts in the MFC system. Molecular biology approaches are effective tools to improve the performance of the biocatalysts for the desired system. In this section, recent topics about the approaches for microbial catalyst development are discussed.

3.1.1 Chemical treatment

Mediators and macromolecular catabolic enzymes, which are used for electron transfer and other metabolic activities, are abundant in the cytoplasm of the microbial cells used in the MFCs. However, it is not easy to transport the mediator molecules to the bacterial outer membrane so as to reach the electrode. The lipopolysaccharide (LPS) layer on the Gram-negative outer membrane is compact and nonconductive; thus, most microbial cells are nonconductive [55]. It was found that chemically perforated pores and channels on the cell membranes accelerated electron transfer, leading to an improved power output for an MFC using *P. aeruginosa* [56]. In their study, polyethyleneimine (PEI)-treated biofilm achieved a doubled power density (41 mW/m^2) compared to the control biofilms. The large pores and channels on the cell membrane created by PEI treatment promoted the diffusion of the self-produced mediators (pyocyanin and pyorubin) of *P. aeruginosa*. The modified cell membrane surface also promoted the adherence of microbial cells to the electrode, which further improved the electron transfer. This method was also applied to *E. coli* [57]. Recently, it has been reported that lysozyme treatment increased 1.75-fold of the MFC performance with *K. rhizophila* P2-A-5 [58]. Thus, chemical treatment is one of the important approaches to modify the microbial cells for the improvement of the MFC performance.

3.1.2 Biosurfactant production by gene modification

To increase the cell permeability of biocatalysts in the MFCs, Zheng et al. proposed a new approach by inducing the biosurfactant production based on a genetic modification [59]. It is true that the efficiency of membrane permeability can be improved with a biosurfactant, which ultimately increases the transport across the membrane. In addition, overexpression of the *rhlA* gene, which is responsible for rhamnolipid (a biosurfactant) production, was also conducted [60]. The biosurfactant directly influenced the overproduction of rhamnolipids from the electrical bacteria, such as *P. aeruginosa*. The electron transport across the membrane was greatly increased as the membrane permeability increased. The power output of the

MFC-catalyzed process by the biosurfactant-producing bacterium was enhanced up to about 2.5 times compared with the wild type.

3.1.3 Introducing the extracellular electron transfer pathway

The sparse availability of genetic tools in manipulating electricity-generating bacteria and the multiple overlapping pathways for extracellular electron transfer make it challenging to modulate electron transfer and/or introduce other functions of interest. In response to this challenge, several studies have taken the complementary approach of engineering portions of the extracellular electron transfer pathways into the well-studied industrial microbe *E. coli* [61]. In these studies, MtrCAB of *S. oneidensis* was successfully expressed in the *E. coli* cells, and the activity of these proteins was confirmed by the metal reduction. Although the introduction of MtrCAB permits extracellular electron transfer in *E. coli*, the low electron flux and the absence of growth in these cells limit their practical application. Recently, in addition to surface-localized cytochromes, it has been further confirmed that CymA, the inner membrane component of *S. oneidensis*, significantly improved the extracellular electron transfer rate or cell viability. This recombinant *E. coli* achieved current generation in an MFC system without the addition of mediators [62, 63]. Our research group is trying to develop an excellent *E. coli* biocatalyst for the anode in an MFC system based on the combination of engineering of central metabolism and introduction of extracellular electron transfer in the presence of an HNQ mediator.

3.2 Extremophilic microorganisms

An extremophilic microorganism thrives in physically or geochemically extreme conditions that are detrimental to most life on Earth. They thrive in extreme hot niches, ice, and salt solutions, as well as acid and alkaline conditions; some may grow in toxic waste, organic solvents, heavy metals, or several other habitats that were previously considered inhospitable for life. Extremophiles can be used to oxidize sulfur compounds in acidic pH to remediate wastewaters and generate electrical energy from marine sediment microbial fuel cells at low temperatures. The MFC performance of these extremophilic microorganisms has been well summarized in several review papers [49, 64]. In this section, the recent advances of MFCs using extremophilic microorganisms as catalysts are briefly introduced and discussed.

3.2.1 Acidophiles and alkaliphiles

An increase of cell voltage is seen at increasing anode pH because of the additional pH gradient representing a source of energy. The practical implication of an elevated cell voltage is that more energy can be gained from MFC systems at higher pH values. By contrast, operating the anode of MFCs at an acidic condition has an advantage that the protons will not cause diffusion limitations in the cathode compartment for the reduction of oxygen, and therefore, it will not limit the current production [65]. However, under a low-pH condition, the microbial cells have to maintain a near-neutral cytoplasm [66] which consumes a portion of the energy derived from the electron transport for other processes, such as proton export, that increases the anode overpotential, leading to decreasing power generation. At pH 2.5, *Acidiphilium* sp. isolated from the environment mediates a direct electron transfer from the glucose metabolism to the anode at a rate of 3 A/m^2 even in the presence of air [67]. This interesting strain produces extracellular polymeric substances and forms a biofilm between the carbon microfibers and in pores on the graphite rod surface [68].

Meanwhile, at high pH, the lower anode potential results in an increased cell voltage. In general, the anode becomes acidified during the MFC operation, and the cathode becomes more alkaline, followed by a reduced cell voltage and power output. *Pseudomonas alcaliphila* is capable of electricity production at high pH, excreting phenazine-1-carboxylic acid that acts as an electron shuttle during the oxidation of citrate [69]. An MFC has also been developed to treat food wastes that comprise 30–55% of all refuse in urban societies. The waste was first treated by anaerobic digestion, the resulting food waste leachate was amended with 100 mM NaCl, and then, electricity was generated in a pH 9 MFC that had a maximum of 63% Coulombic efficiency [70]. Recently, a tubular upflow MFC utilizing seafood processing wastewater has achieved a maximum power density of 105 mW/m^2 at pH 9 [71].

3.2.2 Psychrophiles and thermophiles

Temperature also has a major impact on the output of MFCs. It influences the activity of microorganisms, the electrochemical reactions, and the Gibbs free energy change of the reactions. There exists an optimum temperature for enzymes in the MFCs, and the electrochemical reaction rate increases with increased temperatures. A lower operating temperature adversely influences the output, start-up time, and substrate oxidation rate in the MFC system. This negatively impacts the MFCs for processes such as wastewater treatment because the streams are generally at low temperatures. However, the advantage of low temperature for the MFCs is that they typically produce higher Coulombic efficiencies [72, 73]. The microbial community was enriched from the anaerobic sludge at the anode of an acetate-fed MFC operated at 15°C with psychrophiles *Simplicispira* sp. and *Geobacter* sp. [74]. Another study at 5–10°C enriched the low-temperature microorganisms from the genera *Arcobacter*, *Pseudomonas*, and *Geobacter* [75]. One promising application of the low-temperature MFCs is that for low-power-consuming devices like sensors that are intended to last for an extended period of time.

On the other hand, the advantages of operating at high temperatures are higher microbial activity, better substrate solubility, high mass transfer rate, and lower contamination risk. An example of an improved current generation at a high temperature (60°C) is a marine sediment MFC that generated 209–254 mA/m^2 compared with 10–22 mA/m^2 at 22°C [76]. Recently, an MFC with a higher operating temperature (70°C) has generated 6800 mA/m^2 [77]. Furthermore, the hyperthermophilic MFCs were operated at above 80°C [78]. However, a negative point of thermophilic MFCs is higher rates of evaporation than the MFC system itself. Therefore, a continuous mode of MFCs was proposed to allow a replacement of the anolyte and catholyte [79].

In this section, recent topics of microbial catalysts for MFCs were introduced. There are two approaches in developing the microbial catalysts. One is the modification of existing microorganisms using chemical treatments or biotechnological techniques, including gene editing. The other is exploring new microorganisms from the environment, including extreme conditions. Although new findings and knowledge were obtained from both approaches, a drastic improvement on the MFC performance to achieve a paradigm shift has not appeared yet. In parallel with the improvement of microbial catalysts, the development of the fuel cell system, including the electrodes, was intensively studied to increase the output of MFCs. In particular, the application of graphene-modified electrodes [33] and the investigation of electron acceptors [43] have shown remarkable progress in the past decade. In order to actualize the practical use of MFCs, a synergistic impact from the combination of microbial catalyst and fuel cell system is essential.

4. Conclusion

This chapter focuses on the significance of MFC development, the historical background and fundamental principles of MFCs, and their recent developments, especially on microbial catalysts. MFCs have not reached the desirable level of power generation that supports daily life because of the problems such as scaling-up. On the other hand, developments of technology combining wastewater treatment and power generation, and application for environmental sensors are progressing to a stage close to practical use. If these popularizations continue, it will further develop its application in broader fields. Owing to their limitations, it may be difficult to force MFCs to become the main power supply in our daily life in the future, but it seems possible to use them as an auxiliary power supply. In addition, MFCs may become useful as a power supply in areas where the infrastructure is not well developed, for example, a portable power supply generating electricity if water is added. Regarding microbial catalysts, it is also known that various microorganisms can generate electricity, and if this superior power-generating function of these microorganisms can be integrated into a microbial cell using the synthetic biological method developed recently, the ability of the microbial catalyst will dramatically increase. Soon, its power generation ability could be greatly improved in combination with the progress of other constituents.

Acknowledgements

This work was supported by JSPS KAKENHI Grant Number JP17K06932.

Conflict of interest

The authors declare no conflicts of interest associated with this text.

Author details

Masayuki Azuma* and Yoshihiro Ojima
Department of Applied Chemistry and Bioengineering, Osaka City University, Osaka, Japan

*Address all correspondence to: azuma@bioa.eng.osaka-cu.ac.jp

References

[1] Ivanov I, Vidaković-Koch T, Sundmacher K. Recent advances in enzymatic fuel cells: Experiments and modeling. Energies. 2010;**3**(4):803-846. DOI: 10.3390/en3040803

[2] Santoro C, Arbizzani C, Erable B, Ieropoulos I. Microbial fuel cells: From fundamentals to applications: A review. Journal of Power Sources. 2017;**356**:225-244. DOI: 10.1016/j.jpowsour.2017.03.109

[3] Choudhury P, Prasad Uday US, Bandyopadhyay TK, Ray RN, Bhunia B. Performance improvement of microbial fuel cell (MFC) using suitable electrode and bioengineered organisms: A review. Bioengineered. 2017;**8**(5):471-487. DOI: 10.1080/21655979.2016.1267883

[4] Potter MC. Electrical effects accompanying the decomposition of organic compounds. Proceedings of the Royal Society B: Biological Sciences. 1911;**84**(571):260-276

[5] Cohen B. The bacterial culture as an electrical half-cell. Journal of Bacteriology. 1931;**21**:18-19

[6] Bennetto HP, Delaney GR, Rason JR, Roller SD, Stirling JL, Thurston CF. The sucrose fuel cell: Efficient biomass conversion using a microbial catalyst. Biotechnology Letters. 1985;**7**(10):699-704. DOI: 10.1007/BF01032279

[7] Delaney GM, Bennetto HP, Mason JR, Roller SD, Stirling JL, Thurston CF. Electron-transfer coupling in microbial fuel cells. II. Performance of fuel cells containing selected microorganism-mediator combinations. Journal of Chemical Technology and Biotechnology. 1984;**34**(1):13-27. DOI: 10.1002/jctb.280340104

[8] Kim BH, Kim HJ, Hyun MS, Park DH. Direct electrode reaction of Fe(III)-reducing bacterium, *Shewanella putrefaciens*. Journal of Microbiology and Biotechnology. 1999;**9**(2):127-131. DOI: 10.1002/(SICI)1099-1514(199905/06)20:3<127::AID-OCA650>3.0.CO;2-I

[9] Logan BE, Regan JM. Electricity-producing bacterial communities in microbial fuel cells. Trends in Microbiology. 2006;**14**(12):512-518. DOI: 10.1016/j.tim.2006.10.003

[10] Lovley DR. Microbial fuel cells: Novel microbial physiologies and engineering approaches. Current Opinion in Biotechnology. 2006;**17**(3):327-332. DOI: 10.1016/j.copbio.2006.04.006

[11] Kim HJ, Park HS, Hyun MS, Changa IS, Kim M. A mediator-less microbial fuel cell using a metal reducing bacterium, *Shewanella putrefaciens*. Enzyme and Microbial Technology. 2002;**30**(2):145-152. DOI: 10.1016/S0141-0229(01)00478-1

[12] Reimers CE, Tender LM, Fertig S, Wang W. Harvesting energy from the marine sediment-water interface. Environmental Science & Technology. 2001;**35**(1):192-195

[13] Bond DR, Holmes DE, Tender LM, Lovley DR. Electrode-reducing microorganisms that harvest energy from marine sediments. Science. 2002;**295**(5554):483-485. DOI: 10.1126/science.1066771

[14] Fredrickson JK, Romine MF, Beliaev AS, Auchtung JM, Driscoll ME, Gardner TS, et al. Towards environmental systems biology of *Shewanella*. Nature Reviews Microbiology. 2008;**6**(8):592-603. DOI: 10.1038/nrmicro1947

[15] Shi L, Squier TC, Zachara JM, Fredrickson JK. Respiration of metal (hydr)oxides by *Shewanella* and

Geobacter: A key role for multihaem c-type cytochromes. Molecular Microbiology. 2007;**65**(1):12-20. DOI: 10.1111/j.1365-2958.2007.05783.x

[16] Fan Y, Hu H, Liu H. Sustainable power generation in microbial fuel cells using bicarbonate buffer and proton transfer mechanisms. Environmental Science & Technology. 2007;**41**:8154-8158. DOI: 10.1021/es071739c

[17] Fan Y, Hu H, Liu H. Enhanced Coulombic efficiency and power density of air-cathode microbial fuel cells with an improved cell configuration. Journal of Power Sources. 2007;**17**:348-354. DOI: 10.1016/j.jpowsour.2007.06.220

[18] Ringeisen BR, Henderson E, Wu PK, Pietron J, Ray R, Little B, et al. High power density from a miniature microbial fuel cell using *Shewanella oneidensis* DSP10. Environmental Science & Technology. 2006;**40**:2629-2634. DOI: 10.1021/es052254w

[19] Kaneshiro H, Takano K, Takada Y, Wakisaka T, Tachibana T, Azuma M. A milliliter-scale yeast-based fuel cell with high performance. Biochemical Engineering Journal. 2014;**83**:90-96. DOI: 10.1016/j.bej.2013.12.011

[20] Wang H, Wang G, Ling Y, Qian F, Song Y, Lu X, et al. High power density microbial fuel cell with flexible 3D graphene-nickel foam as anode. Nanoscale. 2013;**5**(21):10283-10290. DOI: 10.1039/c3nr03487a

[21] Nevin KP, Richter H, Covalla SF, Johnson JP, Woodard TL, Orloff AL, et al. Power output and Coulombic efficiencies from biofilms of *Geobacter sulfurreducens* comparable to mixed community microbial fuel cells. Environmental Microbiology. 2008;**10**:2505-2514. DOI: 10.1111/j.1462-2920.2008.01675.x

[22] Choi S, Lee HS, Yang Y, Parameswaran P, Torres CI, Rittmann BE, et al. A μL-scale micromachined microbial fuel cell having high power density. Lab on a Chip. 2011;**11**(6):1110-1117. DOI: 10.1039/c0lc00494d

[23] Liu H, Cheng H, Huang L, Logan BE. Scale-up of membrane-free single-chamber microbial fuel cells. Journal of Power Sources. 2008;**179**:274-279. DOI: 10.1016/j.jpowsour.2007.12.120

[24] Walter XA, Gajda I, Forbes S, Winfield J, Greenman J, Ieropoulos I. Scaling-up of a novel, simplified MFC stack based on a self-stratifying urine column. Biotechnology for Biofuels. 2016;**9**:93. DOI: 10.1186/s13068-016-0504-3

[25] Stoll ZA, Dolfing J, Xu P. Minimum performance requirements for microbial fuel cells to achieve energy-neutral wastewater treatment. Water. 2018;**10**(3):243. DOI: 10.3390/w10030243

[26] Dong Y, Qu Y, He W, Du Y, Liu J, Han X, et al. A 90-liter stackable baffled microbial fuel cell for brewery wastewater treatment based on energy self-sufficient mode. Bioresource Technology. 2015;**195**:66-72. DOI: 10.1016/j.biortech.2015.06.026

[27] Ge Z, He Z. Long-term performance of a 200-liter modularized microbial fuel cell system treating municipal wastewater: Treatment, energy, and cost. Environmental Science: Water Research & Technology. 2016;**2**:274-281. DOI: 10.1039/C6EW00020G

[28] Zhuang L, Yuan Y, Wang Y, Zhou S. Long-term evaluation of a 10-liter serpentine-type microbial fuel cell stack treating brewery wastewater. Bioresource Technology. 2012;**123**:406-412. DOI: 10.1016/j.biortech.2012.07.038

[29] Zhou T, Han H, Liu P, Xiong J, Tian F, Li X. Microbial fuels cell-based biosensor for toxicity retection: A

review. Sensors. 2017;**17**(10):2230. DOI: 10.3390/s17102230

[30] Chae KJ, Choi M, Ajayi FF, Park W, Chang IS, Kim IS. Mass transport through a proton exchange membrane (Nafion) in microbial fuel cells. Energy & Fuels. 2008;**22**:169-176

[31] Ghassemi Z, Slaughter G. Biological fuel cells and membranes. Membranes (Basel). 2017;**7**(1):3. DOI: 10.3390/membranes7010003

[32] Novoselov KS, Geim AK, Morozov SV, Jiang D, Zhang Y, Dubonos SV, et al. Electric field effect in atomically thin carbon films. Science. 2004;**306**:666-669

[33] Yu F, Wang C, Ma J. Applications of graphene-modified electrodes in microbial fuel cells. Materials. 2016;**9**:807. DOI: 10.3390/ma9100807

[34] Yang Y, Liu T, Zhu X, Zhang F, Ye D, Liao Q, et al. Boosting power density of microbial fuel cells with 3D nitrogen-doped graphene aerogel electrode. Advanced Science (Weinh). 2016;**3**(8):1600097. DOI: 10.1002/advs.201600097

[35] Nam JY, Kim HW, Lim KH, Shin HS, Logan BE. Variation of power generation at different buffer types and conductivities in single chamber microbial fuel cells. Biosensors & Bioelectronics. 2010;**25**(5):1155-1159. DOI: 10.1016/j.bios.2009.10.005

[36] Jang JK, Pham TH, Chang IS, Kang KH, Moon H, Cho KS, et al. Construction and operation of a novel mediator- and membrane-less microbial fuel cell. Process Biochemistry. 2004;**39**:1007-1012. DOI: 10.1016/S0032-9592(03)00203-6

[37] Garba NA, Sa'adu L, Balarabe MD. An overview of the substrates used in microbial fuel cells. Greener Journal of Biochemistry and Biotechnology. 2017;**4**(2):007-026. DOI: 10.15580/GJBB.2017.2.051517061.

[38] Reddy LV, Kumar SP, Wee YJ. Microbial fuel cells (MFCs)-a novel source of energy for new millennium. Applied Microbiology and Microbial Biotechnology. 2010;**25**:956-964

[39] McKinlay JB, Zeikus JG. Extracellular iron reduction is mediated in part by neutral red and hydrogenase in *Escherichia coli*. Applied and Environmental Microbiology. 2004;**70**(6):3467-3474. DOI: 10.1128/AEM.70.6.3467-3474.2004

[40] Rosenbaum M, He Z, Angenent LT. Light energy to bioelectricity: Photosynthetic microbial fuel cells. Current Opinion in Biotechnology. 2010;**21**(3):259-264. DOI: 10.1016/j.copbio.2010.03.010

[41] Fischer F. Photoelectrode, photovoltaic and photosynthetic microbial fuel cells. Renewable and Sustainable Energy Reviews. 2018;**90**:16-27. DOI: 10.1016/j.rser.2018.03.053

[42] Tartakovsky B, Guiot SR. A comparison of air and hydrogen peroxide oxygenated microbial fuel cell reactors. Biotechnology Progress. 2006;**22**:241-246. DOI: 10.1021/bp050225j

[43] Ucar D, Zhang Y, Angelidaki I. An overview of electron acceptors in microbial fuel cells. Frontiers in Microbiology. 2017;**8**:643. DOI: 10.3389/fmicb.2017.00643

[44] Oh S, Min B, Logan BE. Cathode performance as a factor in electricity generation in microbial fuel cells. Environmental Science & Technology. 2004;**38**(18):4900-4904. DOI: 10.1016/j.electacta.2015.12.105

[45] Watanabe K. Recent developments in microbial fuel cell technologies

for sustainable bioenergy. Journal of Bioscience and Bioengineering. 2008;**106**(6):528-536. DOI: 10.1263/jbb.106.528

[46] Tharali AD, Sain N, Osborne WJ. Microbial fuel cells in bioelectricity production. Fromtiers in Life Science. 2016;**9**(4):252-266. DOI: 10.1080/21553769.2016.1230787

[47] Schaetzle O, Barrière F, Baronian K. Bacteria and yeasts as catalysts in microbial fuel cells: Electron transfer from micro-organisms to electrodes for green electricity. Energy & Environmental Science. 2008;**1**:607-620. DOI: 10.1039/b810642h

[48] Fan L, Xue S. Overview on electricigens for microbial fuel cell. The Open Biotechnology Journal. 2016;**10**:399. DOI: 10.2174/1874070701610010398

[49] Dopson M, Ni G. Possibilities for extremophilic microorganisms in microbial electrochemical systems. FEMS Microbiology Reviews. 2016;**40**(2):164-181. DOI: 10.1093/femsre/fuv044

[50] Khater D, El-khatib KM, Hazaa M, Hassan RYA. Activated sludge-based microbial fuel cell for bio-electricity generation. Journal of Basic and Environmental Sciences. 2015;**2**:63-73

[51] Clauwaert P, Aelterman P, Pham TH, De Schamphelaire L, Carballa M, Rabaey K, et al. Minimizing losses in bio-electrochemical systems: The road to applications. Applied Microbiology and Biotechnology. 2008;**79**(6):901-913. DOI: 10.1007/s00253-008-1522-2

[52] Kim GT, Webster G, Wimpenny JW, Kim BH, Kim HJ, Weightman AJ. Bacterial community structure, compartmentalization and activity in a microbial fuel cell. Journal of Applied Microbiology. 2006;**101**(3):698-710

[53] Kouzuma A, Kasai T, Hirose A, Watanabe K. Catabolic and regulatory systems in *Shewanella oneidensis* MR-1 involved in electricity generation in microbial fuel cells. Frontiers in Microbiology. 2015;**6**:609. DOI: 10.3389/fmicb.2015.00609

[54] Sure SK, Ackland LM, Torriero AA, Adholeya A, Kochar M. Microbial nanowires: An electrifying tale. Microbiology. 2016;**162**:2017-2028. DOI: 10.1099/mic.0.000382

[55] Vaara M. Agents that increase the permeability of the outer membrane. Microbiological Reviews. 1992;**56**:395-411

[56] Liu J, Qiao Y, Lu ZS, Song H, Li CM. Enhance electron transfer and performance of microbial fuel cells by perforating the cell membrane. Electrochemistry Communications. 2012;**15**(1):50-53. DOI: 10.1016/j.elecom.2011.11.018

[57] Qiao Y, Li CM, Bao SJ, Lu ZS, Hong YH. Direct electrochemistry and electrocatalytic mechanism of evolved *Escherichia coli* cells in microbial fuel cells. Chemical Communications. 2008;**11**:1290-1292. DOI: 10.1039/b719955d

[58] Luo JM, Li M, Zhou MH, Hu YS. Characterization of a novel strain phylogenetically related to *Kocuria rhizophila* and its chemical modification to improve performance of microbial fuel cells. Biosensors & Bioelectronics. 2015;**69**:113-120. DOI: 10.1016/j.bios.2015.02.025

[59] Zheng T, Xu YS, Yong XY, Li B, Yin D, Cheng QW, et al. Endogenously enhanced biosurfactant production promotes electricity generation from microbial fuel cells. Bioresource Technology. 2015;**197**:416-421. DOI: 10.1016/j.biortech.2015.08.136

[60] Van Gennip M, Christensen LD, Alhede M, Phipps R, Jensen PØ, Christophersen L, et al. Inactivation of the *rhlA* gene in *Pseudomonas aeruginosa* prevents rhamnolipid production, disabling the protection against polymorphonuclear leukocytes. APMIS. 2009;**117**:537-546. DOI: 10.1111/j.1600-0463.2009.02466.x

[61] Heather MJ, Aaron EA, Konstantin RM, Yuri YL, Bruce EC, Brett AH, et al. Engineering of a synthetic electron conduit in living cells. Proceedings of the National Academy of Sciences of the United States of America. 2010;**107**(45):19213-19218. DOI: 10.1073/pnas.1009645107

[62] Sturm-Richter K, Golitsch F, Sturm G, Kipf E, Dittrich A, Beblawy S, et al. Unbalanced fermentation of glycerol in *Escherichia coli* via heterologous production of an electron transport chain and electrode interaction in microbial electrochemical cells. Bioresource Technology. 2015;**186**:89-96. DOI: 10.1016/j.biortech.2015.02.116

[63] Jensen HM, TerAvest MA, Kokish MG, Ajo-Franklin CM. CymA and exogenous flavins improve extracellular electron transfer and couple it to cell growth in Mtr-expressing *Escherichia coli*. ACS Synthetic Biology. 2016;**5**(7):679-688. DOI: 10.1021/acssynbio.5b00279

[64] Shrestha N, Chilkoor G, Vemuri B, Rathinam N, Sani RK, Gadhamshetty V. Extremophiles for microbial-electrochemistry applications: A critical review. Bioresource Technology. 2018;**255**:318-330. DOI: 10.1016/j.biortech.2018.01.151

[65] Erable B, Etcheverry L, Bergel A. Increased power from a two-chamber microbial fuel cell with a low-pH air-cathode compartment. Electrochemistry Communications. 2009;**11**(3):619-622. DOI: 10.1016/j.elecom.2008.12.058

[66] Slonczewski JL, Fujisawa M, Dopson M, Krulwich TA. Cytoplasmic pH measurement and homeostasis in bacteria and archaea. Advances in Microbial Physiology. 2009;**55**:1-79. DOI: 10.1016/S0065-2911(09)05501-5

[67] Malki M, De Lacey AL, Rodriguez N, Amils R, Fernandez VM. Preferential use of an anode as an electron acceptor by an acidophilic bacterium in the presence of oxygen. Applied and Environmental Microbiology. 2008;**74**:4472-4476. DOI: 10.1128/AEM.00209-08

[68] Tapia JM, Munoz JA, Gonzalez F, Blazquez ML, Ballester A. Interrelation between cells and extracellular polymeric substances (EPS) from *Acidiphilium* 3.2Sup(5) on carbon surfaces. Advanced Materials Research. 2009;**71-73**:287-290. DOI: 10.4028/www.scientific.net/AMR.71-73.287

[69] Zhang T, Zhang L, Su W, Gao P, Li D, He X, et al. The direct electrocatalysis of phenazine-1-carboxylic acid excreted by *Pseudomonas alcaliphila* under alkaline condition in microbial fuel cells. Bioresource Technology. 2011;**102**:7099-7102. DOI: 10.1016/j.biortech.2011.04.093

[70] Li XM, Cheng KY, Wong JW. Bioelectricity production from food waste leachate using microbial fuel cells: Effect of NaCl and pH. Bioresource Technology. 2013;**149**:452-458. DOI: 10.1016/j.biortech.2013.09.037

[71] Jayashree C, Tamilarasan K, Rajkumar M, Arulazhagan P, Yogalakshmi KN, Srikanth M, et al. Treatment of seafood processing wastewater using upflow microbial fuel cell for power generation and identification of bacterial community in anodic biofilm. Journal of Environmental Management. 2016;**180**:351-358. DOI: 10.1016/j.jenvman.2016.05.050

[72] Jadhav GS, Ghangrekar MM. Performance of microbial fuel cell subjected to variation in pH, temperature, external load and substrate concentration. Bioresource Technology. 2009;**100**:717-723. DOI: 10.1016/j.biortech.2008.07.041

[73] Catal T, Kavanagh P, O'Flaherty V, Leech D. Generation of electricity in microbial fuel cells at sub-ambient temperatures. Journal of Power Sources. 2011;**196**:2676-2681. DOI: 10.1016/j.jpowsour.2010.11.031

[74] Liu L, Tsyganova O, Lee DJ, Su A, Chang JS, Wang A, et al. Anodic biofilm in single chamber microbial fuel cells cultivated under different temperatures. International Journal of Hydrogen Energy. 2012;**37**:15792-15800. DOI: 10.1016/j.ijhydene.2012.03.084

[75] Zhang L, Shen J, Wang L, Ding L, Xu K, Ren H. Stable operation of microbial fuel cells at low temperatures (5-10°C) with light exposure and its anodic microbial analysis. Bioprocess and Biosystems Engineering. 2014;**37**:819-827. DOI: 10.1007/s00449-013-1054-8

[76] Mathis BJ, Marshall CW, Milliken CE, Makkar RS, Creager SE, May HD. Electricity generation by thermophilic microorganisms from marine sediment. Applied Microbiology and Biotechnology. 2008;**78**(1):147-155. DOI: 10.1007/s00253-007-1266-4

[77] Shehab NA, Ortiz-Medin JF, Katuri KP, Hari AR, Amy G, Logan BE, et al. Enrichment of extremophilic exoelectrogens in microbial electrolysis cells using Red Sea brine pools as inocula. Bioresource Technology. 2017;**239**:82-86. DOI: 10.1016/j.biortech.2017.04.122

[78] Fu Q, Fukushima N, Maeda H, Sato K, Kobayashi H. Bioelectrochemical analysis of a hyperthermophilic microbial fuel cell generating electricity at temperatures above 80°C. Bioscience, Biotechnology, and Biochemistry. 2015;**79**(7):1200-1206. DOI: 10.1080/09168451.2015.1015952

[79] Carver SM, Vuoriranta P, Tuovinen OH. A thermophilic microbial fuel cell design. Journal of Power Sources. 2011;**196**:3757-3760. DOI: 10.1016/j.jpowsour.2010.12.088

Chapter 5

Integrated Biologics Manufacturing in Stirred-Suspension Bioreactor: A Stem Cell Perspective

Suman C. Nath and Derrick E. Rancourt

Abstract

Stem cell therapy is garnering attention as several clinical trials have taken place in the recent years by using human pluripotent stem cells (hPSCs). Hundreds of biotechnological companies are investing to find a permanent cure for difficult-to-treat diseases like age-related macular degeneration, Parkinson's disease, diabetes, etc. by using hPSCs. Therefore, clinical-grade cell manufacturing has become an important issue to make cell therapy products safe and effective. Current manufacturing practices are adopted from conventional antibody or protein production in the pharmaceutical industry where cells are used as a vector for producing the desired products. In cell therapy applications, cells are the products that are sensitive to physicochemical parameters and storage conditions anywhere between isolation to patient administration. Moreover, cell-based product manufacturing consists of multi-step processing, including isolation from patients, genetic modification, derivation, expansion, differentiation, purification, characterization, cryopreservation, etc. This can require long processing times and pose high risk of product contamination as well as high production cost. Herein, we discuss the current methods of biologics manufacturing and its limitations. We also review current practices for integrating and automating cell manufacturing facilities. Finally, we propose how to integrate multi-step cell processing in a single bioreactor to make the cell manufacturing practices more direct.

Keywords: biologics, stem cell therapy, genetic modification, integrated manufacturing, bioreactor

1. Introduction

Based on their self-renewal and differentiation capabilities, human pluripotent stem cells (hPSCs) including embryonic stem cells (ESCs) [1] and induced pluripotent stem cells (iPSCs) [2] are attractive tools in the field of regenerative medicine. After the discovery of hiPSCs in 2007, this field expanded vigorously and hundreds of biotechnological companies were established to use these cells for treating degenerative diseases. The most common degenerative diseases treated by the hESCs are age-related macular degeneration (AMD), type I diabetes mellitus, heart failure, Parkinson's disease, and spinal cord injury [3]. Although hiPSCs are a better source for autologous cell therapy applications, they are less preferable for clinical trials because of less genetic stability compared to the hESCs. However, a few clinical trials have already been started using the patient-derived hiPSCs. The Takahashi group from the Riken Center for Developmental Biology has recently conducted a clinical trial for treating wet AMD [4]. Similarly, a Takahashi from Kyoto University is conducting a clinical trial for treating Parkinson's disease by using hiPSCs [5]. A few clinical trials are also ongoing in the USA for treating different diseases like β-thalassemia, liver diseases, diabetes, etc. using hiPSCs and their use is expanding worldwide day by day [6].

As stem cell therapy is garnering increasing attention, a lot of clinical trials are ongoing using both hESCs and hiPSCs cells. About 6849 clinical trials and 1415 stem cell-based therapies were found based upon searches we recently performed on clinicaltrials.gov (October, 2018) [7]. However, the percentage of success is not high enough as speculated from the previous clinical trials. Among the 315 clinical trials conducted (26.0% Phase 1, 40.6% Phase 1/2, 22.5% Phase 2, 3.8% Phase 2/3, and 6.7% Phase 3), only 0.3% went to Phase 4 [3]. The low percentage of completion of clinical trials depends on various factors. One of the major factors is manufacturing practices that can provide high safety and efficacy of cell therapy products. Moreover, production cost of multiple doses also hinders the success rate of clinical trials. As cell therapy revenue exceeded multi-million dollars and has been a profitable business in recent years, but much attention is needed to produce high quality cells for treating incurable diseases [8, 9].

The production of stem cell biologics is adapted from the conventional pharmaceutical protein and vaccine production. Conventional biologics production involves the following basic steps: isolation and identification of raw materials, formulation, filling, packaging, and storage, where the total processing stops at the storage of final products.

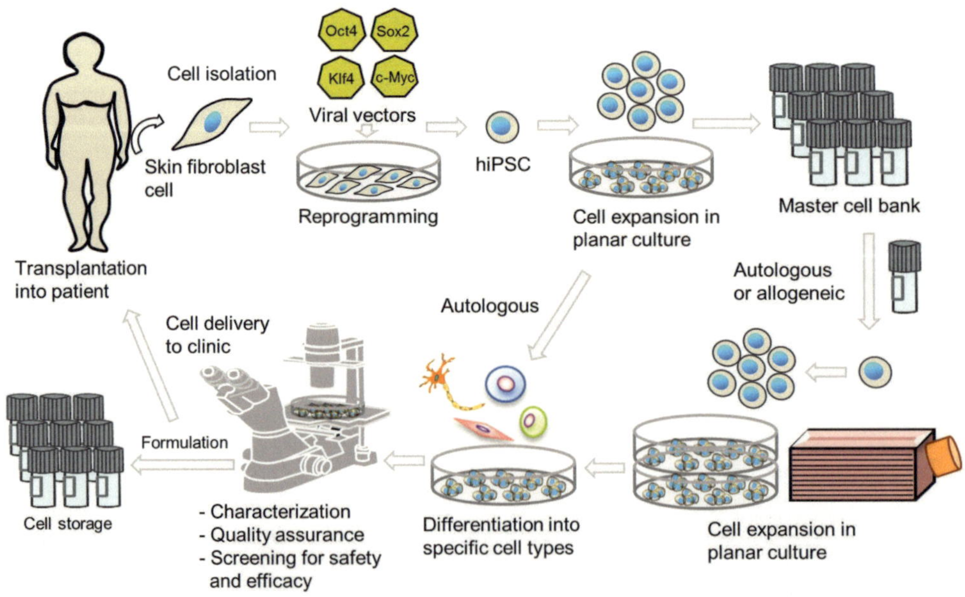

Figure 1.
Schematic illustration of current multi-step cell manufacturing strategies in planar culture for stem cell therapy applications. Skin cells are isolated from the patient and reprogrammed to hiPSCs using viral vectors. After reprogramming, hiPSCs are stored in a master cell bank or differentiated directly in autologous cell therapy applications. In some cases like allogeneic cell therapy applications, cells are expanded in a large amount and then differentiated. After performing characterization, quality assurance, and screening for safety and efficacy, cells are delivered to hospital or stored in a cell bank for future use.

There is a big difference between the production of conventional biologics and cell-based therapy products. For vaccines or pharmaceutical protein production, cells are used as a platform for obtaining desired proteins. After that, cells are discarded. However, in cell-based therapies, cells, which are sensitive to the physical or chemical attributes of the residing environment, are the final products. Therefore, much consideration is needed before translating cell-based products from bench to clinic. This extends to the acquisition of tissue samples and isolation of cells, initial cell purification, selection, activation and transduction, cell expansion in plate or bioreactor culture, differentiation, washing, harvesting and formulation, filling and cryopreservation, and finally, storage and delivery to the clinics (**Figure 1**) [10].

Cumbersome multi-step manufacturing systems can cause batch to batch variability, inefficacy, and low quality of cells for transplantation and need to be simplified and made more direct. In this context, we will discuss current limitations of cell manufacturing strategies and propose how to overcome these by integrating the total process in a single bioreactor to make cell manufacturing straight forward enough to deliver high quality cell therapy products to the clinic. In this review, we will also discuss how to integrate genetic modification—transfection or transduction, reprogramming, differentiation, purification, and formulation of final products in a single bioreactor.

2. Current manufacturing strategies for stem cell therapy

Current manufacturing strategies for cell therapy products are replicated from biologics manufacturing in the pharmaceutical industry. However, the processing of cells is far different from pharmaceutical proteins or vaccines. For pharmaceutical peptide production from microorganisms, the raw materials are extracted from bacteria or fungus [11, 12]. They are then separated, purified, and examined for quality assurance to meet the requirements of regulatory agencies, e.g., Food and Drug Administration (FDA), British Pharmacopeia, etc. The final products are stored or marketed in a dose-dependent manner.

Cell processing is more intensified when the pharmaceutical proteins are produced by using human, animal, or plant cells as a by-product. In this case, high quality products depend on the maintenance of high quality cells, and maintaining a sterile condition is very important. Therefore, good bioprocessing is required to optimize the production of desired proteins. After inoculating from a master cell bank, the cells are cultured for a specific period of time [13, 14]. The supernatant is then collected and the desired proteins are separated, purified, and concentrated. The isolated products then go through quality assurance to meet the criteria of the regulatory agency. Finally, the products are stored and marketed in a dose-dependent manner.

The manufacturing of stem-cell based products is not as straight forward as the production of pharmaceutical proteins or vaccines. This is because cells are the final product in stem cell therapy and are vulnerable to physical or chemical operations from isolation to delivery to patients. Cell manufacturing strategies also vary from source to source and depend on autologous or allogeneic transplantation (**Figure 1**). The major general steps are the acquisition of tissue samples and isolation of cells, initial cell purification, selection, activation and transduction, cell expansion, differentiation, washing, harvesting and formulation, filling and cryopreservation, and finally, storage and delivery to the clinics [10].

For stem-cell based products, cells are isolated from specific tissues of patients, e.g., blood, skin, etc. for autologous transplantation or can be used from cell banks for allogeneic transplantation. Heterogeneity of final products may arise from the cell isolation step because patients' tissues contain various undesired subpopulations. For example, in chimeric antigen receptor T-cell (CAR-T) therapy, cells are isolated from patients' blood tissue, which contains abnormal levels of inhibitory factors and regulatory cells [15, 16] because patients are treated with chemo- and radiotherapies. As a result, heterogeneity occurs in the final products, which need much attention during the cell isolation step. Cells isolated from patients need to be purified by centrifugation, magnetic-activated cell sorting (MACS), or fluorescent-activated cell sorting (FACS). Then, initial cell culture is done for selection, activation, or transduction of specific interest.

After purification, cells are expanded in plate culture or bioreactor. Based on demand, large-scale expansion is required in a sterile condition, which also requires intensive consideration because it is the rate-limiting step for commercialization of cell therapy products. The most important considerations for large-scale expansions are: operational, economic, quality and safety.

Operational design for culture systems (2D or 3D) with manual or automatic (desirable) operation is important before large-scale expansion [17]. Bioreactors are superior to plate culture for obtaining a large number of cells. Online monitoring and control of process parameters (pH, DO, pCO_2, etc.) and considering the shortest possible culture time are also important parameters for operational consideration. A prediction model for medium consumption (glucose and glutamine) and toxic material production (lactic acid and ammonium) is very useful for determining medium feeding regimen. A dedicated single-use vessel is also a big operational consideration before large-scale expansion of cell-based products.

As cell-based products are costlier, economic considerations for medium, efficient cell lines and other indirect utilities are important. However, the most important consideration in large-scale expansion is product quality and safety. For this purpose, dedicated cell manufacturing facilities are required to maintain current manufacturing practices (cGMP) for high product purity and safety.

After large-scale expansion, cells are harvested by detaching them from the culture substrate using enzymatic treatment. Non-enzymatic detachment is also available by changing temperature or pH [18–20]. Aggregate culture in bioreactors may not necessarily need a detachment step for harvesting [21–25]. Next steps are washing and volume reduction, which can be done by centrifugation or tangential flow filtration on a large scale by using automated commercial devices (kSep systems and Terumo BCT).

Purified cells are formulated in a dose-dependent manner and checked for quality assurance. Quality assurance is done in three different stages: microbial contamination, chemical contamination, and quality or potency assurance. Microbial contamination is checked for bacterial, fungal, or viral contamination by sterility tests with various methods [26, 27]. The most commonly used sterility test is a 14-day incubation of cell products for bacterial and fungal contamination [28, 29]. Chemical testing includes checking for molecules accompanying the culture medium or other factors used during isolation, expansion, and storage. One commonly used chemical test is the LAL test for bacterial endotoxin. There is now an automated 15 min test for determining endotoxin in cell therapy products, which was developed following FDA regulations [30]. Other chemical testing concerns are checking for residual proteins of different origins, serum, and other harmful particles originated from cell processing.

In cell therapy products, quality is the major concern, especially because cell growth is a requirement. For that reason, a cell viability assay is done to determine live or dead cells in the product using a variety of staining methods. Colony forming unit (CFU) is also useful for determining biological activity of cell therapy products [31, 32]. Product potency is an important criterion to meet before releasing the product. For example, if a cell therapy product is applied for the chimeric antigen receptor T (CART)-related cancer therapy, it needs to be examined for the secretion of cytotoxic cytokines (IFN-γ) and killing of target cells [33]. However, for hPSCs, the final products are differentiated cells, wherein potency should be checked via transplantation into disease models.

For hPSC-derived products, strict quality control is imperative before transplantation to the patients because there is high risk of oncogene transfer to patients. A clinical trial was halted in 2015 in Japan while treating AMD by autologous hiPSC-derived retinal pigmented epithelial cells because of genetic abnormality [34]. Since genetic abnormalities occur in hiPSC-derived products from reprogramming to finally differentiated cells [35], cells should be strictly screened for epigenetic signatures, karyotyping, telomerase activity, mitochondrial remodeling, etc. [36–38]. Rohani et al. summarized possible molecular cytogenetics for quality control that should be checked before releasing the final products [39]. Some of the proposed quality testings are whole-genome sequencing, single-cell genome sequencing, epigenomic analysis, and mitochondrial DNA integrity testing for maximizing the patient safety.

After passing the product quality assurance, cells need to be delivered to clinics immediately or stored for future use. Cells are shipped generally to the clinics on dry ice (−78°C) or in liquid nitrogen dry shippers (−160°C) if the cells are vitrified. The mostly used technique for cell storage is cryopreservation in liquid nitrogen at −196°C which is adapted from the conventional stem cell banking [40, 41]. For cryopreservation, dimethyl sulfoxide (DMSO), glycerol, sugars, or other polymers are used. Among them, clinical grade DMSO is widely used although it is detrimental and can cause harmful effects to cells [42, 43]. Therefore, removing it from cryopreservation protocols or lowering the concentration is important. However, developing appropriate protocols for freezing and thawing is also important for high recovery of cells. Generally, slow-freezing and quick thawing is highly applicable for better recovery of cryopreserved cells [44, 45]. Since intracellular ice crystal formation is a big obstacle in cryopreservation, using ice recrystallization inhibitors is also an effective process for cryopreservation of clinical cell therapy products [46, 47].

Product delivery is also an important step to consider before administration to the patients. Since the products are carried in an environment where temperature is extreme, the container should be made with such materials that can withstand extreme low temperature and do not cause any leakage compromising the product quality [48]. For autologous cell therapy applications especially for CAR-T cell therapy, a dedicated vessel, which can withstand extreme low temperature, is needed [49].

3. Integrated biologics manufacturing in bioreactors

The conventional production of pharmaceutical proteins or other biologics consists of multiple steps from raw materials to finished products. As biologics need to maintain stringent quality control, multiple steps in production facilities compromise the product quality significantly. They also reduce productivity and become prone to human errors, which decrease product efficacy and safety. Moreover, multiple steps in cell processing consume a lot of time, which indirectly increases production cost. To overcome these drawbacks, integrated pharmaceutical production has been attempted by various pharmaceutical companies. One of the significant attempts was made by the Novartis-MIT Center for Continuous Manufacturing of pharmaceutical products to fully integrate the cell processing system [50, 51]. Another attempt was taken by Genzyme™ for continuous production of pharmaceutical recombinant protein in bioreactors, where cell culture to product isolation and purification was integrated in a single flow [52]. By using this system, they respectively reported successful production of monoclonal antibody as well as highly complex, less stable pharmaceutical protein with consistent product quality, high product output, and low cost. Process integrity is necessary for reducing cumbersome production steps and cutting cost significantly. One such integrated system developed by Johnson & Johnson has recently got FDA approval for large-scale HIV drug production [53] that reduces time and cost by one third compared to the conventional batch processing.

Since biologics production for cell therapies require multiple steps, integration of all of the steps will give high product quality and safety, as well as help overcome stringent regulatory requirements. In this context, we will discuss how to integrate some important basic steps of cell manufacturing especially genetic modification, cellular reprogramming, expansion, and differentiation in bioreactors to promote a single-step approach for cell-based therapies (**Figure 2**).

3.1 Genetic modifications in bioreactor

Genetic modification is one of the biggest steps in producing cell therapy products. In biologics manufacturing, it has been practiced for many years for producing antibodies, proteins, or other biotechnological drugs. It has also been used extensively in the cell therapy industry as various cell-based products have been applied for treating multiple incurable genetic diseases in recent years. Some genetic modifications affect patients directly and some indirectly. For example, in adrenoleukodystrophy (ALD), a neurological disorder occurs due to malfunction of oligodendrocytes and microglia where genetic modification can affect a patient directly. To recover from it, a corrected gene is inserted into the patient-derived hPSCs and transplanted into the

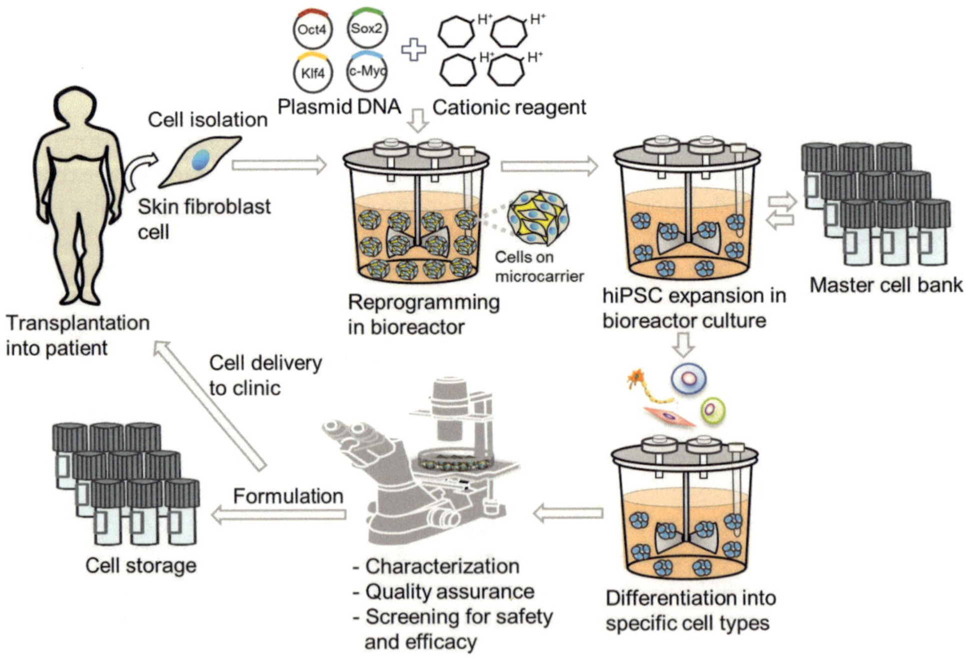

Figure 2.
Schematic illustrations of integrated single-step cell manufacturing strategies in bioreactor culture for stem cell therapy applications. Skin cells are isolated from the patient and reprogrammed to hiPSCs on microcarriers using a nonviral approach. After expansion as aggregates, hiPSCs are stored in a master cell bank or differentiated directly in bioreactor. After performing characterization, quality assurance, and screening for safety and efficacy, cells are delivered to hospital or stored in a cell bank for future use.

patient's brain, which is differentiated into microglia to promote production of myelin in the patient's brain that recovered the ALD [54].

In some gene therapy applications, patients are exposed indirectly to genetic modification. For example, in thalassemia, patient blood cells are extracted from the body and the cells are modified and enriched in *ex vivo* to target the specific antigens of patients' body [55]. Other indirect genetic modifications used for treating CAR or T-cell receptor (TCR) genes to T-cells [56], expression of CD40 ligand in dendritic cells [57], adenosine-deaminase severe-combined immunodeficiency [58], and beta-thalassemia [59], as well as deletion or insertion of desired genes in a specific genomic location. Among them, CAR-T cell therapy has got much attention for treating cancer-related diseases. These genetically modified T-cells can specifically target the antigens and kill the cancer cells efficiently [60]. CARs and TCRs are the mostly used receptors which are engineered to activate the T-cells [61]. Nowadays, a lot of CAR-T cell-based therapies are being established for treating advanced-stage lymphoma [62] and B-cell lymphoma [63] as well as other autoimmune diseases [64].

Viral vectors are commonly used to deliver genetic cargo to cells (**Figure 1**). This involves a two-step process: preparation for viral vectors and transduction for modifying the cells to express desired property. Lentiviral and gamma-retroviral are widely-used for their superior transduction efficiency but their transgenes are integrated with the host genome [65]. Another choice for viral transduction is adenovirus where viral transgenes are not integrated into the host genome but less efficient than lenti- and retro-virus. The major drawbacks in viral vector mediated transduction are concerns for safety of the products [66]. Viral vectors are widely used for reprogramming hiPSCs from skin fibroblasts cells [2].

Other methods for cellular transduction use nonviral approaches, including nucleofection or electroporation, or liposome-mediated delivery of DNA or RNA into cells. Although DNA vectors are easy to scale-up, carry large-size DNA with less immunotoxicity, this process is less efficient than the viral transduction. There are some other methods for skipping the use of viral vectors which are also efficient in doing the transgene expression [67–69]. Hsu et al. reported successful transfection by using commercially available nonviral cationic reagents, for example, TransIT-3D, TransIT-2020, XtremeGENE 9, XtremeGENE HP, JetPrime, Lipofectamine 3000, and Effectene and compared their transfection efficiency [70]. Warren et al. reported efficient reprogramming of hiPSCs from various cell sources by using mRNA and differentiated the cells into three germ layers [71]. hiPSCs were also reprogrammed by using recombinant protein that also maintained all the three germ layers [72].

Since transgene possesses high risk of cancer-causing agents; therefore, removal after transduction is highly desired. There are a few methods developed for the removal of these vectors. One of the methods is the piggyBac transposon system, which has been used to remove tandem Yamanaka reprograming genes Oct4, Sox2, Klf4, and c-Myc from iPSCs following reprogramming [73]. Removal of transgenes after incorporating CAR into T-cells used another transposon system called Sleeping Beauty, which successfully removed any genetic scar from the transduced cells [74, 75]. Likewise, transgene-free iPSCs have also been produced by Cre excision of reprogramming genes via loxP sites [76]. Integration-deficient viral vectors are also good candidates for producing transgene-free cell therapy products by mutating viral integrase [77]. Another approach is to use site-directed integration using targeting nucleases [78–80].

Various genome engineering technologies have been explored for gene addition, deletion, or correction in the cell therapy industry and are increasing day by day [81]. The most widely used targeting nucleases are zinc-finger nucleases (ZFNs), clustered regulatory interspaced short palindromic repeats (CRISPR)/Cas endonucleases, or transcription-activator like effector nucleases (TALENs) [82]. Although the CRISPR/Cas system has recently received much attention due to broad use in genome engineering of patient cells [83], ZFNs are also popular for treating graft-versus-host disease in T-cell therapy [84].

Recently, a nuclease dead variant of Cas9 bearing a transcriptional trans-activator has recently been used in cellular reprogramming by activating the transcription factors Oct4 and Sox2, which maintained pluripotency and expressed the markers for the three germ layers [85].

Although genetic modification is a rate-limiting step in the cell manufacturing industry, the conventional methods make it more complicated because it is a multi-step process. Conventional genetic modification in planar culture is also costly, labor-intensive, and time-consuming. The bioreactor is a better platform for producing large-scale genetically modified cells for commercial purposes because cell expansion is possible in the same vessel which makes the process straightforward (**Figure 2**). For genetic modification in bioreactor, Hsu et al. recently reported how to transfect reprogramming factors in bioreactor where they tried eliminating viral vectors for gene delivery by using cationic reagents [78]. Generally, transfection of reprogramming factors for generating induced pluripotent stem cells (iPSCs) is done in adherent culture and then cells are expanded in 2D or 3D which is a two-step process. By integrating the genetic modification step in bioreactor, it is possible to establish a single-step process which enables cell manufacturing in automated and closed bioreactor system.

Genetic modification is also a challenging step in CAR-T cell therapy-based products. In CAR-T cell therapy, generally cells are isolated from patients' blood sample and then the cells are expanded after selection and activation. Finally, the cells need to be transduced with the CAR or any other antigens depending on target diseases. Conventional methods for genetic transduction are based on planar culture where every step is performed in open culture system. Recently, a few steps are integrated in bag culture system where selection, activation, and expansion can be done in a single step using DynaMag™ CTS™ [86], whereas the Xuri cell expansion System developed by GE Healthcare can expand cells in large numbers [87–89].

Although washing and concentrating the final product are integrated by the COBE® 2991 system developed by Terumo BCT [90], the transduction step is still not integrated in any of the above systems. Integrating the transduction step with the expansion and formulation will make the CAR-T cell therapy straightforward and performing these steps in bioreactor is a good platform since the physiological parameters as well as automated operation is possible in bioreactor culture. Miltenyi Biotec developed a device named CliniMACS Prodigy™ which is based on bag culture for CAR-T cell therapy. This device integrated major steps especially cell preparation, selection, activation, expansion, transduction, washing, and formulation in an automated system [91–93]. Such integration in the bioreactor will pave a straightforward method for producing cell-based products in a closed and automatic method.

3.2 Integrated system for large-scale expansion and differentiation in bioreactor

Current manufacturing practices for stem cell-based products are multi-step: derivation, expansion, and differentiation. In this process, patient-derived skin fibroblast cells are transduced with reprogramming factors in the planar culture. After deriving hiPSCs, cells are expanded in planar or bioreactor culture to obtain a large number of cells. Then cells are differentiated to target cells of interest. The differentiated cells are characterized and transplanted to the patient in a dose-dependent manner. As this process is complicated with multiple steps, it poses high risk of contamination to the final products. Moreover, maintaining cGMP culture platform is also mandatory for cell-therapy products [94–96], which makes the cell manufacturing process more complicated. Therefore, developing an integrated system that can combine all these steps from derivation to final products is required. Here, the bioreactor may be a good platform for doing this (**Figure 2**).

The bioreactor platform is widely used for the large-scale expansion of hPSC-based cell therapy production because bioreactor is easy to operate in an automated mode where various physicochemical parameters can be regulated in a closed-system. Two groups have demonstrated that the bioreactor is conducive to cellular reprogramming [97, 98]. Shafa et al. reported a significantly higher reprogramming efficiency in the bioreactor compared to the planar culture [97]. Since mesenchymal-epithelial transition (MET) is an important early step in cellular reprogramming [99], transformed fibroblasts that are moved into the bioreactor will form aggregates that are efficiently expanded in the bioreactor. Indeed because fibroblasts are substrate-dependent, bioreactor culture may be promoting aggregate formation and therefore cellular reprogramming.

Unfortunately, bioreactor reprogramming methods require genetic modification (retroviral, piggyBAC) prior to bioreactor expansion. It is theoretically possible to pursue cellular reprograming fully and completely in the bioreactor. Recently, for example, Hsu et al. has demonstrated that it is possible to transfect human fibroblasts directly on microcarriers [70]. Reprogrammed cells in theory will leave the microcarrier to form aggregates in the bioreactor via MET.

Following bioreactor derivation of hPSCs, the next big steps are expansion and differentiation. Generally, a large number of cells are required for an effective cell therapy application, which is ranging from 10^8 to 10^{10} cells per 70 kg patient [100]. In the conventional process, cell expansion is performed in planar culture. However, it has many drawbacks and limits the cell expansion in various ways. Planar culture is unable to provide enough growth surfaces for the unlimited expansion.

Another major drawback is surface coating. Extracellular matrix (ECM) is needed for surface coating which is initially derived from animal sources, which poses high risk in clinical-grade manufacturing. Currently, recombinant ECM has been discovered, which can be used efficiently for clinical applications [101]. The advancement in cell coating also stimulated the advancement of integration and automation of cell expansion in adherent culture.

Automated planar culture systems have been established for the expansion of hPSCs for clinical-grade cell manufacturing. One of the notable automated systems for cell manufacturing is CompacT SelecT™ developed by the TAP Biosystems. This system is based on T-flask where 90 T175 flasks can be accommodated for large-scale expansion of cells. All the cell culture steps, cell counting, seeding, medium change, passaging, and plating as well as transient transfection can be done automatically by using this robotic system. However, such systems are not used for differentiation since differentiation is a complicated process, which needs several components to add in the culture medium. As a result, the expansion and differentiation process in planar culture is mostly disintegrated.

Cell expansion in bioreactors need not require surface coating except for microcarrier culture. Bioreactor also provides enough growth surface availability. Generally, a single bioreactor (100 mL working volume) is enough for providing clinically relevant number of cells for autologous cell therapy applications. Several types of bioreactors are employed for the expansion of hPSCs [102]. For anchorage-dependent expansion of hPSCs, microcarriers need to be coated with ECM for cell attachment in the bioreactor [100, 103–105].

After large-scale expansion, cells are harvested by detaching them from the microcarrier using enzymatic treatment. Nonenzymatic detachment is also available by changing temperature or pH [18–20]. Bioreactor expansion of hPSCs on microcarrier is troublesome for clinical application because it needs an extra step for microcarrier separation from the final cell harvest. On the other hand, aggregate culture in bioreactors may not necessarily need a detachment step for harvesting [21–25] and clinically relevant numbers of cells can be produced in a single bioreactor as aggregate [21, 106–108].

A major drawback in aggregate culture is the size limitation. With the increase in aggregate size, the growth potential decreases in the large size aggregate due to diffusion limitation of oxygen and nutrients [109]. Therefore, maintaining aggregate size is an important issue to maintain high growth rate as well as high quality for cell therapy applications [21].

After expansion, cells can be differentiated in the same vessel which makes bioreactor culture a unique choice for integrated biologics manufacturing. Bioreactors were used for differentiation of hPSCs into various cell types, especially for cardiac [110–112], hepatic [113, 114], and neural [115] lineages. To provide straightforward methods for clinical applications, integration of expansion and differentiation is important and there are several reports published recently where expansion and differentiation were integrated [108, 116–118]. However, the integration of derivation with expansion and differentiation is still facing complications and there are a very few reports available.

Steiner et al. reported integration of derivation, propagation and differentiation of hESCs in suspension culture where hESCs were isolated from the inner cell mass in suspension culture that did not involve feeder cells or microcarriers [119]. However, the integration of derivation, expansion, and differentiation is not still realized for personalized medicine especially for autologous or allogenic cell therapy applications. Such integration is needed for overcoming the multi-step cell processing, which will reduce the risk of contamination and save cell processing time as well as reduce manufacturing costs for cell therapy manufacturing.

4. Concluding remarks and future directions

Cell therapy applications utilizing stem cells are increasing day by day and several clinical trials are ongoing to treat incurable diseases. With the growing need for cell-based products, the manufacturing facilities should be compatible for fulfilling the market demand by supplying safe and effective cell-based products. Since the current manufacturing systems are stuck with several drawbacks, especially multi-step processing which poses high risk of contamination as well as long processing time which contributes to increase culture cost, a more straightforward system is required. Bioreactor-based cell manufacturing system can provide a single-step and straightforward processing of cell-based products. Integration of different steps, especially genetic modifications, derivation, and expansion as well as differentiation in bioreactor will pave the future of manufacturing cell-based products. The integrated biologics manufacturing in stirred suspension culture will significantly reduce the risk of contamination of final products, increase product efficacy, and reduce cell processing time and provide a cost-effective platform for cell manufacturing for cell therapy applications.

Acknowledgements

SCN conceptualized, designed, and wrote the manuscript. DER conceptualized and revised the manuscript.

Conflict of interest

The authors declare no conflict of interest.

Author details

Suman C. Nath and Derrick E. Rancourt*
Department of Biochemistry and Molecular Biology, Cumming School of Medicine, University of Calgary, Calgary, AB, Canada

*Address all correspondence to: rancourt@ucalgary.ca

References

[1] Thomson JA, Itskovitz-Eldor J, Shapiro SS, Waknitz MA, Swiergiel JJ, Marshall VS, et al. Embryonic stem cell lines derived from human blastocysts. Science. 1998;**282**:1145-1147. DOI: 10.1126/science.282.5391.1145

[2] Takahashi K, Tanabe K, Ohnuki M, Narita M, Ichisaka T, Tomoda K, et al. Induction of pluripotent stem cells from adult human fibroblasts by defined factors. Cell. 2007;**131**:861-872. DOI: 10.1016/j.cell.2007.11.019

[3] Trounson A, McDonald C. Stem cell therapies in clinical trials: Progress and challenges. Cell Stem Cell. 2015;**17**: 11-22. DOI: 10.1016/j.stem.2015.06.007

[4] Reardon S, Cyranoski D. Japan stem-cell trial stirs envy: Researchers elsewhere can't wait to test iPS cells in humans. Nature. 2014;**513**:278-288. DOI: 10.1038/513287a

[5] Cyranoski D. 'Reprogrammed' stem cells implanted into patient with Parkinson's disease. Nature News. 2018. DOI: 10.1038/d41586-018-07407-9. Available from: https://www.nature.com/articles/d41586-018-07407-9. [News release: 14 November, 2018]

[6] Kimbrel EA, Lanza R. Current status of pluripotent stem cells: Moving the first therapies to the clinic. Nature Reviews. Drug Discovery. 2015;**14**: 681-692. DOI: 10.1038/nrd4738

[7] Available from: https://clinicaltrials.gov/ct2/results?cond=stem+cell+therapy. [Accessed: Jul Nov 11, 2018]

[8] Davie NL, Brindley DA, Culme-Seymour EJ, Mason C. Streaming cell therapy manufacture. Bioprocess International. 2012;**10**:24-29

[9] PR Newswire. Stem cell therapy to redefine regenerative medicine, says Frost & Sullivan. Available from: http://ww2.frost.com/news/press-releases/stem-cell-therapy-redefine-regenerativemedicine-says-frost-sullivan/ [News release: Jul 28, 2015]

[10] Roh KH, Nerem RM, Roy K. Biomanufacturing of therapeutic cells: State of the art, current challenges, and future perspectives. Annual Review of Chemical and Biomolecular Engineering. 2016;**7**:455-478. DOI: 10.1146/annurev-chembioeng-080615-033559

[11] Jozala AF, Geraldes DC, Tundisi LL, Feitosa VA, Breyer CA, Cardoso SL, et al. Biopharmaceuticals from microorganisms: From production to purification. Brazilian Journal of Microbiology. 2016;**47**:51-63. DOI: 10.1016/j.bjm.2016.10.007

[12] Overton TW. Recombinant protein production in bacterial hosts. Drug Discovery Today. 2014;**19**:590-601. DOI: 10.1016/j.drudis.2013.11.008

[13] Tekoah Y, Shulman A, Kizhner T, Ruderfer I, Fux L, Nataf Y, et al. Large-scale production of pharmaceutical proteins in plant cell culture-the Protalix experience. Plant Biotechnology Journal. 2015;**13**: 1199-1208. DOI: 10.1111/pbi.12428

[14] Lai T, Yang Y, Ng SK. Advances in mammalian cell line development technologies for recombinant protein production. Pharmaceuticals. 2013;**6**: 579-603. DOI: 10.3390/ph6050579

[15] Bellone G, Turletti A, Artusio E, Mareschi K, Carbone A, Tibaudi D, et al. Tumor-associated transforming growth factor-beta and interleukin-10 contribute to a systemic Th2 immune phenotype in pancreatic carcinoma patients. The American Journal of Pathology. 1999;**155**:537-547. DOI: 10.1016/S0002-9440(10)65149-8

[16] Gajewski TF, Schreiber H, Fu YX. Innate and adaptive immune cells in the tumor microenvironment. Nature Immunology. 2013;**14**:1014-1022. DOI: 10.1038/ni.2703

[17] Jenkins MJ, Farid SS. Human pluripotent stem cell-derived products: Advances towards robust, scalable and cost-effective manufacturing strategies. Biotechnology Journal. 2015;**10**:83-95. DOI: 10.1002/biot.201400348

[18] Yang HS, Jeon O, Bhang SH, Lee SH, Kim BS. Suspension culture of mammalian cells using thermosensitive microcarrier that allows cell detachment without proteolytic enzyme treatment. Cell Transplantation. 2010;**19**:1123-1132. DOI: 10.3727/096368910X516664

[19] Guillaume-Gentil O, Semenov OV, Zisch AH, Zimmermann R, Voros J, Ehrbar M. pH-controlled recovery of placenta-derived mesenchymal stem cell sheets. Biomaterials. 2013;**2**:4376-4384. DOI: 10.1016/j.biomaterials.2011.02.058

[20] Dou XQ, Yang XM, Li P, Zhang ZG, Schönherr H, Zhanga D, et al. Novel pH responsive hydrogels for controlled cell adhesion and triggered surface detachment. Soft Matter. 2012;**8**: 9539-9544. DOI: 10.1039/C2SM26442K

[21] Nath SC, Tokura T, Kim MH, Kino-Oka M. Botulinum hemagglutinin-mediated in situ break-up of human induced pluripotent stem cell aggregates for high-density suspension culture. Biotechnology and Bioengineering. 2018;**115**:910-920. DOI: 10.1002/bit.26526

[22] Bartosh TJ, Ylöstalo JH, Mohammadipoor A, Bazhanov N, Coble K, Claypool K, et al. Aggregation of human mesenchymal stromal cells (MSCs) into 3D spheroids enhances their antiinflammatory properties. Proceedings of the National Academy of Sciences. 2010;**107**:13724-13729. DOI: 10.1073/pnas.1008117107

[23] Amit M, Laevsky I, Miropolsky Y, Shariki K, Peri M, Itskovitz-Eldor J. Dynamic suspension culture for scalable expansion of undifferentiated human pluripotent stem cells. Nature Protocols. 2011;**6**:572-579. DOI: 10.1038/nprot.2011.325

[24] Larijani MR, Seifinejad A, Pournasr B, Hajihoseini V, Hassani SN, Totonchi M, et al. Long-term maintenance of undifferentiated human embryonic and induced pluripotent stem cells in suspension. Stem Cells and Development. 2011;**20**:1911-1923. DOI: 10.1089/scd.2010.0517

[25] Zweigerdt R, Olmer R, Singh H, Haverich A, Martin U. Scalable expansion of human pluripotent stem cells in suspension culture. Nature Protocols. 2011;**6**:689-700. DOI: 10.1038/nprot.2011.318

[26] Goldring CE, Duffy PA, Benvenisty N, Andrews PW, Ben-David U, Eakins R, et al. Assessing the safety of stem cell therapeutics. Cell Stem Cell. 2011;**8**:618-628. DOI: 10.1016/j.stem.2011.05.012

[27] Rayment EA, Williams DJ. Concise review: Mind the gap: Challenges in characterizing and quantifying cell- and tissue-based therapies for clinical translation. Stem Cells. 2010;**28**: 996-1004. DOI: 10.1002/stem.416

[28] Hocquet D, Sauget M, Roussel S, Malugani C, Pouthier F, Morel P, et al. Validation of an automated blood culture system for sterility testing of cell therapy products. Cytotherapy. 2014;**16**:692-698. DOI: 10.1016/j.jcyt.2013.09.005

[29] Khuu HM, Patel N, Carter CS, Murray PR, Read EJ. Sterility testing of cell therapy products: Parallel comparison of automated methods with a CFR-compliant method. Transfusion. 2006;**46**:2071-2082. DOI: 10.1128/JCM.00302-09

[30] Gee AP, Sumstad D, Stanson J, Watson P, Proctor J, Kadidlo D, et al. A multicenter comparison study between the Endosafe PTS rapid-release testing system and traditional methods for detecting endotoxin in cell-therapy products. Cytotherapy. 2008;**10**:427-435. DOI: 10.1080/14653240802075476

[31] Choi WH, Choi BH, Min BH, Park SR. Low-intensity ultrasound increased colony forming unit-fibroblasts of mesenchymal stem cells during primary culture. Tissue Engineering Part C Methods. 2011;**17**:517-526. DOI: 10.1089/ten.TEC.2010.0231

[32] Schellenberg A, Hemeda H, Wagner W. Tracking of replicative senescence in mesenchymal stem cells by colony-forming unit frequency. Methods in Molecular Biology. 2013;**976**:143-154. DOI: 10.1007/978-1-62703-317-6_11

[33] Dudley ME, Wunderlich JR, Shelton TE, Even J, Rosenberg SA. Generation of tumor-infiltrating lymphocyte cultures for use in adoptive transfer therapy formelanoma patients. Journal of Immunotherapy. 2003;**26**:332-342

[34] Garber K. RIKEN suspends first clinical trial involving induced pluripotent stem cells. Nature Biotechnology. 2015;**33**:890-891. DOI: 10.1038/nbt0915-890

[35] Rohani L, Johnson AA, Arnold A, et al. The aging signature: A hallmark of induced pluripotent stem cells? Aging Cell. 2014;**13**:2-7. DOI: 10.1111/acel.12182

[36] Kim K, Doi A, Wen B, et al. Epigenetic memory in induced pluripotent stem cells. Nature. 2010;**467**:285-290. DOI: 10.1038/nature09342

[37] Feng Q, Lu SJ, Klimanskaya I, et al. Hemangioblastic derivatives from human induced pluripotent stem cells exhibit limited expansion and early senescence. Stem Cells. 2010;**28**:704-712. DOI: 10.1002/stem.321

[38] Vaziri H, Chapman KB, Guigova A, et al. Spontaneous reversal of the developmental aging of normal human cells following transcriptional reprogramming. Regenerative Medicine. 2010;**5**:345-363

[39] Rohani L, Johnson AA, Naghsh P, Rancourt DE, Ulrich H, Holland H. Concise review: Molecular cytogenetics and quality control: Clinical guardians for pluripotent stem cells. Stem Cells Translational Medicine. 2018;**7**:867-875. DOI: 10.1002/sctm.18-0087

[40] Hunt CJ. Cryopreservation of human stem cells for clinical application: A review. Transfusion Medicine and Hemotherapy. 2011;**38**:107-123. DOI: 10.1159/000326623

[41] Thirumala S, Goebel WS, Woods EJ. Clinical grade adult stem cell banking. Organogenesis. 2009;**5**:143-154. DOI: 10.4161/org.5.3.98113

[42] Windrum P, Morris TCM. Severe neurotoxicity because of dimethyl sulphoxide following peripheral blood stem cell transplantation. Bone Marrow Transplantation. 2003;**31**:315

[43] Liseth K, Abrahamsen JF, Bjorsvik S, Grottebo K, Bruserud O. The viability of cryopreserved PBPC depends on the DMSO concentration and the concentration of nucleated cells in the graft. Cytotherapy. 2005;**7**:328-333. DOI: 10.1080/14653240500238251

[44] Moon JH, Lee JR, Jee BC, Suh CS, Kim SH, Lim HJ, et al. Successful vitrification of human amnion-derived mesenchymal stem cells. Human Reproduction. 2008;**23**:1760-1770. DOI: 10.1093/humrep/den202

[45] Kurata H, Takakuwa K, Tanaka K. Vitrification of hematopoietic

progenitor cells obtained from human cord blood. Bone Marrow Transplantation. 1994;**14**:261-263

[46] Capicciotti CJ, Kurach JD, Turner TR, Mancini RS, Acker JP, Ben RN. Small molecule ice recrystallization inhibitors enable freezing of human red blood cells with reduced glycerol concentrations. Scientific Reports. 2015;**5**:9692. DOI: 10.1038/srep09692

[47] Briard JG, Poisson JS, Turner TR, Capicciotti CJ, Acker JP, Ben RN. Small molecule ice recrystallization inhibitors mitigate red blood cell lysis during freezing, transient warming and thawing. Scientific Reports. 2016;**6**:23619. DOI: 10.1038/srep23619

[48] Khuu HM, Cowley H, David-Ocampo V, Carter CS, Kasten-Sportes C, Wayne AS, et al. Catastrophic failures of freezing bags for cellular therapy products: Description, cause, and consequences. Cytotherapy. 2002;**4**:539-549. DOI: 10.1080/146532402761624700

[49] Röllig C, Babatz J, Wagner I, Maiwald A, Schwarze V, Ehninger G, et al. Thawing of cryopreserved mobilized peripheral blood—Comparison between water bath and dry warming device. Cytotherapy. 2002;**4**:551-555. DOI: 10.1080/14653240276162471

[50] Bisson W. Continuous Manufacturing–The Ultra-Lean Way of Manufacturing. ISPE Innovations in Process Technology for Manufacture of APIs and BPCs. Copenhagen; 2008. pp. 7-11

[51] Schaber SD, Gerogiorgis DI, Ramachandran R, Evans JM, Barton PI, Trout BL. Economic analysis of integrated continuous and batch pharmaceutical manufacturing: A case study. Industrial & Engineering Chemistry Research. 2011;**50**:10083-10092

[52] Warikoo V. Feasibility Study to Integrate Perfusion Cell Culture Processes to Continuous Downstream Processing. Anaheim, CA: American Chemical Society—Biotechnology Division; 2011

[53] FDA urges companies to get on board with continuous manufacturing. Available from: http://www.fiercepharma.com/manufacturing/fda-urges-companies-to-get-on-board-continuous-manufacturing. [News release: Apr 14, 2016]

[54] Cartier N, Hacein-Bey-Abina S, Bartholomae CC, Veres G, Schmidt M, Kutschera I, et al. Hematopoietic stem cell gene therapy with a lentiviral vector in Xlinked adrenoleukodystrophy. Science. 2009;**326**:818-823. DOI: 10.1126/science.1171242

[55] Naldini L. Ex vivo gene transfer and correction for cell-based therapies. Nature Reviews. Genetics. 2011;**12**: 301-315. DOI: 10.1038/nrg2985

[56] Wang X, Riviere I. Manufacture of tumor- and virus-specific T lymphocytes for adoptive cell therapies. Cancer Gene Therapy. 2015;**22**:85-94. DOI: 10.1038/cgt.2014.81

[57] Kikuchi T, Worgall S, Singh R, Moore MA, Crystal RG. Dendritic cells genetically modified to express CD40 ligand and pulsed with antigen can initiate antigen-specific humoral immunity independent of CD4+ T cells. Nature Medicine. 2000;**6**:1154-1159. DOI: 10.1038/80498

[58] Aiuti A, Cattaneo F, Galimberti S, Benninghoff U, Cassani B, Callegaro L, et al. Gene therapy for immunodeficiency due to adenosine deaminase deficiency. The New England Journal of Medicine. 2009;**360**:447-458. DOI: 10.1056/NEJMoa0805817

[59] Payen E, Colomb C, Negre O, Beuzard Y, Hehir K, Leboulch P. Lentivirus vectors in beta-thalassemia.

Methods in Enzymology. 2012;**507**:109-124. DOI: 10.1016/B978-0-12-386509-0.00006-5

[60] Song XT. Combination of virotherapy and T-cell therapy: Arming oncolytic virus with T-cell engagers. Discovery Medicine. 2013;**16**:261-266

[61] Kerkar SP. "Model T" cells: A time-tested vehicle for gene therapy. Frontiers in Immunology. 2013;**4**:304. DOI: 10.3389/fimmu.2013.00304

[62] Kochenderfer JN, Rosenberg SA. Treating B-cell cancer with T cells expressing anti-CD19 chimeric antigen receptors. Nature Reviews. Clinical Oncology. 2013;**10**:267-276. DOI: 10.1038/nrclinonc.2013.46

[63] Kochenderfer JN, Wilson WH, Janik JE, Dudley ME, Stetler-Stevenson M, Feldman SA, et al. Eradication of B-lineage cells and regression of lymphoma in a patient treated with autologous T cells genetically engineered to recognize CD19. Blood. 2010;**116**:4099-4102. DOI: 10.1182/blood-2010-04-281931

[64] Jethwa H, Adami AA, Maher J. Use of gene-modified regulatory T-cells to control autoimmune and alloimmune pathology: Is now the right time? Clinical Immunology. 2014;**150**:51-63. DOI: 10.1016/j.clim.2013.11.004

[65] Sakuma T, Barry MA, Ikeda Y. Lentiviral vectors: Basic to translational. The Biochemical Journal. 2012;**443**:603-618. DOI: 10.1042/BJ20120146

[66] Hacein-Bey-Abina S, Hauer J, Lim A, Picard C, Wang GP, Berry CC, et al. Efficacy of gene therapy for X-linked severe combined immunodeficiency. The New England Journal of Medicine. 2010;**363**:355-364. DOI: 10.1056/NEJMoa1000164

[67] Hsu CYM, Uludag H. Nucleic-acid based gene therapeutics: Delivery challenges and modular design of non-viral gene carriers and expression cassettes to overcome intracellular barriers for sustained targeted expression. Journal of Drug Targeting. 2012;**20**:301-328. DOI: 10.3109/1061186X.2012.655247

[68] Djurovic S, Iversen N, Jeansson S, Hoover F, Christensen G. Comparison of nonviral transfection and adeno-associated viral transduction on cardiomyocytes. Molecular Biotechnology. 2004;**28**:21-32. DOI: 10.1385/MB:28:1:21

[69] Dullaers M, Breckpot K, Van Meirvenne S, Bonehill A, Tuyaerts S, Michiels A, et al. Side-by-side comparison of lentivirally transduced and mRNA-electroporated dendritic cells: Implications for cancer immunotherapy protocols. Molecular Therapy. 2004;**10**:768-779. DOI: 10.1016/j.ymthe.2004.07.017

[70] Hsu CYM, Walsh T, Borys B, Kallos M, Rancourt DE. An integrated approach towards the bio-manufacturing of engineered cell therapy products in a continuous stirred suspension bioreactor. Molecular Therapy-Methods & Clinical Development. 2018;**9**:376-389. DOI: 10.1016/j.omtm.2018.04.007

[71] Warren L, Manos PD, Ahfeldt T, Loh YH, Li H, Lau F, et al. Highly efficient reprogramming to pluripotency and directed differentiation of human cells with synthetic modified mRNA. Cell Stem Cell. 2010;**7**:618-630. DOI: 10.1016/j.stem.2010.08.012

[72] Kim D, Kim CH, Moon JI, Chung YG, Chang MY, Han BS, et al. Generation of human induced pluripotent stem cells by direct delivery of reprogramming proteins. Cell Stem Cell. 2009;**4**:472-476. DOI: 10.1016/j.stem.2009.05.005

[73] Woltjen K, Michael IP, Mohseni P, Desai R, Mileikovsky M, Hämäläinen R,

et al. *piggyBac* transposition reprograms fibroblasts to induced pluripotent stem cells. Nature. 2009;**458**:766-770. DOI: 10.1038/nature07863

[74] Singh H, Figliola MJ, Dawson MJ, Olivares S, Zhang L, et al. Manufacture of clinical-grade CD19-specific T cells stably expressing chimeric antigen receptor using sleeping beauty system and artificial antigen presenting cells. PLoS One. 2013;**8**:e64138. DOI: 10.1371/journal.pone.0064138

[75] Huls MH, Figliola MJ, Dawson MJ, Olivares S, Kebriaei P, et al. Clinical application of *Sleeping Beauty* and artificial antigen presenting cells to genetically modify T cells from peripheral and umbilical cord blood. Journal of Visualized Experiments. 2013;**72**:e50070. DOI: 10.3791/50070

[76] Sommer CA, Sommer AG, Longmire TA, Christodoulou C, Thomas DD, et al. Excision of reprogramming transgenes improves the differentiation potential of iPS cells generated with a single excisable vector. Stem Cells. 2010;**28**:64-74. DOI: 10.1002/stem.255

[77] Yáñez-Muñoz RJ, Balaggan KS, MacNeil A, Howe SJ, Schmidt M, Smith AJ, et al. Effective gene therapy with nonintegrating lentiviral vectors. Nature Medicine. 2006;**12**:348-353. DOI: 10.1038/nm1365

[78] Lombardo A, Genovese P, Beausejour CM, Colleoni S, Lee YL, Kim KA, et al. Gene editing in human stem cells using zinc finger nucleases and integrase-defective lentiviral vector delivery. Nature Biotechnology. 2007;**25**:1298-1306. DOI: 10.1038/nbt1353

[79] Provasi E, Genovese P, Lombardo A, Magnani Z, Liu PQ, Reik A, et al. Editing T cell specificity towards leukemia by zinc finger nucleases and lentiviral gene transfer. Nature Medicine. 2012;**18**:807-815. DOI: 10.1038/nm.2700

[80] Seki T, Yuasa S, Oda M, Egashira T, Yae K, et al. Generation of induced pluripotent stem cells from human terminally differentiated circulating T cells. Cell Stem Cell. 2010;**7**:11-14. DOI: 10.1016/j.stem.2010.06.003

[81] Mullard A. Novartis secures first CRISPR pharma collaborations. Nature Reviews. Drug Discovery. 2015;**14**:82. DOI: 10.1038/nrd4546

[82] Gaj T, Gersbach CA, Barbas CF. ZFN, TALEN, and CRISPR/Cas-based methods for genome engineering. Trends in Biotechnology. 2013;**31**:397-405. DOI: 10.1016/j.tibtech.2013.04.004

[83] Jamebozorgi K, Taghizadeh E, Rostami D, Pormasoumi H, Barreto GE, Hayat SMG, et al. Cellular and molecular aspects of Parkinson treatment: Future therapeutic perspectives. Molecular Neurobiology. 2018:1-13. DOI: 10.1007/s12035-018-1419-8

[84] Torikai H, Reik A, Liu PQ, Zhou Y, Zhang L, et al. A foundation for universal T-cell based immunotherapy: T cells engineered to express a CD19-specific chimeric-antigen-receptor and eliminate expression of endogenous TCR. Blood. 2012;**119**:5697-5705. DOI: 10.1182/blood-2012-01-405365

[85] Liu P, Chen M, Liu Y, Qi LS, Ding S. CRISPR-based chromatin Remodeling of the endogenous Oct4 or Sox2 locus enables reprogramming to pluripotency. Cell Stem Cell. 2018;**22**:252-261. DOI: 10.1016/j.stem.2017.12.001

[86] Hollyman D, Stefanski J, Przybylowski M, Bartido S, Borquez-Ojeda O, et al. Manufacturing validation of biologically functional T cells targeted to CD19 antigen for autologous adoptive cell therapy. Journal of Immunotherapy. 2009;**32**:169-180. DOI: 10.1097/CJI.0b013e318194a6e8

[87] Somerville RP, Devillier L, Parkhurst MR, Rosenberg SA, Dudley

ME. Clinical scale rapid expansion of lymphocytes for adoptive cell transfer therapy in the WAVE(R) bioreactor. Journal of Translational Medicine. 2012;**10**:69. DOI: 10.1186/1479-5876-10-69

[88] Jin J, Sabatino M, Somerville R, Wilson JR, Dudley ME, Stroncek DF, et al. Simplified method of the growth of human tumor infiltrating lymphocytes in gas permeable flasks to numbers needed for patient treatment. Journal of Immunotherapy. 2012;**35**:283-292. DOI: 10.1097/CJI.0b013e31824e801f

[89] Bajgain P, Mucharla R, Wilson J, Welch D, Anurathapan U, Liang B, et al. Optimizing the production of suspension cells using the G-rex 'M' series. Molecular Therapy. 2014;**1**:14015. DOI: 10.1038/mtm.2014.15

[90] Acker JP, Marks DC, Sheffield WP. Quality assessment of established and emerging blood components for transfusion. Journal of Blood Transfusion. Vol. 2016. Article ID: 4860284. DOI: 10.1155/2016/4860284

[91] Casati A, Varghaei-Nahvi A, Feldman SA, Assenmacher M, Rosenberg SA, Dudley ME, et al. Clinical-scale selection and viral transduction of human naive and central memory CD8+ T cells for adoptive cell therapy of cancer patients. Cancer Immunology, Immunotherapy. 2013;**62**:1563-1573. DOI: 10.1007/s00262-013-1459-x

[92] Terakura S, Yamamoto TN, Gardner RA, Turtle CJ, Jensen MC, Riddell SR. Generation of CD19-chimeric antigen receptor modified CD8+ T cells derived from virus-specific central memory T cells. Blood. 2012;**119**:72-82. DOI: 10.1182/blood-2011-07-366419

[93] Apel M, Brüning M, Granzin M, Essl M, Stuth J, Blaschke J, et al. Integrated clinical scale manufacturing system for cellular products derived by magnetic cell separation, centrifugation and cell culture. Chemie Ingenieur Technik. 2013;**85**:103-110. DOI: 10.1002/cite.201200175

[94] Heathman TRJ, Nienow AW, McCall MJ, Coopman K, Kara B, Hewitt CJ. The translation of cell-based therapies: Clinical landscape and manufacturing challenges. Regenerative Medicine. 2015;**10**:49-64. DOI: 10.2217/rme.14.73

[95] Sensebe L, Bourin P, Tarte K. Good manufacturing practices production of mesenchymal stem/stromal cells. Human Gene Therapy. 2011;**22**:19-26. DOI: 10.1089/hum.2010.197

[96] Wuchter P, Bieback K, Schrezenmeier H, Bornhauser M, Muller LP, et al. Standardization of good manufacturing practice-compliant production of bone marrow-derived human mesenchymal stromal cells for immunotherapeutic applications. Cytotherapy. 2015;**17**:128-139. DOI: 10.1016/j.jcyt.2014.04.002

[97] Shafa M, Day B, Yamashita A, Meng G, Liu S, Krawetz R, et al. Derivation of iPSCs in stirred suspension bioreactors. Nature Methods. 2012;**9**:465-466. DOI: 10.1038/nmeth.1973

[98] Fluri DA, Tonge PD, Song H, Baptista RP, Shakiba N, Shukla S, et al. Derivation, expansion and differentiation of induced pluripotent stem cells in continuous suspension cultures. Nature Methods. 2012;**9**:509-516. DOI: 10.1038/nmeth.1939

[99] Samavarchi-Tehrani P, Golipour A, David L, Sung HK, Beyer TA, Datti A, et al. Functional genomics reveals a BMP-driven mesenchymal-to-epithelial transition in the initiation of somatic cell reprogramming. Cell Stem Cell. 2010;7:64-77. DOI: 10.1016/j.stem.2010.04.015

[100] Serra M, Brito C, Correia C, Alves PM. Process engineering of

human pluripotent stem cells for clinical application. Trends in Biotechnology. 2012;**30**:350-359. DOI: 10.1016/j.tibtech.2012.03.003

[101] Miyazaki T, Futaki S, Suemori H, Taniguchi Y, Yamada M, Kawasaki M, et al. Laminin E8 fragments support efficient adhesion and expansion of dissociated human pluripotent stem cells. Nature Communications. 2012;**3**:1236. DOI: 10.1038/ncomms2231

[102] Wang Y, Cheng L, Gerecht S. Efficient and scalable expansion of human pluripotent stem cells under clinically compliant settings: A view in 2013. Annals of Biomedical Engineering. 2014;**42**:1357-1372. DOI: 10.1007/s10439-013-0921-4

[103] Fan Y, Wu J, Ashok P, Hsiung M, Tzanakakis ES. Production of human pluripotent stem cell therapeutics under defined xeno-free conditions: Progress and challenges. Stem Cell Reviews. 2015;**11**:96-109. DOI: 10.1007/s12015-014-9544-x

[104] Lam AT, Chen AK, Li J, Birch WR, Reuveny S, Oh SK. Conjoint propagation and differentiation of human embryonic stem cells to cardiomyocytes in a defined microcarrier spinner culture. Stem Cell Research & Therapy. 2014;**5**:110. DOI: 10.1186/scrt498

[105] Badenes SM, Fernandes TG, Rodrigues CAV, Diogo MM, Cabral JMS. Microcarrier-based platforms for in vitro expansion and differentiation of human pluripotent stem cells in bioreactor culture systems. Journal of Biotechnology. 2016;**234**:71-82. DOI: 10.1016/j.jbiotec.2016.07.02

[106] Kropp C, Massai D, Zweigerdt R. Progress and challenges in large-scale expansion of human pluripotent stem cells. Process Biochemistry. 2017;**59**:244-254. DOI: 10.1016/j.procbio.2016.09.032

[107] Olmer R, Lange A, Selzer S, Kasper C, Haverich A, Martin U, et al. Suspension culture of human pluripotent stem cells in controlled, stirred bioreactors. Tissue Engineering Part C. 2012;**18**:772-784. DOI: 10.1089/ten.TEC.2011.0717

[108] Haraguchi Y, Matsuura K, Shimizu T, Yamato M, Okano T. Simple suspension culture system of human iPS cells maintaining their pluripotency for cardiac cell sheet engineering. Journal of Tissue Engineering and Regenerative Medicine. 2015;**9**:1363-1375. DOI: 10.1002/term.1761

[109] Nath SC, Horie M, Nagamori E, Kino-Oka M. Size- and time-dependent growth properties of human induced pluripotent stem cells in the culture of single aggregate. Journal of Bioscience and Bioengineering. 2017;**124**:469-475. DOI: 10.1016/j.jbiosc.2017.05.006

[110] Rungarunlert S, Ferreira JN, Dinnyes A. Novel bioreactor platform for scalable cardiomyogenic differentiation from pluripotent stem cell-derived embryoid bodies. Methods in Molecular Biology. 2016;**1502**: 169-179. DOI: 10.1007/7651_2016_341

[111] Kempf H, Kropp C, Olmer R, Martin U, Zweigerdt R. Cardiac differentiation of human pluripotent stem cells in scalable suspension culture. Nature Protocols. 2015;**10**:1345-1361. DOI: 10.1038/nprot.2015.089

[112] Matsuura K, Wada M, Shimizu T, Haraguchi Y, Sato F, Sugiyama K, et al. Creation of human cardiac cell sheets using pluripotent stem cells. Biochemical and Biophysical Research Communications. 2012;**425**:321-327. DOI: 10.1016/j.bbrc.2012.07.089

[113] Park Y, Chen Y, Ordovas L, Verfaillie CM. Hepatic differentiation of human embryonic stem cells on

microcarriers. Journal of Biotechnology. 2014;**174**:39-48. DOI: 10.1016/j.jbiotec.2014.01.025

[114] Vosough M, Omidinia E, Kadivar M, Shokrgozar MA, Pournasr B, Aghdami N, et al. Generation of functional hepatocyte-like cells from human pluripotent stem cells in a scalable suspension culture. Stem Cells and Development. 2013;**22**:2693-2705. DOI: 10.1089/scd.2013.0088

[115] Yan Y, Song L, Tsai AC, Ma T, Li Y. Generation of neural progenitor spheres from human pluripotent stem cells in a suspension bioreactor. Methods in Molecular Biology. 2016;**1502**:119-128. DOI: 10.1007/7651_2015_310

[116] Lam AT, Chen AK, Ting SQ, Reuveny S, Oh SK. Integrated processes for expansion and differentiation of human pluripotent stem cells in suspended microcarriers cultures. Biochemical and Biophysical Research Communications. 2016;**473**:764-768. DOI: 10.1016/j.bbrc.2015.09.079

[117] Ting S, Chen A, Reuveny S, Oh SK. An intermittent rocking platform for integrated expansion and differentiation of human pluripotent stem cells to cardiomyocytes in suspended microcarrier cultures. Stem Cell Research. 2014;**13**:202-213. DOI: 10.1016/j.scr.2014.06.002

[118] Fonoudi H, Ansari H, Abbasalizadeh S, Larijani MR, Kiani S, Hashemizadeh S, et al. A universal and robust integrated platform for the scalable production of human cardiomyocytes from pluripotent stem cells. Stem Cells Translational Medicine. 2015;**4**:1482-1494. DOI: 10.5966/sctm.2014-0275

[119] Steiner D, Khaner H, Cohen M, Even-Ram S, Gil Y, Itsykson P, et al. Derivation, propagation and controlled differentiation of human embryonic stem cells in suspension. Nature Biotechnology. 2010;**28**:361-364. DOI: 10.1038/nbt.1616

Chapter 6

A Simple Way to Produce Gold Nanoshells for Cancer Therapy

Rosa Isela Ruvalcaba Ontiveros, José Alberto Duarte Moller, Anel Rocío Carrasco Hernandez, Hilda Esperanza Esparza-Ponce, Erasmo Orrantia Borunda, Cynthia Deisy Gómez Esparza and Juan Manuel Olivares Ramírez

Abstract

Gold nanoshells (GNSs), formed by a silica core surrounded by a gold shell, present a shift on their surface plasmon resonance (SPR) to the near-infrared (NIR) part of the electromagnetic spectrum when synthesized with specific dimensions. This chapter presents a simple method to prepare the nanoshells, a step-by-step characterization, as well as their absorbance spectrum. For the synthesis, silica spheres, with approximately 190 ± 5 nm in diameter, were prepared using the Stöber method and then functionalized with 3-aminopropyltriethoxysilane (APTES). The gold nanoparticles (GNPs), with a diameter of 7 ± 3 nm, were produced by the reduction of chloroauric acid. Then, the silica was seeded with the GNPs to later grow a gold shell with the help of $Au(OH)_4^-$ ions and formaldehyde. UV-Vis spectroscopy results showed an increase of absorbance starting at 520 nm. It reached its maximum around 600 nm and kept absorbing all through 1200 nm. Transmission electron microscope (TEM) and scanning electron microscope (SEM) images suggest that the absorption peak movement coincided with the completion of the shell. Furthermore, when the sample was irradiated with an 820 nm wavelength/3.1 mW laser, its temperatures increased by 6.3°C in 2 min, showing its absorbance in the NIR.

Keywords: gold, nanoshells, surface plasmon resonance, near-infrared absorption, silica core, core-shell particles

1. Introduction

There are hundreds of types of cancer, and each one has different characteristics [1]. Therefore, science utilizes the most innovative discoveries in an effort to find new treatments, and nanotechnology offers a wide variety of options. One example of this is the nanoparticle colloids. They can be designed to concentrate on specific organs (passive targeting), or their surfaces can be modified by an antibody or ligand to get attached to a specific target (active targeting) [2]. Furthermore, metallic nanoparticles, like GNS, present interesting optical properties. The shell, formed by GNPs, confines the plasmons to the surface of the particle, changing the plasmon frequency of the gold. Therefore, the GNSs absorb different wavelengths than gold in bulk. Moreover, when the wavelength of the incident light is larger than the size of the nanoparticle exciting the plasmons at their natural resonance frequency, light is absorbed more strongly causing an increase in temperature. When the GNSs are synthesized with specific geometry and dimensions, their SPR changes causing their absorption to shift to the NIR region of the electromagnetic spectrum [3]. This shift offers a great potential for applications in the medical field because GNPs are bio-inert [4], and the cytotoxicity of the silica has been widely studied [5]. Additionally, the wavelengths of the NIR spectrum are considered the optical window of the human body. As a result, while most biological soft tissues have low absorption of these wavelengths [6], GNSs absorb them causing them to increase their temperature.

GNSs have been synthesized over different templates. Polystyrene cores claim to offer a narrower plasmon resonance absorption peak due to their higher reflective index [7]; iron oxide nanoparticles present a superparamagnetic template useful for magnetic resonance imaging [8]; silver nanoparticles have also been used as a mold for hollow gold nanoshells [9]. However, the functionalization of the polystyrene takes more time, reactants, and supervision which increases the chances of error as compared with the functionalization of the silica. Besides, the cytotoxicity of the silica nanoparticles makes them a good option for medical applications. Moreover, once the GNS is produced, the silica core can be diluted with hydrochloric acid to obtain hollow gold nanoparticles [10] that can be used for the controlled release of drugs [11] due to their capacity for encapsulating sensitive materials and their low thermal expansion coefficient. Therefore, providing a simpler and more efficient method of synthesis of GNS on silica templates provides a more promising variety of applications like for photothermal therapy [12], optical imaging [13], and drug release [2], as well as providing a near instantaneous in situ whole blood assay [14].

The synthesis of the GNSs has been extensively explored. Different methods, like reflux systems [15] or flow micro-reactors [16], can be used as well as procedures involving high temperatures [17]. But most of those methods last over 30 h [18]. In this chapter, we present a simple and effective method of preparation that shortens the time of the traditional procedures published before and uses only a magnetic stirrer with heating for the synthesis.

The reductions of the time were obtained by first modifying the Stöber method of synthesis of silica particles from 2 h to 30 min Samples were obtained at 30, 60, 90, and 120 min throughout the reaction to determine the minimum time of reaction needed. Also, the seeding process can be shortened from 2 h to 30 min. During the seeding process, where the silica is decorated with GNPs, a sample was obtained using only 30 min of resting time and compared with another sample obtained after the full 2 h of the resting time previously suggested. In both cases, SEM images were obtained showing that 30 min were sufficient to accomplish the synthesis of the silica as well as their seeding. In consequence, the total time of the process was reduced by 3 h.

2. Background

The "Birth of Nanotechnology" was the title used by David Thompson [19] on his article acknowledging Michael Faraday's synthesis of gold nanoparticles in 1857. What Faraday called "Colloidal Ruby Gold" [20] was, in fact, a solution of dispersed GNPs so small that no microscope of that time was able to observe them. It wasn't until 1985 that Turkevich et al. [21] used an electron microscope to corroborate that Faraday's ruby gold was formed by GNPs with an average size of 6 ± 2 nm. Separately, in 1967 Werner Stöber et al. developed a method of synthesizing silica spheres in the micron size range [22] to be used especially in the medical field due to its known cytotoxicity, and in 1998 they were used by Naomi Halas et al. as the templates of GNS [23].

3. Experimental

3.1 Materials

Ethanol (100%), tetraethyl orthosilicate (TEOS) (98%), 3-aminopropyl-triethoxysilane (APTES) (99%), trisodium citrate dihydrate, gold (III) chloride trihydrate ($HAuCl_4$, 49%), formaldehyde (37%), and sodium borohydride ($NaBH_4$, 98%) were purchased from Sigma-Aldrich. Potassium carbonate (K_2CO_3, 99%) and ammonium hydroxide (28%) were purchased from J.T. Baker. All the solutions were prepared with deionized water.

3.2 Characterization

Images were obtained using the field-emission scanning electron microscope (SEM, JEOL JSM-7401F) and the transmission electron microscope (TEM, HT7700 Hitachi). For the ultraviolet-visible (UV-Vis) spectra, the Evolution 220 spectrophotometer UV-Vis (Thermo Scientific) was used. The FTIR spectra were obtained with an IRAffinity-1S Fourier transform infrared spectrophotometer (Shimadzu). The sample was irradiated with an 820 nm wavelength/3.1 mW laser (Multi-Channel Fiber-Coupled Laser Source, Thorlabs), and the infrared images were taken with a Non-contact Digital IR Thermometer (TrueIR Agilent Keysight U5855A). Measurement of the particles and histograms were acquired with the Image J® software [24].

3.3 Preparation of silica spheres

Silica particles were prepared by modifying the Stöber method [22]. About 50 ml of ethanol, 2.5 ml of deionized water, and 4.25 ml of ammonium hydroxide were magnetically stirred in an 80 ml glass flask for 5 min. Then, 0.75 ml of TEOS was added dropwise. The solution was heated at 40°C. Temperature and agitation were kept for 2 h. The color of the solution changed from transparent to opaque white approximately 10 min after adding the TEOS as shown in **Figure 1**. This time corresponds to the induction period needed to form the SiO_2 nucleus from the concentration used of the TEOS monomer [25]. Samples were obtained at 30, 60, 90, and 120 min after adding the TEOS to observe the evolution of the process.

3.4 Functionalization of the silica spheres

In order to create open links over the silica to attach the GNPs, the silica was functionalized with APTES on a 1 ml:1 μl silica/APTES volume ratio. About 50 ml

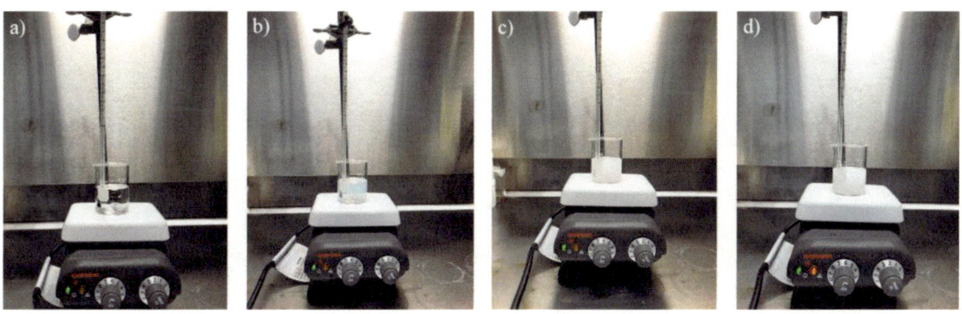

Figure 1.
Images of the synthesis of SiO_2 particles: (a) right after adding TEOS, (b) at 10 min of reaction, (c) at 30 min of reaction, and (d) at 2 h of reaction.

of the silica template solution was magnetically stirred for 5 min with 50 μl of APTES in an 80 ml glass flask. The solution was left still overnight at room temperature. The opaque white functionalized silica particles precipitated in the solution leaving a clear fluid at the top. To separate the functionalized silica, the mixture was centrifuged at 6000 rpm for 1.5 min and washed in deionized water three times. Finally, they were sonicated in 20 ml of deionized water final volume.

3.5 Synthesis of gold nanoparticles

The method presented by Abdollahi et al. [10] was followed to elaborate the GNPs. First, 100 ml of deionized water at room temperature was placed in a 140 ml flask under magnetic agitation. Then 1 ml of 1% $HAuCl_4$ solution, 2 ml of 1% trisodium citrate, and 1 ml of freshly made 0.075% $NaBH_4$ in 1% trisodium citrate were added in that order. The mixture was stirred for 10 min and used immediately to avoid the agglomeration. The GNP may also be stored at 4°C in an amber glass bottle for later use.

Throughout the synthesis, the gold solution changed its color from light yellow (**Figure 2a**) to wine red (**Figure 2b**). This is a characteristic of the GNP formation [26].

3.6 Seeding process

For the seeding process, 100 ml of GNPs and 10 ml of functionalized silica were magnetically stirred in a 140 ml glass flask for 5 min as shown in **Figure 3a**. Then, it was left still for 2 h. **Figure 3b** presents how the seeded silica spheres precipitated and changed their color from opaque white to lavender, while the mother solution changed from wine red to transparent. The mixture was centrifuged at 6000 rpm for 2 min and washed in deionized water three times. Finally, it was sonicated in 20 ml of deionized water final volume. The same procedure was followed, but the solution was left still for only 30 min to observe the development of the seeding process through time.

3.7 Gold hydroxide solution

For the shell growth process, a gold hydroxide solution was prepared by mixing 100 ml of 2 mM K_2CO_3 solution and 1.5 ml of 1% $HAuCl_4$ in a 140 ml glass flask for

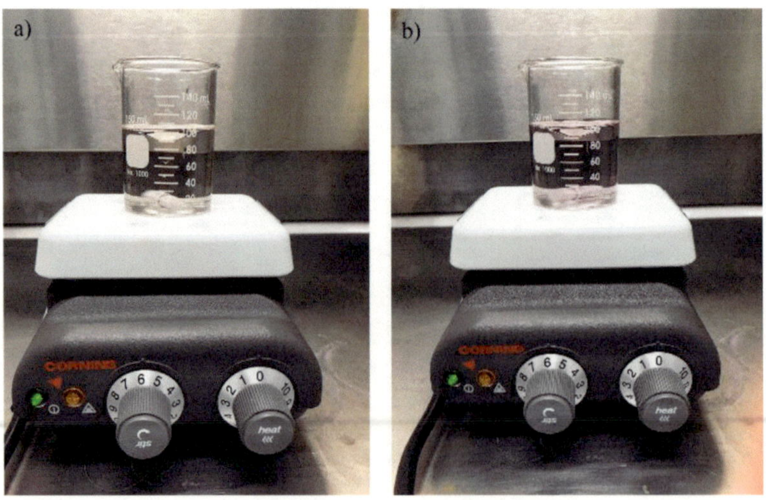

Figure 2.
Synthesis of GNPs at (a) the beginning of the reaction and (b) after 10 min of reaction.

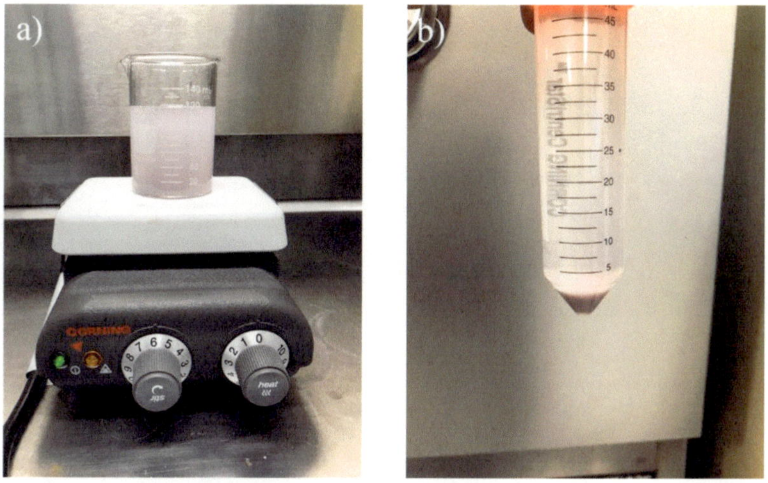

Figure 3.
Images at (a) the beginning of the seeding process and (b) after 2 h of resting time.

30 min. The color of the solution changed from light yellow (**Figure 4a**) to transparent (**Figure 4b**). It was left still overnight at room temperature in an amber glass bottle to facilitate the formation of $Au(OH)_4^-$ ions [18].

3.8 Shell growth

The shell was developed from the gold seeds deposited over the functionalized silica particles with the help of the $Au(OH)_4^-$ ions. About 100 ml of the gold hydroxide solution (**Figure 5a**) and 5 ml of seeded silica were magnetically stirred in a 140 ml glass flask for 5 min (**Figure 5b**). Next, 5 ml of formaldehyde was added to the solution (**Figure 5c**) and stirred for 10 min (**Figure 5d**). The solution was left still for 50 min. Finally, it was centrifuged at 6000 rpm for 2 min, washed, and dispersed in 10 ml of deionized water final volume.

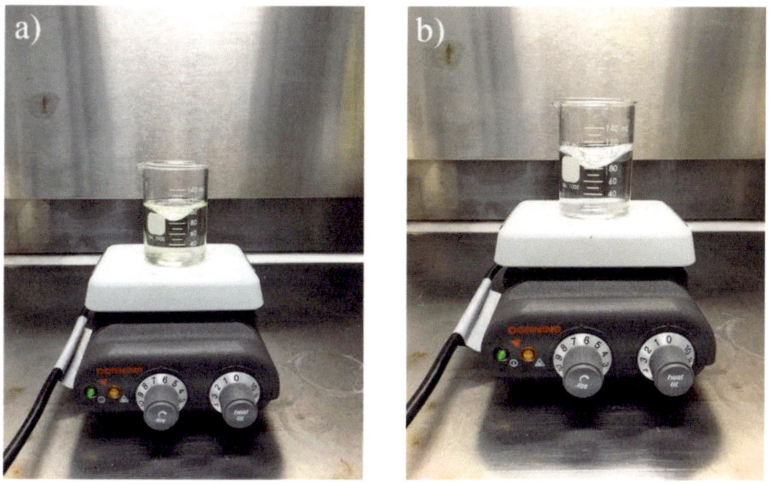

Figure 4.
Images illustrating the change of color of the gold hydroxide solution at (a) the beginning of the synthesis (light yellow) and (b) 30 min of reaction (transparent).

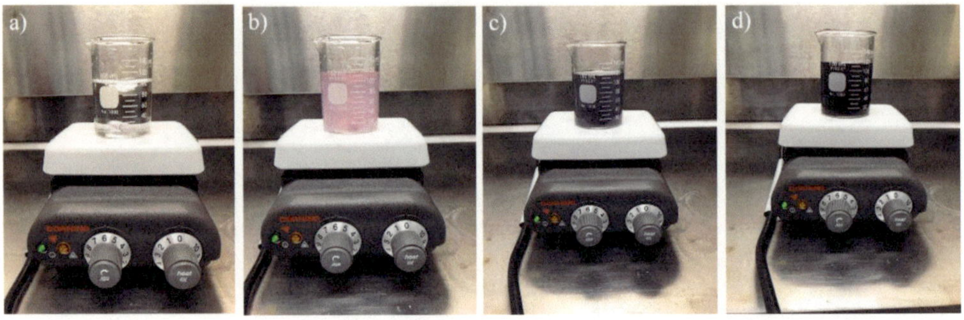

Figure 5.
Images of the shell growing process. (a) Gold hydroxide solution, (b) gold hydroxide + seeded silica, (c) gold hydroxide + seeded silica + formaldehyde, and (d) solution after 10 min of reaction.

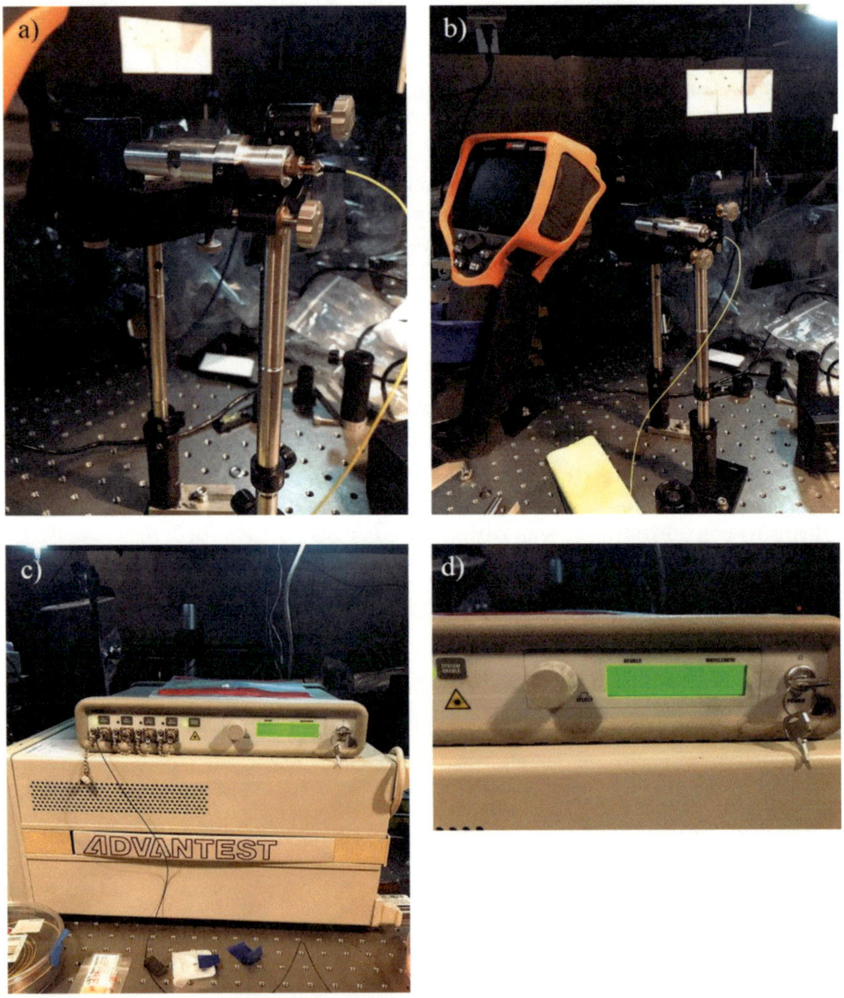

Figure 6.
Images of the installation of the (a) 820 nm wavelength laser, (b) non-contact digital IR thermometer, (c) connection to the multi-channel laser source, and (d) selection of the channel with the desired wavelength.

3.9 Thermography

To obtain the IR images, first, the 820 nm wavelength laser was fastened to the support for it to aim directly to the sample (**Figure 6a**). Then the Digital IR thermometer was also secured and directed to the GNS (**Figure 6b**). Next, the laser was connected to the Multi-channel laser source (**Figure 6c**). Finally, the channel with the desired wavelength was selected (**Figure 6d**), and the irradiation was started.

4. Results and discussion

4.1 Characterization of the silica templates

TEM images obtained from the silica samples taken at 30, 60, 90, and 120 min after adding the TEOS are presented in **Figure 7**. When comparing the images, no significant variation in the size of the silica particles is noticeable.

To corroborate that the silica particles do not change substantially when the reaction time is over 30 min, the images were studied with the software Image J®, and the diameter distribution of the particles was analyzed. Over 1000 particles from the different samples were measured to obtain the histograms presented in **Figure 8** where samples A, B, C, and D correspond to 30, 60, 90, and 120 min of reaction time, respectively. They illustrate that the diameter distribution of the silica spheres throughout the synthesis oscillates around the 190 ± 5 nm on all the samples.

To have a better understanding of the information, **Table 1** contains useful statistic information from the samples.

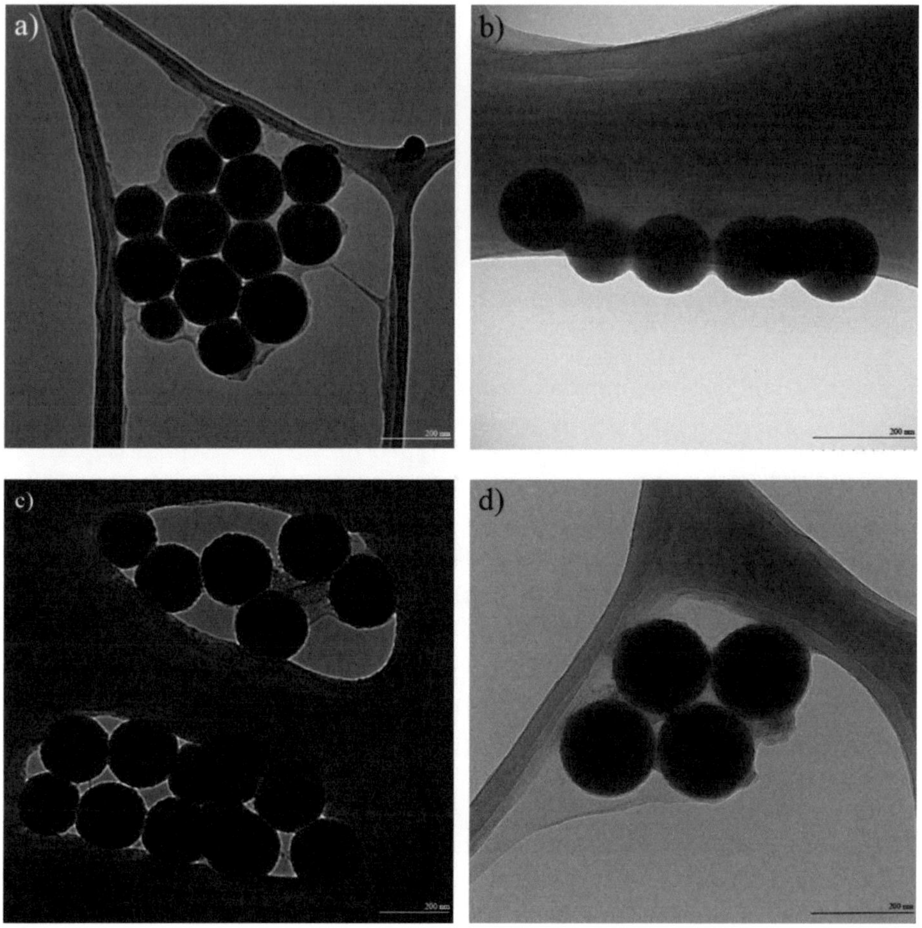

Figure 7.
TEM images of silica particles at (a) 30, (b) 60, (c) 90, and (d) 120 min after adding TEOS.

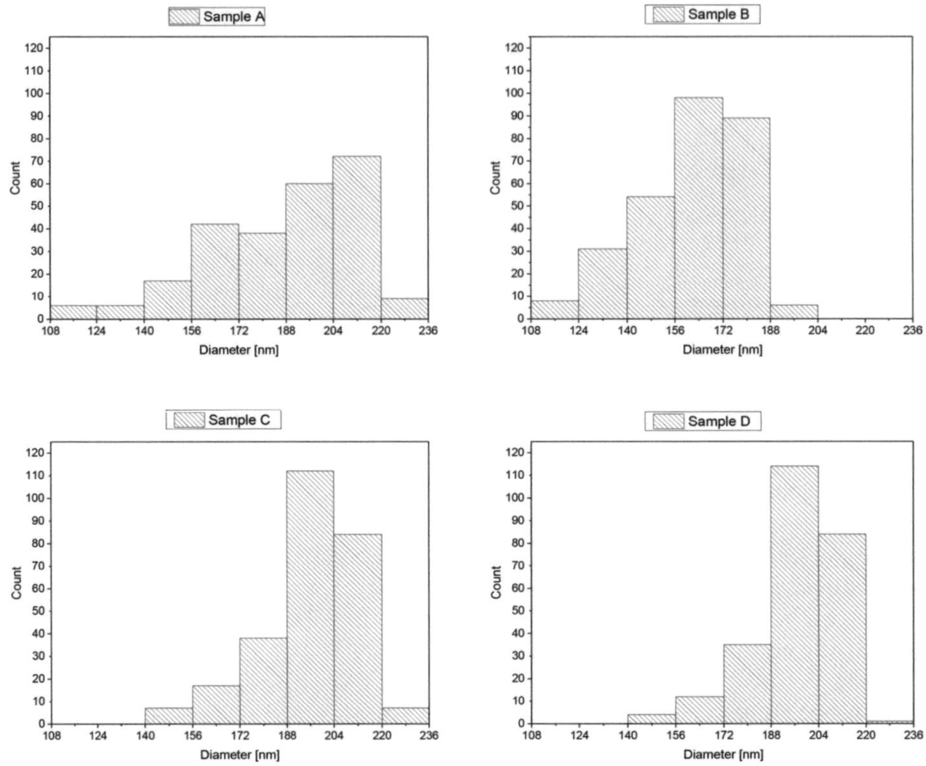

Figure 8.
Histograms illustrating the diameter distribution of the silica particles throughout their synthesis.

	Count	Mean	Std dev.	Mode
Sample A	250	187	25	197 (79)
Sample B	287	162	18	162 (80)
Sample C	265	196	16	201 (92)
Sample D	250	197	13	198 (105)

Table 1.
Statistic information obtained by measuring the diameters of silica particles from the different samples.

The mean, standard deviation, and mode obtained after analyzing the samples show that, in general, the silica templates keep their size and shape after 30 min of synthesis. Therefore, the objective of synthesizing silica particles with diameters of 190 ± 5 nm was achieved within 30 min of reaction time. More than 30 min of synthesis does not result in any relevant change in the sample. For this reason, the total process time can be reduced from 2 h to 30 min, shortening the reaction time by 1 h and 30 min when compared with similar published works where the synthesis time is at least 2 h [10, 17, 27, 28].

4.2 Characterization of the functionalized silica templates

The functionalization of the silica with a primary amine group ($-NH_2$) was accomplished by the use of APTES which changed the superficial charge of the

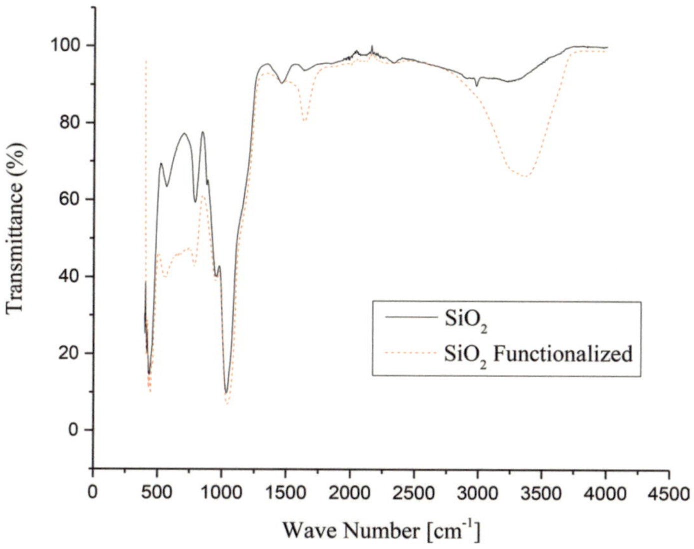

Figure 9.
FTIR spectrum of silica particles and silica particles functionalized with APTES.

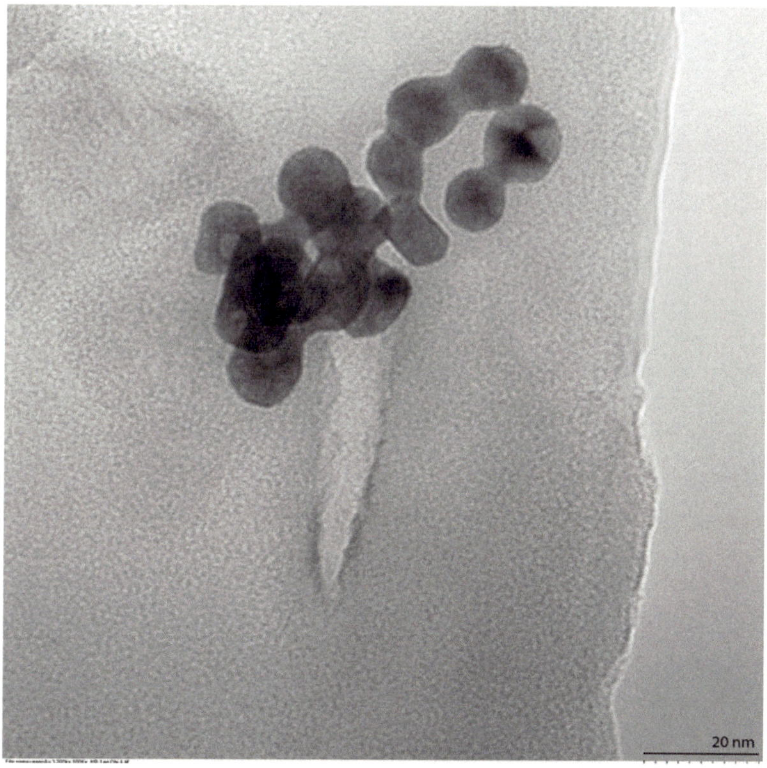

Figure 10.
TEM images of GNPs.

silica providing an electrostatic link for the GNPs to attach [29]. This superficial modification was verified by the FTIR spectrum shown in **Figure 9** where the vibrations of primary amines are found between 3550 and 3330 cm^{-1} which correspond to the vibrations of a primary amine group [30].

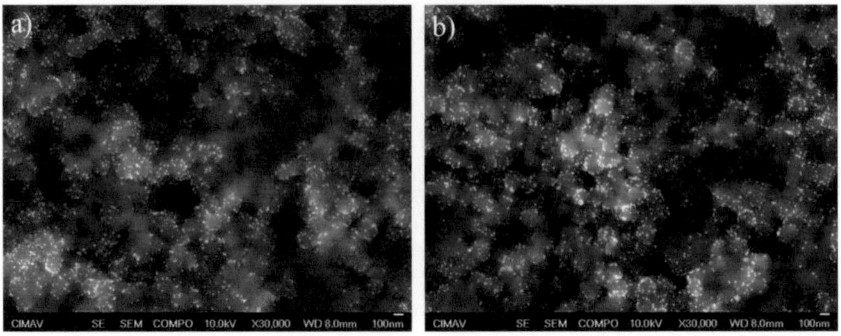

Figure 11.
SEM images illustrating the seeding process with (a) 30 min of resting time and (b) 2 h of resting time.

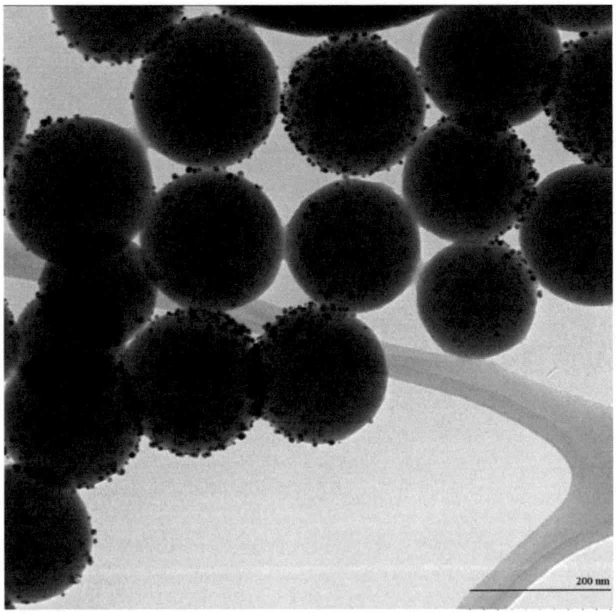

Figure 12.
TEM image of a gold decorated silica particle.

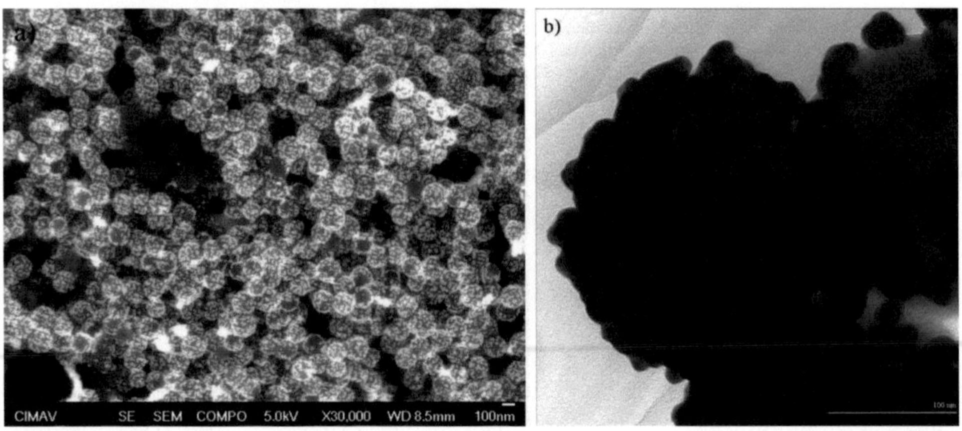

Figure 13.
(a) SEM and (b) TEM images of GNS.

4.3 Characterization of the gold nanoparticles and seeded silica

The GNPs were analyzed under a TEM. **Figure 10** illustrates the GNPs with a diameter of 7 ± 3 nm and spherical shape overall.

The seeding process was followed with 2 h of still time as well as with 30 min of still time. The first and second samples were observed under the microscope. The samples were taken with the purpose of observing the development of the seeds. **Figure 11a** presents an SEM image of seeded silica with 30 min of resting time, while **Figure 11b** presents an SEM image of seeded silica with 2 h of resting time. The images show that 30 min is enough time to create the seeds because both images display approximately the same number of nucleus per silica particle.

Even though a complete shell was not formed, the seeds are ready to grow the gold shell on the next step. A TEM image of a seeded silica particle is presented in **Figure 12**. This image corroborates the seeding process as well as the silica functionalization.

4.4 Characterization of the gold nanoshells

Figure 13a and **b** presents SEM and TEM images of the synthesized GNS, respectably. They illustrate that the silica particles are almost surrounded by gold. The higher density of gold, the separation of the GNPs [31], and the dielectric properties of the silica [3] contribute to the absorption of the NIR wavelength, which causes the increase in temperature.

4.5 The UV-Vis spectrum

Figure 14 presents the UV-Vis spectrum of the particles through the process. Silica particles, as well as functionalized silica particles, do not show significant

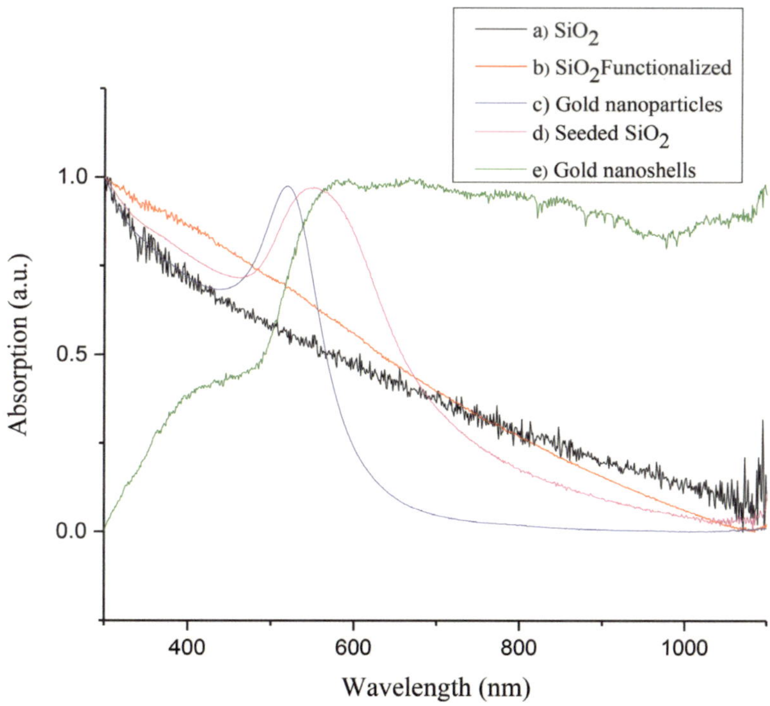

Figure 14.
UV-Vis spectrum of (a) silica, (b) functionalized silica, (c) gold nanoparticles, (d) gold seeded silica, and (e) gold nanoshells.

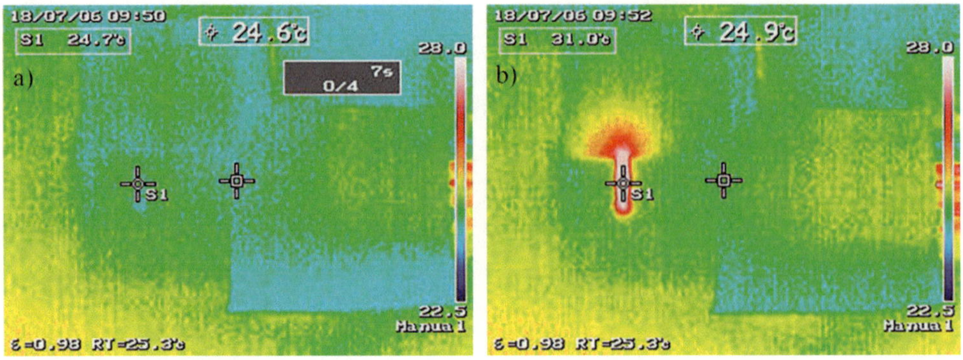

Figure 15.
Thermography images of the GNS taken (a) before and (b) after being irradiated with 840 nm wavelength laser for 2 min.

absorption on the NIR. As for the GNPs, they exhibit their characteristic absorption between 520 and 530 nm [29]. However, on the seeded silica particles, the slight shift to the NIR is noticeable. While on the GNS, the peak not only shifted to the NIR, but it kept a high absorbance all the way to 1100 nm. This range is part of the optical window of the human body [5]. The absorbance of the GNSs is due to the SPR that creates an electric field on the surface increasing the absorption of these wavelengths. SPR happens when metal nanoparticles are irradiated with a wavelength bigger than their size, exciting the electrons of the conducting band [2].

4.6 Heat generation of GNS from light energy

To verify the absorbance of the GNS, they were irradiated with an 840 nm laser with a power of 3.1 mW. **Figure 15a** and **b** displays the thermography images of the sample while being irradiated at time zero and 2 min later. The temperature of the sample increased from 24.7 to 31.0°C. This confirms that GNSs are able to absorb light from the NIR and convert it in heat.

5. Conclusions

Synthesizing GNS by seeding and growing a gold shell over silica spheres with GNPs showed to be an effective method to tune their absorption to the NIR. The SEM and TEM images show the evolution of the process, while the absorbance spectrum displays the GNS shifting over the NIR. Therefore, we obtained a simple technique of producing GNS that can be used for medical applications thanks to the bio-inert GNPs [3] and the widely studied cytotoxicity of the silica [5]. This method does not require long periods of time, when compared with previously published mechanisms, and does not need sophisticated equipment.

Acknowledgements

We thank Consejo Nacional de Ciencia y Tecnologia (Conacyt) and Centro de Investigacion en Materiales Avanzados (CIMAV) for the financial support, Dr. Jose Guadalupe Murillo Ramirez for his help with the use of 852 nm wavelength laser, Dr. Pedro Piza for lending us the thermographic camera, Ing. Wilber Antunez Flores, and M.C. Karla Campos Venegas for helping us obtain the TEM and SEM images.

Nomenclature

GNSs	gold nanoshells
SPR	surface plasmon resonance
NIR	near-infrared
APTES	3-aminopropyltriethoxysilane
GNP	gold nanoparticles
TEM	transmission electron microscope
SEM	scanning electron microscope
TEOS	tetraethyl orthosilicate

Author details

Rosa Isela Ruvalcaba Ontiveros[1], José Alberto Duarte Moller[1*],
Anel Rocío Carrasco Hernandez[1], Hilda Esperanza Esparza-Ponce[1],
Erasmo Orrantia Borunda[1], Cynthia Deisy Gómez Esparza[1] and
Juan Manuel Olivares Ramírez[2]

1 Centro de Investigación en Materiales Avanzados (CIMAV), Chihuahua, México

2 Universidad Tecnológica de San Juan del Río, San Juan del Río, México

*Address all correspondence to: alberto.duarte@cimav.edu.mx

References

[1] Instituto Nacional del Cáncer. Tipos de Cancer [Internet]. 2015. Available from: https://www.cancer.gov/espanol/tipos

[2] Pissuwan D, Valenzuela SM, Cortie MB. Therapeutic possibilities of plasmonically heated gold nanoparticles. Trends in Biotechnology. 2006;**24**(2):62-67. DOI: 10.1016/j.tibtech.2005.12.004

[3] Petryayeva E, Krull UJ. Localized surface plasmon resonance: Nanostructures, bioassays and biosensing—A review. Analytica Chimica Acta. 2011;**706**(1):8-24. Available from: http://linkinghub.elsevier.com/retrieve/pii/S0003267011011196

[4] Jenkins JT, Halaney DL, Sokolov KV, Ma LL, Shipley HJ, Mahajan S, et al. Excretion and toxicity of gold-iron nanoparticles. Nanomedicine: Nanotechnology, Biology and Medicine. 2013;**9**(3):356-365. DOI: 10.1016/j.nano.2012.08.007

[5] Spyrogianni A, Sotiriou GA, Brambilla D, Leroux J-C, Pratsinis SE. The effect of settling on cytotoxicity evaluation of SiO_2 nanoparticles. Journal of Aerosol Science. 2017;**108**(January):56-66. Available from: https://linkinghub.elsevier.com/retrieve/pii/S0021850216302166

[6] Anderson RR, Parrish JA. The optics of human skin. The Journal of Investigative Dermatology. 1981;**77**(1):13-19. DOI: 10.1111/1523-1747.ep12479191

[7] Yong K-T, Sahoo Y, Swihart MT, Prasad PN. Synthesis and plasmonic properties of silver and gold nanoshells on polystyrene cores of different size and of gold-silver core-shell nanostructures. Colloids and Surfaces A: Physicochemical and Engineering Aspects. 2006;**290**(1-3): 89-105. Available from: http://linkinghub.elsevier.com/retrieve/pii/S0927775706003542

[8] Mohammad F, Al-Lohedan HA. Luteinizing hormone-releasing hormone targeted superparamagnetic gold nanoshells for a combination therapy of hyperthermia and controlled drug delivery. Materials Science and Engineering: C. 2017;**76**:692-700. DOI: 10.1016/j.msec.2017.03.162

[9] Sun Y, Xia Y. Increased sensitivity of surface plasmon resonance of gold nanoshells compared to that of gold solid colloids in response to environmental changes. Analytical Chemistry. 2002;**74**(20):5297-5305. DOI: 10.1021/ac0258352

[10] Abdollahi SN, Naderi M, Amoabediny G. Synthesis and characterization of hollow gold nanoparticles using silica spheres as templates. Colloids and Surfaces A: Physicochemical and Engineering Aspects. 2013;**436**:1069-1075. DOI: 10.1016/j.colsurfa.2013.08.028

[11] David Lou XW, Archer LA, Yang Z. Hollow micro-/nanostructures: Synthesis and applications. Advanced Materials. 2008;**20**(21):3987-4019. DOI: 10.1002/adma.200800854

[12] Wang L, Yuan Y, Lin S, Huang J, Dai J, Jiang Q, et al. Photothermo-chemotherapy of cancer employing drug leakage-free gold nanoshells. Biomaterials. 2016;**78**:40-49. DOI: 10.1016/j.biomaterials.2015.11.024

[13] Tuersun P, Han X. Optimal design of gold nanoshells for optical imaging and photothermal therapy. Optik. 2014;**125**(14):3702-3706. DOI: 10.1016/j.ijleo.2014.03.007

[14] Hirsch LR, Jackson JB, Lee A, Halas NJ, West JL. A whole blood immunoassay using gold nanoshells. Analytical Chemistry. 2003;**75**(10):2377-2381. DOI: 10.1021/ac0262210

[15] Brito-Silva AM, Sobral-Filho RG, Barbosa-Silva R, de Araújo CB, Galembeck A, Brolo AG. Improved synthesis of gold and silver nanoshells. Langmuir. 2013;**29**(13):4366-4372. DOI: 10.1021/la3050626

[16] Watanabe S, Asahi Y, Omura H, Mae K, Miyahara MT. Flow microreactor synthesis of gold nanoshells and patchy particles. Advanced Powder Technology. 2016;**27**(6):2335-2341. DOI: 10.1016/j.apt.2016.08.013

[17] Phonthammachai N, Kah JCY, Jun G, Sheppard CJR, Olivo MC, Mhaisalkar SG, et al. Synthesis of contiguous silica-gold core-shell structures: Critical parameters and processes. Langmuir. 2008;**24**(9):5109-5112. DOI: 10.1021/la703580r

[18] Abdollahi SN, Naderi M, Amoabediny G. Synthesis and physicochemical characterization of tunable silica-gold nanoshells via seed growth method. Colloids and Surfaces A: Physicochemical and Engineering Aspects. 2012;**414**:345-351. DOI: 10.1016/j.colsurfa.2012.08.043

[19] Thompson D. Michael Faraday's recognition of ruby gold: The birth of modern nanotechnology. Gold Bulletin. 2007;**40**(4):267-269. DOI: 10.1007/BF03215598

[20] Faraday M. The Bakerian lecture: Experimental relations of gold (and other metals) to light. Philosophical Transactions. Royal Society of London. 1857;**147**:145-181. DOI: 10.1098/rstl.1857.0011

[21] Turkevich J. Colloidal gold. Part II. Gold Bulletin. 1985;**18**(4):125-131. DOI: 10.1007/BF03214694

[22] Stöber W, Fink A, Bohn E. Controlled growth of monodisperse silica spheres in the micron size range. Journal of Colloid and Interface Science. 1968;**26**(1):62-69. Available from: http://linkinghub.elsevier.com/retrieve/pii/0021979768902725

[23] Westcott SL, Oldenburg SJ, Lee TR, Halas NJ. Formation and adsorption of clusters of gold nanoparticles onto functionalized silica nanoparticle surfaces. Langmuir. 1998;**14**(19):5396-5401. DOI: 10.1021/la980380q

[24] ImageJ [Internet]. 2018. Avaliable from: https://imagej.nih.gov/ij/

[25] Pham T, Jackson JB, Halas NJ, Lee TR. Preparation and characterization of gold nanoshells coated with self-assembled monolayers. Langmuir. 2002;**18**(12):4915-4920. DOI: 10.1021/la015561y

[26] Pandey PC, Pandey G, Walcarius A. 3-Aminopropyltrimethoxysilane mediated solvent induced synthesis of gold nanoparticles for biomedical applications. Materials Science and Engineering: C. 2017;**79**:45-54. DOI: 10.1016/j.msec.2017.05.009

[27] Lee SH, Jamison AC, Hoffman DM, Jacobson AJ, Lee TR. Preparation and characterization of polymeric thin films containing gold nanoshells via electrostatic layer-by-layer self-assembly. Thin Solid Films. 2014;**558**:200-207. DOI: 10.1016/j.tsf.2014.02.021

[28] Elbialy N, Mohamed N, Monem AS. Synthesis, characterization and application of gold nanoshells using mesoporous silica core. Microporous and Mesoporous Materials. 2014;**190**:197-207. DOI: 10.1016/j.micromeso.2014.02.003

[29] Wei X, Liu Z, Jin X, Huang L, Gurav DD, Sun X, et al. Plasmonic nanoshells enhanced laser desorption/ionization

mass spectrometry for detection of serum metabolites. Analytica Chimica Acta. 2017;**950**:147-155. DOI: 10.1016/j.aca.2016.11.017

[30] Socrates G. Infrared and Raman characteristic group frequencies. 3rd ed. West, Sussex: John Wiley & Sons. Journal of Raman Spectroscopy. 2004;**35**:905-905. DOI: 10.1002/jrs.1238

[31] Huang X, El-Sayed MA. Gold nanoparticles: Optical properties and implementations in cancer diagnosis and photothermal therapy. Journal of Advanced Research. 2010;**1**(1):13-28. Available from: http://linkinghub.elsevier.com/retrieve/pii/S2090123210000056

Chapter 7

Engineering of Surface Proteins in Extracellular Vesicles for Tissue-Specific Targeting

Stefan Vogt, Gerhard Stadlmayr, Johannes Grillari, Florian Rüker and Gordana Wozniak-Knopp

Abstract

Extracellular vesicles (EVs) have in the recent decades gained an important stand as vehicles enabling cell-to-cell transport and communication. With the advanced development towards their clinical use and increasing versatility of potential applications, improving their tissue-specific targeting in order to enhance their functionality in drug delivery opened as a challenging engineering field. In the past, the question of specific intercellular contact has been addressed by decoration of the EV surface with agents able of specific target recognition. An attractive possibility here is the modification of strongly overexpressed EV surface marker proteins towards recognition of target cells. As these proteins are involved in a plethora of biological functions in EV biogenesis, cargo targeting and intercellular transfer, a minimal impact on protein architecture upon modifications is desirable, which would also increase the stability of the exosomal preparation intended for therapeutic use. This chapter focuses on the possibilities of engineering of the EV marker proteins towards antigen-recognition units broadly applicable to endow EVs with tissue-targeting functionality.

Keywords: extracellular vesicles, exosomes, tetraspanin, tissue-specific targeting, exosomal drug delivery

1. Introduction

The transfer of extracellular vesicles (EVs) has emerged in the last two decades as a novel mechanism for intercellular communication. Nomenclature of EVs has by now been agreed to be based on biogenesis pathway. Therefore, EVs budding from plasma membranes are termed ectosomes or microvesicles, while exosomes are formed via the endosomal compartment within multivesicular bodies (MVB) which then are released by their fusion with the plasma membrane. Finally, when blebbing from apoptotic cells, the term apoptotic bodies has been retained [1]. Biomarkers to differentiate these EVs are still questionable, and in general, size is used to differentiate exosomes from ectosomes, while apoptotic bodies are considered to present phosphatidylserine on the outside and thus can be stained by annexin V. Exosomes (40–150 nm in diameter) are produced by formation of endosomal intraluminal vesicles (ILVs) in multivesicular bodies (MVBs) and are secreted by fusion of these vesicles with the plasma membrane [2, 3] (**Figure 1**).

 IntechOpen

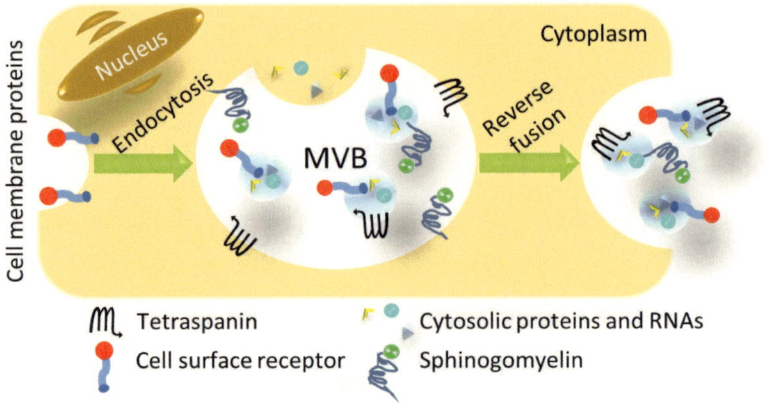

Figure 1.
Biogenesis of EVs.

Since they have been known to mediate directional intercellular transfer of their content, typically proteins, mRNAs, microRNAs (miRNAs) and a variety of non-coding RNAs [4], several studies were oriented towards their application as drugs *per se* or as therapeutic tool for drug delivery [5]. Their membrane is strongly enriched in sphingomyelin and cholesterol, which contributes to their unique buoyancy coefficient and enables practical isolation from other particles of cellular origin using differential centrifugation methods. Another discernible feature of exosome membranes is that it reflects the composition of the MVB membrane and has a high density of endosomal membrane proteins, such as proteins involved in MVB biogenesis (Alix and TSG101), membrane transport, in particular the components of the endosomal sorting complex required for transport (ESCRT) and most prominently tetraspanin proteins (CD9, CD37, CD53, CD63, CD81, CD82 and CD151) [6, 7]. Some tetraspanins such as CD9, CD81 and CD151 are more broadly expressed, while others are restricted to specific EV subsets [8, 9]. The overexpression of tetraspanins, as shown for CD9, can in several production cell lines enhance the exosome production and, in addition, cause a reduced overall size of the vesicles [10].

Tetraspanins play an important role in several physiological processes [11] and have been discovered to cooperate in states of health and disease in signal transduction, cellular activation, polarization, motility, adhesion, tissue differentiation, angiogenesis, tumorigenesis and metastasis [12–14], both by regulating cellular interactions as cell-membrane bound molecules and indirectly through exosomes. They are involved in each step of the metastatic cascade due to their ability to interact with cell surface receptors, adhesion molecules, matrix-remodeling proteases and signaling molecules. In this pathological state, they hence regulate cell proliferation, participate in epithelial-mesenchymal transition, modulate integrin-mediated cell adhesion and mediate the invasion through modulation of angiogenesis, tumor-endothelial cell interactions and regulation of cancer cell migration through the regulation of tumor microenvironment, as well as direct influence on extracellular matrix [15]. In this chapter, we introduce the characteristics of EVs and the engineering approaches aimed at their surface proteins to achieve tissue-specific targeting.

2. Structure of the tetraspanin proteins

The rod-shaped structure of a tetraspanin consists of four transmembrane helices that connect the two extracellular loops [16] (**Figure 2**). The short extracellular loop (SEL), which has not yet been reported to contain any element of a

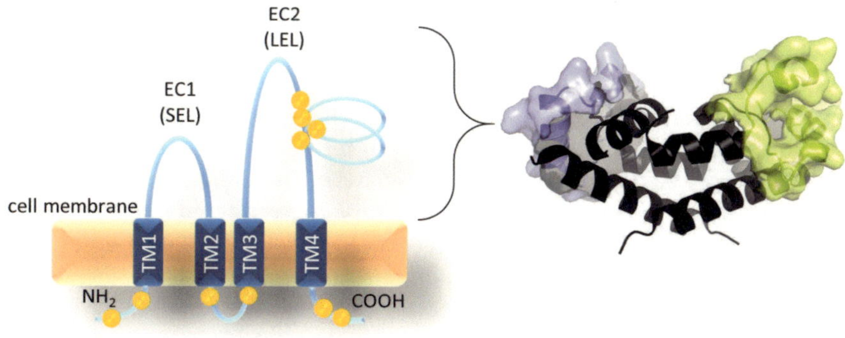

Figure 2.
Organization of tetraspanin CD81: 4 transmembrane helices (TM) span 2 extracellular loops. Positions of cysteine residues are indicated with yellow dots. Crystal structure of EC2 domain is presented as a cartoon diagram of the dimeric form (PDB: 1G8Q); the conserved three-helix bundle is black and the variable domains C and D are surfaced in blue or green for each protomer. The figure was prepared using PyMOL (PyMOL molecular graphics system, version 1.3 Schrödinger, LLC).

particular secondary structure, is of 12–31 amino acids in length. In the studies oriented towards tetraspanin engineering, the EC1 domain has until now been less addressed due to its lower degree of structural organization. Its replacement for a strand of glycine-serine residues has led to an unusual cell surface expression and a distribution dissimilar to the wild-type parental molecule, which may suggest the importance of this structural subunit for the stability of the membrane-bound tetraspanin [17].

The large extracellular loop (LEL) or extracellular domain 2 (EC2) has on the other hand been well characterized. The crystal structure of CD81 [18] demonstrated the four invariant cysteine resides within EC2 to form the 2 disulfide bridges, as a hallmark stability feature of the conserved tetraspanin fold [8, 19] (**Figure 2**). Moreover, the topology of several other tetraspanins was predicted by molecular modeling studies in the absence of an available crystal structure, using the CD81 EC2 structure as a template: these were the tetraspanins CD37, CD53, CD82 and CD151 [20, 21]. It has been early established that binding of CD81-specific antibodies depends on the formation of the disulfide bonds [22]. One class of tetraspanins harbors the EC2 composed of five α-helical elements, forming stalk and head elements of a mushroom-like structure, and the second class of tetraspanins is postulated to contain an additional helical element, stabilized with another pair of cysteine residues. The EC2 fold structure is evolutionarily conserved across species for any particular tetraspanin in spite of a high degree of variability at the amino acid level [23], with a subdomain consisting of a three-helix-bundle fold and a second subdomain variable in size among members of the tetraspanin family [18, 20].

The presence of an energetically unfavorable hydrophobic patch and the dimeric form of CD81 LEL in a crystal hinted at the high likelihood of tetraspanin assembly into dimers or multimers in the cell membrane. As shown later, clustering of tetraspanins leads to the organization of tetraspanin-associated proteins in a tetraspanin web or tetraspanin-enriched microdomain (TEM) [24, 25]. Biologically, such domains induce partitioning of the complexes into lipid rafts and clustering of the lipid rafts [26].

Apart from the hydrophobic interactions of the extracellular domains, the membrane-proximal cysteines residing in transmembrane domains contribute to the dimer formation and stability of tetraspanins. For the CD9 tetraspanin, it has been demonstrated that these residues are positioned close to the dimerization interface and influence homotypic and heterotypic tetraspanin association depending on their reversible palmitoylation status [27]. Similarly, the mutation of N-terminal and C-terminal transmembrane cysteines to serine eliminated palmitoylation of CD151, which turned out to be deleterious for the assembly with other cell surface proteins, including tetraspanins CD9 and CD63, their organization to TEMs and subsequently their subcellular distribution and cell morphology [28]. At the same time, it had minimal influence on the density of tetraspanin protein complexes and was dispensable for CD151-$\alpha_3\beta_1$ integrin association. Depalmitoylation of CD81 did not impact its surface expression and stability, but rendered it less available for contact with its natural interaction partner CD9 and the relevant epitopes less accessible for binding of structurally dependent antibodies [29].

The four transmembrane helices of tetraspanin proteins form two largely separated pairs of antiparallel helices: one pair comprises TM1/TM2 and the other TM3/TM4. The two pairs of helices only converge close to the cytoplasmic side of the membrane through contacts between TM2 and TM3. In the recently solved crystal structure of full-length CD81, this cone-like structure has been shown to harbor a binding pocket for cholesterol [30], and mutations within transmembrane domains in certain tetraspanins have been connected with pathological states [31]. The short cytoplasmic tails show no obvious functional significance in signaling processes, suggesting that their signaling competence relies on association with other molecules [32]; nevertheless, its mutation can lead to different assembly with association partners as shown for CD9 [33].

3. Natural ligands of tetraspanin proteins

3.1 The tetraspanin web and intertetraspanin contacts

The ability of members of the tetraspanin family to assemble into a unique biological feature known as tetraspanin-enriched microdomain (TEM) is due to their mutual interactions; however, these structures include also receptors, integrins and signaling molecules such as phosphatidyl-kinase C (PKC) and phosphatidylinositol-4-kinase (PI4K) [9]. These interactions are fundamental for cellular functions such as cell adhesion, proliferation and motility. Interactions between tetraspanin members are important in maintaining the integrity and stability of the tetraspanin web and providing binding sites for different ligands. The multimers of newly synthesized proteins are formed in the Golgi apparatus. The predominantly cross-linked tetraspanin species are homodimers, but also higher order complexes and low amounts of heterodimeric tetraspanins (CD81/CD9, CD9/CD151, CD81/CD151) were identified [27]. It has been suggested that tetraspanin homodimers, formed in the Golgi and present at the cell surface, serve as building blocks in the assembly of higher organized tetraspanin protein complexes. Interestingly, the exosomes originating from cell lines overexpressing CD9 are believed to be enriched in more stable TEMs [10]. Overall, although most tetraspanins can interact with most other tetraspanins, and similarly engage with several other proteins, the nature of these interactions has been until recently classified only according to their stability in the presence of detergents of different stringency, which does not necessarily reflect their significance in the cellular milieu [34]. A thorough characterization of strength and abundance of the interactions between the members participating in a tetraspanin web in a particular cell and physiological situation is therefore needed and will support the understanding of its mediated biological effects. Similarly, most data on tetraspanin functionality come from studies on their localization on cell membranes, while functional data in vesicles are still scarce. Therefore, we here summarize the known cellular functions, while speculating how this might translate to EVs.

An important step towards the understanding of the specificity of the tetraspanin interactions in TEM has been achieved by delineation of the involved tetraspanin regions by dissecting the model tetraspanin into domains, differently amenable for modification. Early experiments that addressed the relatively unstructured and at the same time antigen-binding competent regions that appeared attractive for mutagenesis resulted in a protein that showed aberrant clustering involving both homo- and heterodimerization of resulting full-length tetraspanins [35], albeit the mutagenesis method employed in this study was a complete deletion of targeted domains. The CD81 D-region was studied in more detail: the CD9 and CD151 tetraspanins were more competent of clustering with CD81 when homologously engrafted with CD81 D-region [36]. When the mutagenized CD81 EC2 molecular subunits were transplanted to other tetraspanins, the extremely flexible conformation of the solvent-exposed D-segment of CD81 EC2 was sufficient to overcome the orientational restrictions to initiate the homotypic contact for dimerization, and this finding has been corroborated with both wet-lab data and the insights from molecular dynamic simulation of the cell membrane-embedded protein [37].

3.2 Interaction of tetraspanins with integrins and matrix-degrading enzymes: role of tetraspanins in cancer and metastasis

Important association partners of tetraspanins are the integrins. The role of such complexes in invasive growth *in vivo* as well as the effect of integrin-mediated binding events on cell proliferation and invasion is well established. Especially, the laminin-binding integrins ($\alpha_6\beta_4$, $\alpha_3\beta_1$, $\alpha_6\beta_1$ and $\alpha_7\beta_1$) exhibit extensive interactions with tetraspanin proteins [12, 38]. The functionality of integrins may depend critically on their interaction of tetraspanins: it has early been described that the remarkably stable association of the tetraspanin CD151 and the integrin $\alpha_3\beta_1$ leads to a high level of activation of cellular PI4K [39]. Further, CD151 interacts directly with the α_3 subunit and links it to other tetraspanins, CD9 and CD81. Loss of CD151 abrogates the $\alpha_3\beta_1$ mediated mobility on its ligands, laminin-332 and laminin-551. CD9/CD81 complex may even regulate the integrin-mediated functions independently of CD151 by forming a complex with the integrin and directing the PKCα-$\alpha_3\beta_1$ association [40]. Another example of tetraspanin-integrin association reveals its proangiogenic role through VEGF induction, mediated by cooperation between TM4SF5 and integrin α_5 of epithelial cells [41]. Interestingly, removal of CD151 palmitoylation sites did not disrupt the CD151–$\alpha_6\beta_4$ complex in epithelial cells but strongly influenced $\alpha_6\beta_4$-integrin–dependent cell morphology [42]. The rat tetraspanin D6.1A (human homolog is CO-029) was able to induce systemic angiogenesis by initiation of an angiogenic loop that reached organs distant from the tumor, probably due to the abundance of D6.1A in tumor-derived exosomes. This is in line with reports claiming that EVs prepare niches for metastatic tumor cells at tissues distant from the primary tumor [43]. This tetraspanin associates with integrins $\alpha_3\beta_1$, $\alpha_6\beta_1$ and $\alpha_6\beta_4$, as well as with tetraspanins CD9 and CD81, and is similarly to CD151 linked to tumor-promoting activities [44].

Active complexes of tetraspanins and integrins influence biological processes other than cellular signaling by interacting with cellular metalloproteinases, important players in the remodeling of extracellular matrix. A study of MDA-MB-231 cells, a breast cancer cell line, has indicated that the $\alpha_3\beta_1$-tetraspanin protein complex may be linked to an invasive phenotype of tumor cells via modulation of various signaling pathways, including activation of membrane metalloproteinase-2 (MMP-2), an enzyme associated with invasive migration of the cells, and affecting phosphatidylinositol-3-kinase (PI3K) signaling pathways, which control actin cytoskeleton dynamics [45]. By the incorporation of the members of a disintegrin and metalloproteinase (ADAM) family members the tetraspanins are able to influence the cellular ectodomain cleavage and release activity of these enzymes [46]. The tetraspanins of the TspanC8 group (tetraspanins with 8 cysteines) have a significant impact on the cellular exit and catalytic activity of ADAM10 [47], in particular the activity of Tspan15/ADAM10 promoted N-cadherin cleavage [48–50]. Different TspanC8/ADAM10 complexes seem to have different substrate specificities [51]. The silencing of CD9 enhanced shedding of ADAM17-substrates TNF-α and ICAM-1 [52].

Important discovery of the possible consequences of fine differences in composition of TEMs has been delivered by the study of exosomes enriched in Tspan8-α_4 complex that were preferentially taken up by the endothelial and pancreatic cells [53]. The fact that such modifications can allow selective targeting *in vitro* and *in vivo* holds promise to achieve improved exosomal delivery by engineering of their membrane components.

3.3 The role of tetraspanins in immune complexes

In antigen-presenting cells (APCs), tetraspanins integrate into TEMs protein-recognition receptors binding to conserved repeated motifs of microbes, such as Toll-like receptors, and MHCII molecules into tetraspanin web platforms, as well as Fc_γ receptor I in phagocytic cells, Fc_γ receptor IIb and III upon the activation of macrophages and Fc_ε receptor I in monocytes and skin-derived dendritic cells [54].

The particular role of CD81 protein in the formation of specialized microdomains in the plasma membrane of the cells of the immune system was discovered by elucidating its function of recruiting various adhesion molecules, receptors and signaling proteins to the central zone of the immune synapse in T-lymphocytes and APCs [55]. Therefore, it has early been proposed for CD81 to play a key role during antigenic presentation, since it colocalizes with the T-cell receptor/CD3 [56], and CD81 indeed turned out to be a regulator of CD3 clustering and sustained CD3 signaling [57].

Further, the T-cell side of the immune synapse is densely populated by tetraspanins CD9 and CD151. The abolishment of their expression reduces markers of activation of T-lymphocytes conjugated to the APCs, such as IL-2 secretion and expression of CD69 [58].

Another role in the immune response of tetraspanin CD81 is amplifying and sustaining B-cell receptor signaling from lipid rafts by ligation to the co-receptor CD19/CD21 complex. The signaling through a variety of cell surface protein complexes implies a role of lipid rafts, again highlighting the ability of tetraspanin to facilitate raft association [9].

3.4 Tetraspanins in pathogen infection

Several studies have been oriented towards the research of tetraspanins as ligand molecules for pathogen entry. CD81 has been identified as a ligand for hepatitis C virus (HCV) recognition and viral entry [59]. The ligand for viral glycoprotein E2 is the D-domain of the LEL, a dynamic region positioned within the triple-bundle helix, whose conformation in solution differs substantially from the one suggested by crystal structure. Challenging for structure-based design, this region nevertheless presents an attractive target for design of therapeutically relevant ligands with methods such as NMR [60]. Apart of the EC2 domain, other regions of CD81 have proven important for virus infection. Experimental evidence here was based on the exchange of the structural domains of the molecule with the ones of tetraspanins of different degrees of homology, and it was found that closely related substitutions were more efficient at functional complementation of CD81. Viral entry has been shown to correlate with surface expression of the chimeric protein and to depend on the presence of the cholesterol-coordinating glutamate residue [61]. EWI-2wint, a cleavage form of EWI-2, a member of the immunoglobulin superfamily, has an inhibitory effect on HCV infection by obstructing the interaction between CD81 and HCV E2 [62]. The related factor EWI-F inhibits *Plasmodium* infection, whereas its silencing increases infection efficiency [63].

Tetraspanin microdomains have been described to regulate HIV-1 entry, assembly and transfer between the cells [64, 65]. CD81 influences importantly the early stages of virus replication by controlling the stability of HIV-1 restriction factor and consequently the activity of viral reverse transcriptase [66], while CD63 facilitates endocytosis of the HIV receptor CXCR4 [67] as well as supports the replication steps in macrophages [68, 69]. Also Coronaviruses and low-pathogenicity Influenza A viruses utilize TEM domains as entry portals to co-engage with cellular receptors and proteases, which enable viral proteolytic priming [70]. As shown in an *in vivo* model with CD151 null-mice, this tetraspanin is a critical novel host factor of nuclear export signaling of Influenza A virus, used complementary to the viral nuclear export proteins [71].

Tspan9 modulates the early endosome compartment to make it more permissive for membrane fusion of early-penetrating viruses, and its depletion strongly inhibits infection by alphaviruses that fuse in early endosomes but does not alter the delivery of virus to early endosomes or change their pH or protease activity [72]. It is unclear, what function then EV-based tetraspanins might have in the context of viral infection, and it might be speculated that cells use EVs to titrate away virus into membrane structures that are unable to provide replication and protein synthesis machineries for the virus. This is supported by the findings that EVs might have anti-influenza infection activity *in vitro* [73].

4. Extracellular vesicles as mediators of cell-cell interaction

4.1 Biological basis for therapeutic applications: EVs as mediators of intercellular interactions

EVs are secreted by most cell types and are taken up by recipient cells, where their cargo consisting of a cocktail of proteins, mRNAs and non-coding RNAs alters the behavior of the recipient cells in a way that might be even considered similar to hormones or cytokines [74], e.g. in the context of skin or bone cell paracrine signaling [75–77]. siRNAs (small interfering RNAs) and miRNA-based inhibitors have been recognized as potent novel drug candidates for many years. As EVs can be loaded with different drugs *in vitro*, they qualify as an attractive drug delivery system. The specificity of the recipient cell targeting *in vivo* is understood in a limited way only, although there is evidence of accumulation of specific EVs [43]. For example, EVs from human mesenchymal stem cells accumulated in the liver, spleen and sites of acute kidney injury [78]. Such tropism for a specific cell type, a requirement for targeted drug delivery, appears to be determined by surface proteins of the source cells. The composition of EV membrane reflecting the one of their source cell makes these particles non-immunogenic, and their small size allows them to pass the immune surveillance of the host organism [79, 80]. The reported engagement of exosomes in physiological processes in normal and diseased central nervous system makes them attractive vehicles for delivering neurotherapeutic agents across the blood-brain barrier [81–83]. Nevertheless, their delivery in humans seems so far limited to liver and kidney as they are reported not to reach therapeutic amounts in brain, heart and other tissues due to lack of specific targeting and thus low enrichment of the intended therapeutic ingredient in the target tissue. Modifications of the EV surface membrane to achieve enhanced targeting of a specific cell type are hence a common strategy embodied in several different engineering approaches.

4.2 Mechanisms of EV entry into the target cell

When an EV attaches to the target cell surface, it can in some cases activate the cognate receptors without internalization or transfer of the content to the recipient cell via its fusion with the target cell membrane or via endocytosis [84]. Endocytosis is an active process that requires cytoskeletal remodeling dependent on actin dynamics and includes clathrin-dependent endocytosis, phagocytosis and macropinocytosis. The clathrin dependent endocytosis has been established as cellular entry for EVs based on the experiments with specific inhibitors of this pathway. Additionally, endocytic uptake of EVs can involve lipid rafts, sometimes dependent on caveolin proteins. The size of EVs may be a limiting factor for cellular entry via endocytosis [85]. The EV uptake by phagocytosis was monitored by their high level of accumulation in phagocytic cells and localization into the phagolysosome [86], as well as the identification of the crucial role of the phosphatidylserine binding T-cell immunoglobulin and mucin domain containing (TIM4) receptor for the uptake of exosomes into macrophages [87, 88]. The contribution of macropinocytosis pathway was revealed with studies where exosomal uptake was decreased by the inhibition of cytoskeletal rearrangements that normally lead to membrane ruffles [89], as well as with its promotion caused by the activation of the agonistically acting epidermal growth factor [90].

5. Molecular engineering to facilitate EV labeling and delivery to target cells

5.1 Modifications of EV membrane with non-covalent and chemical modifications

The role of EVs as encapsulated intercellular messengers makes them attractive for development into nanoscale therapeutic agents [91], and therefore, the need of augmenting the interaction with the recipient cell has been widely recognized. The higher rigidity of the membrane of EVs in comparison with their source cells does not appear to obstruct the efficient application of common hydrophobic insertion strategies to EVs. The introduction of small lipophilic ligands, such as membrane dyes, works effectively for their labeling and aids in monitoring in *in vitro* and *in vivo* experiments. Furthermore, hydrophobic loading is used for encapsulation of certain drugs and leads to their increased stability and therapeutic effect [92, 93].

Due to the negative charge on the recipient cells, binding and uptake of EVs were enhanced by increasing the charge interactions with an artificially introduced positive surface potential, as exemplified in their derivatization with cationic lipids [94]. A downside of this method can be that the extremely charged cationic reagents can cause cytotoxicity and the cellular uptake of the modified particles can proceed differently from the usual pathways, which results in an unpredictable cellular fate of the particle and its cargo and possibly its undesired degradation. The innate slightly negative electrostatic potential exhibited by unmodified EVs should contribute to the longer half-life of such particles *in vivo* as inferred from liposomal studies [95]. Further, the extremes of unbalanced positive charge may negatively impact the storage stability and the application of such reagents in high concentrations.

As an alternative to other methods, labeling of EV surfaces with fluorescent probes has been achieved using click chemistry without an apparent effect on the size and function of the particles [96]. Nevertheless, there is an immense complexity behind the engineering and downstream methods developed for other biologicals conjugated with small molecules, especially when intended for therapeutic purposes [97, 98].

The idea of decoration of EV membranes with functional ligands further led to the consideration of receptor binding strategies. Such EV functionalization can readily be achieved by modification of the source cell. A resulting opportunity harvesting the specificity of this approach could also be used as a strategy to eliminate vesicles implicated in pathological processes, such as cancer metastasis, or to neutralize the undesired activity of therapeutically applied vesicles. Transferrin-conjugated superparamagnetic nanoparticles, reactive with surface-expressed transferrin receptor of exosomes, enabled their isolation from blood and endowed the vesicles with superior targeting properties [99]. A robust labeling approach of microvesicles was the expression of biotin acceptor peptide-transmembrane domain (BAP-TM) receptor on the source cells in combination with biotinylation *in vivo* [100].

Another example of employing an EV-localizing protein for efficient presentation on the EV surface involved a peptide or a protein fused with the C1C2 domain of lactadherin, which binds to EV outer membrane due to its affinity to phosphatidylserine, strongly enriched in exosome membranes. This method was used to generate antibodies against tumor biomarkers [101] and to increase the host immune response to tumor-associated antigens [102, 103].

Strong binding of lactadherin to the exosome membrane was also the basis of an efficient labeling protocol utilizing overexpression of a fusion protein composed of *Gaussia* luciferase and a truncated lactadherin in source melanoma cells. After harvest by ultracentrifugation, the labeled exosomes were successfully used for *in vivo* biodistribution studies [104].

5.2 Engineering of tetraspanin proteins for detection and monitoring of EVs

Regarding the tetraspanins not only as very abundant, but as prominent structural and stability elements of the exosomal membrane, it is sensible to engineer genetic fusions of these proteins enriched in EVs to ensure a high density of the expressed product and optimize the derivatization for the different purposes of engineering: cognate ligand binding, labeling for visualization or isolation, modifying cargo uptake and transfer and stability (**Figure 3**). Most practically, transgene

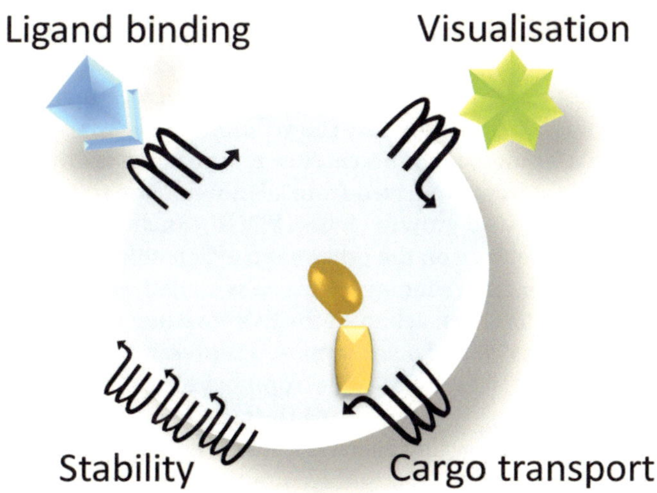

Figure 3.
Main engineering goals utilizing the modification of tetraspanins.

expression in the parent cell would be induced to deliver EVs enriched in the modified protein. The success of such set-up depends on the understanding of both molecular biology and protein architecture of the targeted species.

In one of the pioneering studies, certain sites on the tetraspanin CD63 have been chosen to allow the integration of fluorescent fusion proteins on the extra- and intravesicular side of the exosomal membrane [105]. CD63-GFP fusions have proven valuable reporters in the elucidation of the role of immune synapse in secretion of exosomes from T-cells to APCs and their fusion with recipient cells [106], in determining differential uptake properties of different immune cell subsets for EVs originating from a cancer cell line [107] and in an *in vivo* study imaging the fate of EVs produced by breast cancer cell line in *nude* mice [108].

5.3 Engineering of EV surface proteins for enhanced cell-type specific targeting: introduction of target specificity, stability and improved cargo delivery

In the pioneering example of derivatization of a protein enriched in exosomal membrane to achieve specific targeting [109], the rabies virus glycoprotein (RVG)-peptide fused with lysosome-associated membrane glycoprotein 2b (Lamp-2b) was used to target exosomes to the central nervous system in an *in vivo* mouse model. Immature dendritic cells-derived exosomes, enriched in an N-terminal fusion of an α_v integrin-specific internalizing RGD-sequence containing peptide with this scaffold, could internalize efficiently into target-positive breast cancer cells [110]. The display of such constructs could be efficiently enhanced by introducing a protective glycosylation motive, which improved the surface expression of exosome-bound N-terminally fused peptides by preventing their acid-mediated proteolysis during endosomal passage and indeed led to a more efficient specific cellular uptake of exosomes engineered in this way [111]. Specific targeting of IL3-receptor overexpressing chronic myeloid leukemia (CML) and acute myeloid leukemia (AML) cells has been shown for exosomes armed with a fusion protein of IL3 and Lamp2b and loaded with Imatinib, a tyrosine kinase inhibitor, and led to a the reduction of tumor size *in vivo* [112]. Exosomes engineered in this way achieved a higher abundance at the tumor site and were hence able to inhibit xenograft growth more efficiently than the active ingredient alone or than loaded control exosomes.

Glycosylphosphatidylinositol (GPI)-mediated anchoring of the specific targeting unit was demonstrated to be a stable alternative to Lamp2b fusions. EVs expressing GPI-anchored nanobodies, specific to epidermal growth factor receptor (EGFR), displayed enhanced binding to the EGFR-overexpressing cancer cells. This, however, did not lead to an increased uptake, and it was suggested that in this particular biological system not only the affinity but also the density of the targeting ligand must be high enough to induce receptor clustering and subsequent internalization [113]. Exosomes derived from a HEK cell line transfected with a construct of platelet-derived growth factor (PDGF)-anchoring sequence and an EGFR-binding peptide were on the other hand efficient in targeting EGFR-expressing tumors *in vivo* and reducing their size with delivery of microRNA Let7 [114]. Enhanced uptake has been achieved for EVs enriched in a fusion protein of tetraspanin CD63 and stearylated octaarginine, a representative cell-penetrating peptide, by their ability to induce active macropinocytosis [115].

The concept of enhancing the stability of the targeted exosomal surface protein intended for fusion with a targeting agent to allow a higher degree of versatility for their modification for an improved target recognition has raised interest in their engineering at the protein level. A significant increase in thermal stability has been achieved by introduction of additional disulfide bonds in the EC2 of tetraspanin CD81, with the best variant exhibiting a positive shift in the melting temperature for 45°C comparing with the wild-type protein [116]. When engrafted with a human transferrin-receptor specific peptide, the stabilized scaffold exhibited significantly better biophysical properties than the analogously engrafted wild-type protein. In the same study, a mutant has been discovered that exhibits reversible unfolding behavior up to a temperature of 110°C in contrast to the wild-type CD81 EC2, which presents another option for extensive engineering required for directed evolution of tetraspanin proteins towards novel antigen binding.

Not only can the overexpression of tetraspanins in source cell lines increase the production and stability of exosomes, but also the exosomes engineered to contain adeno-associated viruses (AAVs), designed for improved delivery of genetic material to target cell [100], were superior in their yield and functionality when CD9 was overexpressed in AAV-producer cells [10]. CD9 overexpression has also increased the speed and transduction efficiency of lentiviral gene delivery into numerous cell lines, confirming the important role of this tetraspanin in gene transfer [117].

The current scope of EV engineering reaches beyond receptor targeting systems and aspires towards modifications with complex modules, assigning them simultaneously with multiple novel functionalities, such as specific recognition as well as enzymatic activity, to enhance their potential therapeutic effect. Recently, regression of orthotopic Her2-positive tumors has been achieved by applying exosomes, able of specific targeting via their surface decoration with a fusion protein of high-affinity anti-Her2 single-chain Fv, and containing mRNA encoding a bacterial enzyme Hchr6, which in the strongly Her2-positive cells catalyzed the conversion of the prodrug CNOB to cytotoxic MCHB [118]. Tetraspanin engineering could support a sophisticated concept to aid intracellular delivery of exosomal cargo proteins, directionally incorporated during vesicle biogenesis [119]. Genes encoding two recombinant proteins, a fusion of photoreceptor cryptochrome 2 (CRY2) and the protein of interest and a fusion of tetraspanin protein CD9 and CRY-interacting basic-helix–loop–helix 1 (CIB1), were co-transfected into a single cell line. The blue light-induced binding of CRY2 and CIB1 enabled docking of CRY2-fused target proteins into nascent exosomes, and in the absence of the blue light, the cargo protein was released into the exosomal lumen. Transfer of Cre recombinase confirmed the efficiency of this system *in vitro* and *in vivo*.

6. Conclusions

The tetraspanins, well established as the biomarker proteins of extracellular vesicles, have been addressed for increasing EV stability and improving their function as delivery vehicles, both by assigning them with target recognition properties and modulating their cargo transfer. From the current point of view, the complexity of the tetraspanin-mediated interactions and signaling networks formed in a cell is yet to be discerned. Tetraspanins are known to interact naturally with a plethora of cell surface-bound ligands, which results in potent biological effects conveyed through different pathways; however, systematic evaluation of the affinity of the association with the interaction partners would assist in prediction of the consequential cellular processes as well as in determining optimal choice of the tetraspanin targeted for modification. There are recent reports describing modified tetraspanins mediating both surface protein interactions and an intracellular fusion-mediated enzymatic activity, which underline the feasibility of engineering versatile functions into tetraspanin proteins as fusion partners. The structural details on tetraspanins modified in this way, as well as the read-out revealing their actual behavior in the foreseen role and the influence of such modifications on the fate of an EV preparation, will pave the way into the design and production of EV-based reagents as therapeutics.

Acknowledgements

The financial support by the Austrian Federal Ministry for Digital and Economic Affairs and the National Foundation for Research, Technology and Development is gratefully acknowledged. S.V. was also supported by the PhD program BioToP (Biomolecular Technology of Proteins) funded by the Austrian Science Fund (FWF W1224).

Conflict of interest

The authors declare no conflict of interest.

Author details

Stefan Vogt[1], Gerhard Stadlmayr[2], Johannes Grillari[3,4], Florian Rüker[1,2] and Gordana Wozniak-Knopp[1,2]*

1 acib GmbH (Austrian Centre of Industrial Biotechnology), Graz, Austria

2 Christian Doppler Laboratory for Innovative Immunotherapeutics, Department of Biotechnology, University of Natural Resources and Life Sciences (BOKU), Vienna, Austria

3 Christian Doppler Laboratory for Biotechnology of Skin Aging, Department of Biotechnology, University of Natural Resources and Life Sciences (BOKU), Vienna, Austria

4 Evercyte GmbH, Wien, Austria

*Address all correspondence to: gordana.wozniak@boku.ac.at

References

[1] Wickman G, Julian L, Olson MF. How apoptotic cells aid in the removal of their own cold dead bodies. Cell Death & Differentiation. 2012;**19**:735-742. DOI: 10.1038/cdd.2012.25

[2] Harding C, Heuser J, Stahl P. Receptor-mediated endocytosis of transferrin and recycling of the transferrin receptor in rat reticulocytes. The Journal of Cell Biology. 1983;**97**(2):329-339. DOI: 10.1083/jcb.97.2.329

[3] Pan BT, Teng K, Wu C, Adam M, Johnstone RM. Electron microscopic evidence for externalization of the transferrin receptor in vesicular form in sheep reticulocytes. The Journal of Cell Biology. 1985;**101**(3):942-948. DOI: 10.1083/jcb.101.3.942

[4] Keerthikumar S, Chisanga D, Ariyaratne D, Al Saffar H, Anand S, Zhao K, et al. ExoCarta: A web-based compendium of exosomal cargo. Journal of Molecular Biology. 2016;**428**(4):688-692. DOI: 10.1016/j.jmb.2015.09.019

[5] Lässer C, Jang SC, Lötvall J. Subpopulations of extracellular vesicles and their therapeutic potential. Molecular Aspects of Medicine. 2018;**60**:1-14. DOI: 10.1016/j.mam.2018.02.002

[6] Escola JM, Kleijmeer MJ, Stoorvogel W, Griffith JM, Yoshie O, Geuze HJ. Selective enrichment of tetraspan proteins on the internal vesicles of multivesicular endosomes and on exosomes secreted by human B-lymphocytes. The Journal of Biological Chemistry. 1998;**273**(32):20121-20127. DOI: 10.1074/jbc.273.32.20121

[7] Kalra H, Drummen GPC, Mathivanan S. Focus on extracellular vesicles: Introducing the next small big thing. International Journal of Molecular Sciences. 2016;**17**:170. DOI: 10.3390/ijms17020170

[8] Maecker HT, Todd SC, Levy S. The tetraspanin superfamily: Molecular facilitators. The FASEB Journal. 1997;**11**(6):428-442. DOI: 10.1109/TNN.2011.2169808

[9] Wright MD, Moseley GW, Van Spriel AB. Tetraspanin microdomains in immune cell signalling and malignant disease. Tissue Antigens. 2004;**64**:533-542. DOI: 10.1111/j.1399-0039.2004.00321.x

[10] Schiller LT, Lemus-Diaz N, Rinaldi Ferreira R, Böker KO, Gruber J. Enhanced production of exosome-associated AAV by overexpression of the Tetraspanin CD9. Molecular Therapy–Methods & Clinical Development. 2018;**9**:278-287. DOI: 10.1016/j.omtm.2018.03.008

[11] Yáñez-Mó M, Mittelbrunn M, Sánchez-Madrid F. Tetraspanins and intercellular interactions. Microcirculation. 2001;**8**(3):153-168. DOI: 10.1038/sj.mn.7800076

[12] Berditchevski F. Complexes of tetraspanins with integrins: More than meets the eye. Journal of Cell Science. 2001;**114**:4143-4151. DOI: 10.1091/MBC.12.2.351

[13] Ang J, Fang BL, Ashman LK, Frauman AG. The migration and invasion of human prostate cancer cell lines involves CD151 expression. Oncology Reports. 2010;**24**(6):1593-1597. DOI: 10.3892/or-00001022

[14] Yáñez-Mó M, Siljander PRM, Andreu Z, Zavec AB, Borràs FE, Buzas EI, et al. Biological properties of extracellular vesicles and their physiological functions. Journal of Extracellular Vesicles. 2015;**4**:1-60. DOI: 10.3402/jev.v4.27066

[15] Detchokul S, Williams ED, Parker MW, Frauman AG. Tetraspanins as regulators of the tumour microenvironment: Implications for metastasis and therapeutic strategies. British Journal of Pharmacology. 2014;**171**:5462-5490. DOI: 10.1111/bph.12260

[16] Min G, Wang H, Sun TT, Kong XP. Structural basis for tetraspanin functions as revealed by the cryo-EM structure of uroplakin complexes at 6-Å resolution. The Journal of Cell Biology. 2006;**173**(6):975-983. DOI: 10.1083/jcb.200602086

[17] Masciopinto F, Campagnoli S, Abrignani S, Uematsu Y, Pileri P. The small extracellular loop of CD81 is necessary for optimal surface expression of the large loop, a putative HCV receptor. Virus Research. 2001;**80**(1-2):1-10. DOI: 10.1016/S0168-1702(01)00245-3

[18] Kitadokoro K, Bordo D, Galli G, Petracca R, Falugi F, Abrignani S, et al. CD81 extracellular domain 3D structure: Insight into the tetraspanin superfamily structural motifs. The EMBO Journal. 2001;**20**(1-2):12-18. DOI: 10.1093/emboj/20.1.12

[19] Tomlinson MG, Hanke T, Hughes DA, Barclay AN, Scholl E, Hünig T, et al. Characterization of mouse CD53: Epitope mapping, cellular distribution and induction by T cell receptor engagement during repertoire selection. European Journal of Immunology. 1995;**25**(8):2201-2205. DOI: 10.1002/eji.1830250813

[20] Seigneuret M, Delaguillaumie A, Lagaudrière-Gesbert C, Conjeaud H. Structure of the tetraspanin main extracellular domain: A partially conserved fold with a structurally variable domain insertion. The Journal of Biological Chemistry. 2001;**276**(43):40055-40064. DOI: 10.1074/jbc.M105557200

[21] Seigneuret M. Complete predicted three-dimensional structure of the facilitator transmembrane protein and hepatitis C virus receptor CD81: Conserved and variable structural domains in the tetraspanin superfamily. Biophysical Journal. 2006;**90**(1):212-227. DOI: 10.1529/biophysj.105.069666

[22] Imai T, Yoshie O. C33 antigen and M38 antigen recognized by monoclonal antibodies inhibitory to syncytium formation by human T cell leukemia virus type 1 are both members of the transmembrane 4 superfamily and associate with each other and with CD4 or CD8 in T cells. Journal of Immunology. 1993;**151**(11):6470-6481

[23] Bienstock RJ, Carl Barrett J. KAI1, a prostate metastasis suppressor: Prediction of solvated structure and interactions with binding partners; integrins, cadherins, and cell-surface receptor proteins. Molecular Carcinogenesis. 2001;**32**(3):139-153. DOI: 10.1002/mc.1073

[24] Rubinstein E, Le NF, Lagaudrière-Gesbert C, Billard M, Conjeaud H, Boucheix C. CD9, CD63, CD81, and CD82 are components of a surface tetraspan network connected to HLA-DR and VLA integrins. European Journal of Immunology. 1996;**26**(11):2657-2665. DOI: 10.1002/eji.1830261117

[25] Hemler ME. Tetraspanin functions and associated microdomains. Nature Reviews Molecular Cell Biology. 2005;**6**:801-811. DOI: 10.1038/nrm1736

[26] Claas C, Stipp CS, Hemler ME. Evaluation of prototype Transmembrane 4 superfamily protein complexes and their relation to lipid rafts. The Journal of Biological Chemistry. 2001;**276**(11):7974-7984. DOI: 10.1074/jbc.M008650200

[27] Kovalenko OV, Yang X, Kolesnikova TV, Hemler ME. Evidence for specific

tetraspanin homodimers: Inhibition of palmitoylation makes cysteine residues available for cross-linking. The Biochemical Journal. 2004;**377**(Pt 2): 407-417. DOI: 10.1042/BJ20031037

[28] Yang X, Kovalenko OV, Tang W, Claas C, Stipp CS, Hemler ME. Palmitoylation supports assembly and function of integrin-tetraspanin complexes. The Journal of Cell Biology. 2004;**167**(6):1231-1240. DOI: 10.1083/jcb.200404100

[29] Delandre C, Penabaz TR, Passarelli AL, Chapes SK, Clem RJ. Mutation of juxtamembrane cysteines in the tetraspanin CD81 affects palmitoylation and alters interaction with other proteins at the cell surface. Experimental Cell Research. 2009;**315**(11):1953-1963. DOI: 10.1016/j.yexcr.2009.03.013

[30] Zimmerman B, Kelly B, McMillan BJ, Seegar TCM, Dror RO, Kruse AC, et al. Crystal structure of a full-length human tetraspanin reveals a cholesterol-binding pocket. Cell. 2016;**167**(4):1041-1051.e11. DOI: 10.1016/j.cell.2016.09.056

[31] Sung CH, Davenport CM, Nathans J. Rhodopsin mutations responsible for autosomal dominant retinitis pigmentosa. Clustering of functional classes along the polypeptide chain. The Journal of Biological Chemistry. 1993;**268**(35):26645-26649

[32] Fitter S, Sincock PM, Jolliffe CN, Ashman LK. Transmembrane 4 superfamily protein CD151 (PETA-3) associates with beta 1 and alpha IIb beta 3 integrins in haemopoietic cell lines and modulates cell-cell adhesion. The Biochemical Journal. 1999;**338**(Pt 1):61-70. DOI: 10.1042/0264-6021:3380061

[33] Wang H-X, Kolesnikova TV, Denison C, Gygi SP, Hemler ME. The C-terminal tail of tetraspanin protein CD9 contributes to its function and molecular organization. Journal of Cell Science. 2011;**124**(16):2702-2710. DOI: 10.1242/jcs.085449

[34] van Deventer SJ, Dunlock V-ME, van Spriel AB. Molecular interactions shaping the tetraspanin web. Biochemical Society Transactions. 2017;**45**(3):741-750. DOI: 10.1042/BST20160284

[35] Homsi Y, Schloetel JG, Scheffer KD, Schmidt TH, Destainville N, Florin L, et al. The extracellular δ-domain is essential for the formation of CD81 tetraspanin webs. Biophysical Journal. 2014;**107**(1):100-113. DOI: 10.1016/j.bpj.2014.05.028

[36] Homsi Y, Lang T. The specificity of homomeric clustering of CD81 is mediated by its δ-loop. FEBS Open Bio. 2017;7(2):274-283. DOI: 10.1002/2211-5463.12187

[37] Schmidt TH, Homsi Y, Lang T. Oligomerization of the tetraspanin CD81 via the flexibility of its δ-loop. Biophysical Journal. 2016;**110**(11):2463-2474. DOI: 10.1016/j.bpj.2016.05.003

[38] Hemler ME. Integrin associated proteins. Current Opinion in Cell Biology. 1998;**10**:578-585. DOI: 10.1016/S0955-0674(98)80032-X

[39] Yauch RL, Berditchevski F, Harler MB, Reichner J, Hemler ME. Highly stoichiometric, stable, and specific association of integrin alpha 3beta 1 with CD151 provides a major link to phosphatidylinositol 4-kinase, and may regulate cell migration. Molecular Biology of the Cell. 1998;**9**(10):2751-2765. DOI: 10.1091/mbc.9.10.2751

[40] Gustafson-Wagner E, Stipp CS. The CD9/CD81 Tetraspanin complex and Tetraspanin CD151 regulate α3β1 integrin-dependent tumor cell behaviors by overlapping but distinct mechanisms. PLoS One. 2013;**8**(4):e61834. DOI: 10.1371/journal.pone.0061834

[41] Choi S, Lee SA, Kwak TK, Kim HJ, Lee MJ, Ye SK, et al. Cooperation between integrin α 5 and tetraspan TM4SF5 regulates VEGF-mediated angiogenic activity. Blood. 2009;**113**(8):1845-1855. DOI: 10.1182/blood-2008-05-160671

[42] Yang X, Claas C, Kraeft SK, Chen LB, Wang Z, Kreidberg JA, et al. Palmitoylation of tetraspanin proteins: Modulation of CD151 lateral interactions, subcellular distribution, and integrin-dependent cell morphology. Molecular Biology of the Cell. 2002;**13**(3):767-781. DOI: 10.1091/mbc.01-05-0275

[43] Hoshino A, Costa-Silva B, Shen TL, Rodrigues G, Hashimoto A, Tesic Mark M, et al. Tumour exosome integrins determine organotropic metastasis. Nature. 2015;**527**(7578):329-335. DOI: 10.1038/nature15756

[44] Gesierich S, Paret C, Hildebrand D, Weitz J, Zgraggen K, Schmitz-Winnenthal FH, et al. Colocalization of the tetraspanins, CO-029 and CD151, with integrins in human pancreatic adenocarcinoma: Impact on cell motility. Clinical Cancer Research. 2005;**11**(8):2840-2852. DOI: 10.1158/1078-0432.CCR-04-1935

[45] Sugiura T, Berditchevski F. Function of α3β1-tetraspanin protein complexes in tumor cell invasion. Evidence for the role of the complexes in production of matrix metalloproteinase 2 (MMP-2). The Journal of Cell Biology. 1999;**146**(6):1375-1389. DOI: 10.1083/jcb.146.6.1375

[46] Yáñez-Mó M, Gutiérrez-López MD, Cabañas C. Functional interplay between tetraspanins and proteases. Cellular and Molecular Life Sciences. 2011;**68**:3323-3335. DOI: 10.1007/s00018-011-0746-y

[47] Dornier E, Coumailleau F, Ottavi JF, Moretti J, Boucheix C, Mauduit P, et al. Tspanc8 tetraspanins regulate ADAM10/Kuzbanian trafficking and promote notch activation in flies and mammals. The Journal of Cell Biology. 2012;**199**(3):481-496. DOI: 10.1083/jcb.201201133

[48] Prox J, Willenbrock M, Weber S, Lehmann T, Schmidt-Arras D, Schwanbeck R, et al. Tetraspanin15 regulates cellular trafficking and activity of the ectodomain sheddase ADAM10. Cellular and Molecular Life Sciences. 2012;**69**(17):2919-2932. DOI: 10.1007/s00018-012-0960-2

[49] Jouannet S, Saint-Pol J, Fernandez L, Nguyen V, Charrin S, Boucheix C, et al. TspanC8 tetraspanins differentially regulate the cleavage of ADAM10 substrates, notch activation and ADAM10 membrane compartmentalization. Cellular and Molecular Life Sciences. 2016;**73**(9):1895-1915. DOI: 10.1007/s00018-015-2111-z

[50] Noy PJ, Yang J, Reyat JS, Matthews AL, Charlton AE, Furmston J, et al. TspanC8 tetraspanins and a disintegrin and metalloprotease 10 (ADAM10) interact via their extracellular regions: Evidence for distinct binding mechanisms for different TspanC8 proteins. The Journal of Biological Chemistry. 2016;**291**(7):3145-3157. DOI: 10.1074/jbc.M115.703058

[51] Matthews AL, Szyroka J, Collier R, Noy PJ, Tomlinson MG. Scissor sisters: Regulation of ADAM10 by the TspanC8 tetraspanins. Biochemical Society Transactions. 2017;**45**(3):719-730. DOI: 10.1042/BST20160290

[52] Gutiérrez-López MD, Gilsanz A, Yàñez-Mó M, Ovalle S, Lafuente EM, Domínguez C, et al. The sheddase activity of ADAM17/TACE is regulated by the tetraspanin CD9. Cellular and Molecular Life Sciences. 2011;**68**(19):3275-3292. DOI: 10.1007/s00018-011-0639-0

[53] Rana S, Yue S, Stadel D, Zöller M. Toward tailored exosomes: The exosomal tetraspanin web contributes to target cell selection. The International Journal of Biochemistry & Cell Biology. 2012;**44**(9):1574-1584. DOI: 10.1016/j.biocel.2012.06.018

[54] Saiz ML, Rocha-Perugini V, Sánchez-Madrid F. Tetraspanins as organizers of antigen-presenting cell function. Frontiers in Immunology. 2018;**9**:1074. DOI: 10.3389/fimmu.2018.01074

[55] Levy S, Shoham T. The tetraspanin web modulates immune-signalling complexes. Nature Reviews Immunology. 2005;**5**:136-148. DOI: 10.1038/nri1548

[56] Mittelbrunn M, Yanez-Mo M, Sancho D, Ursa A, Sanchez-Madrid F. Cutting edge: Dynamic redistribution of Tetraspanin CD81 at the central zone of the immune synapse in both T lymphocytes and APC. Journal of Immunology. 2002;**169**(12):6691-6695. DOI: 10.4049/jimmunol.169.12.6691

[57] Rocha-Perugini V, Zamai M, Gonzalez-Granado JM, Barreiro O, Tejera E, Yanez-Mo M, et al. CD81 controls sustained T cell activation signaling and defines the maturation stages of cognate immunological synapses. Molecular and Cellular Biology. 2013;**33**(18):3644-3658. DOI: 10.1128/MCB.00302-13

[58] Rocha-Perugini V, González-Granado JM, Tejera E, López-Martín S, Yañez-Mó M, Sánchez-Madrid F. Tetraspanins CD9 and CD151 at the immune synapse support T-cell integrin signaling. European Journal of Immunology. 2014;**44**(7):1967-1975. DOI: 10.1002/eji.201344235

[59] Pileri P, Uematsu Y, Campagnoli S, Galli G, Falugi F, Petracca R, et al. Binding of hepatitis C virus to CD81. Science. 1998;**282**(5390):938-941. DOI: 10.1126/science.282.5390.938

[60] Rajesh S, Sridhar P, Tews BA, Feneant L, Cocquerel L, Ward DG, et al. Structural basis of ligand interactions of the large extracellular domain of tetraspanin CD81. Journal of Virology. 2012;**86**(18):9606-9616. DOI: 10.1128/JVI.00559-12

[61] Banse P, Moeller R, Bruening J, Lasswitz L, Kahl S, Khan AG, et al. CD81 receptor regions outside the large extracellular loop determine hepatitis C virus entry into hepatoma cells. Viruses. 2018;**10**(4):207. DOI: 10.3390/v10040207

[62] Rocha-Perugini V, Montpellier C, Delgrange D, Wychowski C, Helle F, Pillez A, et al. The CD81 partner EWI-2wint inhibits hepatitis C virus entry. PLoS One. 2008;**3**(4):e1866. DOI: 10.1371/journal.pone.0001866

[63] Charrin S, Yalaoui S, Bartosch B, Cocquerel L, Franetich JF, Boucheix C, et al. The Ig domain protein CD9P-1 down-regulates CD81 ability to support plasmodium yoelii infection. The Journal of Biological Chemistry. 2009;**284**(46):31572-31578. DOI: 10.1074/jbc.M109.057927

[64] Florin L, Lang T. Tetraspanin assemblies in virus infection. Frontiers in Immunology. 2018;**9**:1140. DOI: 10.3389/fimmu.2018.01140

[65] Sims B, Farrow AL, Williams SD, Bansal A, Krendelchtchikov A, Matthews QL. Tetraspanin blockage reduces exosome-mediated HIV-1 entry. Archives of Virology. 2018;**163**(6):1683-1689. DOI: 10.1007/s00705-018-3737-6

[66] Rocha-Perugini V, Suárez H, Álvarez S, López-Martín S, Lenzi GM, Vences-Catalán F, et al. CD81 association with SAMHD1 enhances HIV-1 reverse transcription by increasing dNTP levels. Nature Microbiology. 2017;**2**(11):1513-1522. DOI: 10.1038/s41564-017-0019-0

[67] Yoshida T, Kawano Y, Sato K, Ando Y, Aoki J, Miura Y, et al. A CD63 mutant inhibits T-cell tropic human immunodeficiency virus type 1 entry by disrupting CXCR4 trafficking to the plasma membrane. Traffic. 2008;**9**(4):540-558. DOI: 10.1111/j.1600-0854.2007.00700.x

[68] Chen H, Dziuba N, Friedrich B, von Lindern J, Murray JL, Rojo DR, et al. A critical role for CD63 in HIV replication and infection of macrophages and cell lines. Virology. 2008;**379**(2):191-196. DOI: 10.1016/j.virol.2008.06.029

[69] Li G, Endsley MA, Somasunderam A, Gbota SL, Mbaka MI, Murray JL, et al. The dual role of tetraspanin CD63 in HIV-1 replication. Virology Journal. 2014;**11**(1):23. DOI: 10.1186/1743-422X-11-23

[70] Earnest JT, Hantak MP, Park J-E, Gallagher T. Coronavirus and influenza virus proteolytic priming takes place in tetraspanin-enriched membrane microdomains. Journal of Virology. 2015;**89**(11):6093-6104. DOI: 10.1128/JVI.00543-15

[71] Qiao Y, Yan Y, Sen TK, Tan SSL, Seet JE, Arumugam TV, et al. CD151, a novel host factor of nuclear export signaling in influenza virus infection. The Journal of Allergy and Clinical Immunology. 2018;**141**(5):1799-1817. DOI: 10.1016/j.jaci.2017.11.032

[72] Stiles KM, Kielian M. Role of TSPAN9 in alphavirus entry and early endosomes. Journal of Virology. 2016;**90**(9):4289-4297. DOI: 10.1128/JVI.00018-16

[73] Jantaratrirat S, Boonarkart C, Ruangrung K, Suptawiwat O, Auewarakul P. Microparticle release from cell lines and its anti-influenza activity. Viral Immunology. 2018;**31**(6):447-456. DOI: 10.1089/vim.2017.0201

[74] Bayraktar R, Van Roosbroeck K, Calin GA. Cell-to-cell communication: microRNAs as hormones. Molecular Oncology. 2017;**11**:1673-1686. DOI: 10.1002/1878-0261.12144

[75] Weilner S, Schraml E, Wieser M, Messner P, Schneider K, Wassermann K, et al. Secreted microvesicular miR-31 inhibits osteogenic differentiation of mesenchymal stem cells. Aging Cell. 2016;**15**(4):744-754. DOI: 10.1111/acel.12484

[76] Weilner S, Keider V, Winter M, Harreither E, Salzer B, Weiss F, et al. Vesicular Galectin-3 levels decrease with donor age and contribute to the reduced osteo-inductive potential of human plasma derived extracellular vesicles. Aging (Albany NY). 2016;**8**(1):16-33. DOI: 10.18632/aging.100917

[77] Terlecki-Zaniewicz L, Lämmermann I, Latreille J, Bobbili MR, Pils V, Schosserer M, et al. Small extracellular vesicles and their miRNA cargo are anti-apoptotic members of the senescence-associated secretory phenotype. Aging (Albany NY). 2018;**10**(5):1103-1132. DOI: 10.18632/aging.101452

[78] Bruno S, Grange C, Deregibus MC, Calogero RA, Saviozzi S, Collino F, et al. Mesenchymal stem cell-derived microvesicles protect against acute tubular injury. Journal of the American Society of Nephrology. 2009;**20**(5):1053-1067. DOI: 10.1681/ASN.2008070798

[79] György B, Hung ME, Breakefield XO, Leonard JN. Therapeutic applications of extracellular vesicles: Clinical promise and open questions. Annual Review of Pharmacology and Toxicology. 2015;**55**(1):439-464. DOI: 10.1146/annurev-pharmtox-010814-124630

[80] Zhu X, Badawi M, Pomeroy S, Sutaria DS, Xie Z, Baek A, et al. Comprehensive toxicity and

immunogenicity studies reveal minimal effects in mice following sustained dosing of extracellular vesicles derived from HEK293T cells. Journal of Extracellular Vesicles. 2017;**6**(1):1324730. DOI: 10.1080/20013078.2017.1324730

[81] Wood MJA, O'Loughlin AJ, Lakhal S. Exosomes and the blood-brain barrier: Implications for neurological diseases. Therapeutic Delivery. 2011;**2**:1095-1099. DOI: 10.1002/star.19940460805

[82] Robbins PF, Kassim SH, Tran TLN, Crystal JS, Morgan RA, Feldman SA, et al. A pilot trial using lymphocytes genetically engineered with an NY-ESO-1-reactive T-cell receptor: Long-term follow-up and correlates with response. Clinical Cancer Research. 2015;**21**(5):1019-1027. DOI: 10.1158/1078-0432.CCR-14-2708

[83] EL Andaloussi S, Lakhal S, Mäger I, Wood MJA. Exosomes for targeted siRNA delivery across biological barriers. Advanced Drug Delivery Reviews. 2013;**65**:391-397. DOI: 10.1016/j.addr.2012.08.008

[84] Mulcahy LA, Pink RC, Carter DRF. Routes and mechanisms of extracellular vesicle uptake. Journal of Extracellular Vesicles. 2014;**3**:24641. DOI: 10.3402/jev.v3.24641

[85] Conner SD, Schmid SL. Regulated portals of entry into the cell. Nature. 2003;**422**:37-44. DOI: 10.1038/nature01451

[86] Feng D, Zhao WL, Ye YY, Bai XC, Liu RQ, Chang LF, et al. Cellular internalization of exosomes occurs through phagocytosis. Traffic. 2010;**11**(5):675-687. DOI: 10.1111/j.1600-0854.2010.01041.x

[87] Zakharova L, Svetlova M, Fomina AF. T cell exosomes induce cholesterol accumulation in human monocytes via phosphatidylserine receptor. Journal of Cellular Physiology. 2007;**212**(1):174-181. DOI: 10.1002/jcp.21013

[88] Miyanishi M, Tada K, Koike M, Uchiyama Y, Kitamura T, Nagata S. Identification of Tim4 as a phosphatidylserine receptor. Nature. 2007;**450**(7168):435-439. DOI: 10.1038/nature06307

[89] Fitzner D, Schnaars M, van Rossum D, Krishnamoorthy G, Dibaj P, Bakhti M, et al. Selective transfer of exosomes from oligodendrocytes to microglia by macropinocytosis. Journal of Cell Science. 2011;**124**(3):447-458. DOI: 10.1242/jcs.074088

[90] Nakase I, Kobayashi NB, Takatani-Nakase T, Yoshida T. Active macropinocytosis induction by stimulation of epidermal growth factor receptor and oncogenic Ras expression potentiates cellular uptake efficacy of exosomes. Scientific Reports. 2015;**5**:10300. DOI: 10.1038/srep10300

[91] Armstrong JPK, Stevens MM. Strategic design of extracellular vesicle drug delivery systems. Advanced Drug Delivery Reviews. 2018;**130**:12-16. DOI: 10.1016/j.addr.2018.06.017

[92] Sun D, Zhuang X, Xiang X, Liu Y, Zhang S, Liu C, et al. A novel nanoparticle drug delivery system: The anti-inflammatory activity of curcumin is enhanced when encapsulated in exosomes. Molecular Therapy. 2010;**18**(9):1606-1614. DOI: 10.1038/mt.2010.105

[93] Zhuang X, Xiang X, Grizzle W, Sun D, Zhang S, Axtell RC, et al. Treatment of brain inflammatory diseases by delivering exosome encapsulated anti-inflammatory drugs from the nasal region to the brain. Molecular Therapy. 2011;**19**(10):1769-1779. DOI: 10.1038/mt.2011.164

[94] Nakase I, Futaki S. Combined treatment with a pH-sensitive

fusogenic peptide and cationic lipids achieves enhanced cytosolic delivery of exosomes. Scientific Reports. 2015;**5**:10112. DOI: 10.1038/srep10112

[95] Zhao W, Zhuang S, Qi XR. Comparative study of the in vitro and in vivo characteristics of cationic and neutral liposomes. International Journal of Nanomedicine. 2011;**6**:3087-3098. DOI: 10.2147/IJN.S25399

[96] Smyth T, Petrova K, Payton NM, Persaud I, Redzic JS, Graner MW, et al. Surface functionalization of exosomes using click chemistry. Bioconjugate Chemistry. 2014;**25**(10):1777-1784. DOI: 10.1021/bc500291r

[97] Drake PM, Rabuka D. Recent developments in ADC technology: Preclinical studies signal future clinical trends. BioDrugs. 2017;**31**:521-531. DOI: 10.1007/s40259-017-0254-1

[98] Vezina HE, Cotreau M, Han TH, Gupta M. Antibody–drug conjugates as Cancer therapeutics: Past, present, and future. Journal of Clinical Pharmacology. 2017;**57**:S11-S25. DOI: 10.1002/jcph.981

[99] Qi H, Liu C, Long L, Ren Y, Zhang S, Chang X, et al. Blood exosomes endowed with magnetic and targeting properties for Cancer therapy. ACS Nano. 2016;**10**(3):3323-3333. DOI: 10.1021/acsnano.5b06939

[100] Maguire CA, Balaj L, Sivaraman S, Crommentuijn MHW, Ericsson M, Mincheva-Nilsson L, et al. Microvesicle-associated AAV vector as a novel gene delivery system. Molecular Therapy. 2012;**20**(5):960-971. DOI: 10.1038/mt.2011.303

[101] Delcayre A, Estelles A, Sperinde J, Roulon T, Paz P, Aguilar B, et al. Exosome display technology: Applications to the development of new diagnostics and therapeutics. Blood Cells, Molecules, and Diseases. 2005;**35**(2):158-168. DOI: 10.1016/j.bcmd.2005.07.003

[102] Zeelenberg IS, Ostrowski M, Krumeich S, Bobrie A, Jancic C, Boissonnas A, et al. Targeting tumor antigens to secreted membrane vesicles in vivo induces efficient antitumor immune responses. Cancer Research. 2008;**68**(4):1228-1235. DOI: 10.1158/0008-5472.CAN-07-3163

[103] Zeelenberg IS, van Maren WWC, Boissonnas A, Van Hout-Kuijer MA, Den Brok MHMGM, Wagenaars JAL, et al. Antigen localization controls T cell-mediated tumor immunity. Journal of Immunology. 2011;**187**(3):1281-1288. DOI: 10.4049/jimmunol.1003905

[104] Takahashi Y, Nishikawa M, Takakura Y. In vivo tracking of extracellular vesicles in mice using fusion protein comprising Lactadherin and Gaussia luciferase. Methods in Molecular Biology. 2017;**1660**:245-254. DOI: 10.1007/978-1-4939-7253-1_20

[105] Stickney Z, Losacco J, McDevitt S, Zhang Z, Lu B. Development of exosome surface display technology in living human cells. Biochemical and Biophysical Research Communications. 2016;**472**(1):53-59. DOI: 10.1016/j.bbrc.2016.02.058

[106] Mittelbrunn M, Gutiérrez-Vázquez C, Villarroya-Beltri C, González S, Sánchez-Cabo F, González MÁ, et al. Unidirectional transfer of microRNA-loaded exosomes from T cells to antigen-presenting cells. Nature Communications. 2011;**2**(1):282. DOI: 10.1038/ncomms1285

[107] Sadovska L, Zandberga E, Sagini K, Jekabsons K, Riekstiņa U, Kalniņa Z, et al. A novel 3D heterotypic spheroid model for studying extracellular vesicle-mediated tumour and immune cell communication. Biochemical and Biophysical Research Communications. 2017;**495**(2):1930-1935. DOI: 10.1016/j.bbrc.2017.12.072

[108] Suetsugu A, Honma K, Saji S, Moriwaki H, Ochiya T, Hoffman RM. Imaging exosome transfer from breast cancer cells to stroma at metastatic sites in orthotopic nude-mouse models. Advanced Drug Delivery Reviews. 2013;**65**:383-390. DOI: 10.1016/j.addr.2012.08.007

[109] Alvarez-Erviti L, Seow Y, Yin H, Betts C, Lakhal S, Wood MJA. Delivery of siRNA to the mouse brain by systemic injection of targeted exosomes. Nature Biotechnology. 2011;**29**(4):341-345. DOI: 10.1038/nbt.1807

[110] Tian Y, Li S, Song J, Ji T, Zhu M, Anderson GJ, et al. A doxorubicin delivery platform using engineered natural membrane vesicle exosomes for targeted tumor therapy. Biomaterials. 2014;**35**(7):2383-2390. DOI: 10.1016/j.biomaterials.2013.11.083

[111] Hung ME, Leonard JN. Stabilization of exosome-targeting peptides via engineered glycosylation. The Journal of Biological Chemistry. 2015;**290**(13):8166-8172. DOI: 10.1074/jbc.M114.621383

[112] Bellavia D, Raimondo S, Calabrese G, Forte S, Cristaldi M, Patinella A, et al. Interleukin 3- receptor targeted exosomes inhibit in vitro and in vivo chronic myelogenous leukemia cell growth. Theranostics. 2017;**7**(5):1333-1345. DOI: 10.7150/thno.17092

[113] Kooijmans SAA, Aleza CG, Roffler SR, van Solinge WW, Vader P, Schiffelers RM. Display of GPI-anchored anti-EGFR nanobodies on extracellular vesicles promotes tumour cell targeting. Journal of Extracellular Vesicles. 2016;**5**(1):31053. DOI: 10.3402/jev.v5.31053

[114] Ohno SI, Takanashi M, Sudo K, Ueda S, Ishikawa A, Matsuyama N, et al. Systemically injected exosomes targeted to EGFR deliver antitumor microrna to breast cancer cells. Molecular Therapy. 2013;**21**(1):185-191. DOI: 10.1038/mt.2012.180

[115] Nakase I, Noguchi K, Aoki A, Takatani-Nakase T, Fujii I, Futaki S. Arginine-rich cell-penetrating peptide-modified extracellular vesicles for active macropinocytosis induction and efficient intracellular delivery. Scientific Reports. 2017;**7**(1):1991. DOI: 10.1038/s41598-017-02014-6

[116] Vogt S, Stadlmayr G, Stadlbauer K, Sádio F, Andorfer P, Grillari J, et al. Stabilization of the CD81 large extracellular loop with De novo disulfide bonds improves its amenability for peptide grafting. Pharmaceutics. 2018;**10**(3):138. DOI: 10.3390/pharmaceutics10030138

[117] Böker KO, Lemus-Diaz N, Rinaldi Ferreira R, Schiller L, Schneider S, Gruber J. The impact of the CD9 Tetraspanin on Lentivirus infectivity and exosome secretion. Molecular Therapy. 2018;**26**(2):634-647. DOI: 10.1016/j.ymthe.2017.11.008

[118] Wang J-H, Forterre AV, Zhao J, Frimannsson DO, Delcayre A, Antes TJ, et al. Anti-HER2 scFv-directed extracellular vesicle-mediated mRNA-based gene delivery inhibits growth of HER2-positive human breast tumor xenografts by prodrug activation. Molecular Cancer Therapeutics. 2018;**17**(5):1133-1142. DOI: 10.1158/1535-7163.MCT-17-0827

[119] Yim N, Ryu SW, Choi K, Lee KR, Lee S, Choi H, et al. Exosome engineering for efficient intracellular delivery of soluble proteins using optically reversible protein-protein interaction module. Nature Communications. 2016;**7**:12277. DOI: 10.1038/ncomms12277